Electronic Troubleshooting

DON MATSUDA

Prentice Hall
Englewood Cliffs, New Jersey 07632

Library of Congress Cataloging-in-Publication Data

Matsuda, Don, (date)
Electronic troubleshooting / Don Matsuda.
p. cm.
Includes index.
ISBN 0-13-248055-7
1. Electronic apparatus and appliances—Maintenance and repair.
I. Title.
TK7870.2M38 1992
621.381—dc20 91-35761
CIP

Acquisition Editor: *Susan Willig*
Editorial Assistant: *Shirley Chlopak*
Supplements Editor: *Judith Casillo*
Manufacturing Buyer: *Ed O'Dougherty*
Prepress Buyer: *Ilene Levy*
Cover Designer: *Bruce Kenselaar*
Production Editor: *Patrick Walsh*

Printed in the United States of America
10 9 8 7 6 5 4 3 2 1

ISBN 0-13-248055-7

Prentice-Hall International (UK) Limited, *London*
Prentice-Hall of Australia Pty. Limited, *Sydney*
Prentice-Hall Canada Inc., *Toronto*
Prentice-Hall Hispanoamericana, S.A., *Mexico*
Prentice-Hall of India Private Limited, *New Delhi*
Prentice-Hall of Japan, Inc., *Tokyo*
Simon & Schuster Asia Pte. Ltd., *Singapore*
Editora Prentice-Hall do Brasil, Ltda., *Rio de Janeiro*

Contents

Chapter 4 Troubleshooting Branching Paths 79

Chapter 5 In-Circuit Checks on Basic Electronic Parts 105

Chapter 6 Basic Power Supplies 148

Chapter 7 Checking Signal Circuits 186

Chapter 8 Troubleshooting Basic Transistor Circuits 214

Chapter 9 Troubleshooting Basic Amplifier Stages 250

Chapter 10 Regulated Power Supplies 280

Chapter 11 Power Amplifiers and High-Voltage Circuits 296

Chapter 12 High-Frequency, Wideband, and Related Circuits 326

Preface

In many years of teaching, writing, and troubleshooting, I have yet to come across a textbook devoted to troubleshooting a circuit down to the defective part. Most books will show you how the circuit works. They might show you how to find the defective circuit, but not the defective part in it. They assume that if you're smart enough to learn how it works, you're smart enough to fix it. Nothing could be further from the truth. Troubleshooting is a special field of work. It has its own rules. These rules are derived from everyday practice, and they must be taught from the point of view of everyday practice.

In tech school, you learn theory and design. They give you the right-value voltages and it is your job to find the right-value parts. This book does the opposite. From wrong-value voltages, you find wrong-value parts. Troubleshooting is different. This book is different. It has ideas that are used by every experienced technician. It shows how troubleshooters work their way into the circuits by starting from the outside. Once inside the circuits, they often make simple voltmeter checks. The mere fact that the voltages are too high or too low is usually enough for locating the bad part. Hence there is usually no need to get bogged down in a morass of mathematical calculations.

Don Matsuda

1

The Game Plan: The Troubleshooter's Approach

EXPERIENCE REQUIRED

If you want to learn to troubleshoot electronic circuits, this chapter will help you get started. You should at least have seen or heard about the following widely known electronic applications, devices, and words:

1. Radios, stereo systems, television sets, transmitters, satellite communications, radar, computers, industrial controls, medical equipment, and robotics, to mention a few.
2. Input and output devices such as microphones, speakers, phono pickups, cassette recorders, video cameras, video display tubes, switches, keyboards, and digital displays.
3. Wires, connections, circuits, lines, ac cords, plugs, switches, fuses, electric meters, transistors, and integrated-circuit chips (ICs).
4. Voltage and power as types of electrical measurements.
5. Dc as a type of electrical current that is different from ac.
6. Signals of various kinds, including AM, FM, TV, VHF, UHF, telephone, telegraph, and code.

It would be helpful if you understand games, riddles, percentages, odds, detective stories, or similar activities. A knowledge of basic electricity and electronics would be very helpful but is not absolutely necessary at this point.

1-1 STARTING FROM THE OUTSIDE AND WORKING YOUR WAY IN

All electronic systems, such as radios, transmitters, VCRs, and computers, are connected to input and output devices. We are all familiar with microphones, speakers, video cameras, display tubes, keyboards to printers, pushbuttons, and indicator lights. One of the first steps in troubleshooting is to check these devices. When something goes wrong with a system, these outer devices may exhibit symptoms that point to the circuit that is causing the trouble.

A second source of clues is the system power supply. Practically all electronic systems require electrical power from batteries or ac outlets that feed their power-supply sections. One of the first things you will learn to check when a system fails completely is the power supply. You will check it for partial failures, too.

Another source of clues is the set of knobs and pushbuttons on the system control panels. With experience you will be able to manipulate these controls to find out what might be wrong within the system.

> *Note:* One advantage of checking the clues from these outer sources is that you may not have to tear the system apart to fix it. It is easier to start from the outside.

To gain entry to the system circuits, you must know how it is divided up into sections. Figure 1-1 shows how the system might be divided. In this case we have only three sections. There may be many more. Each of these has a particular job or function. Each modifies the output in a specific way.

Inside each section you'll find electronic circuits. Often a section consists of a single board on which the circuits are mounted. This will make it easier to identify and locate a section.

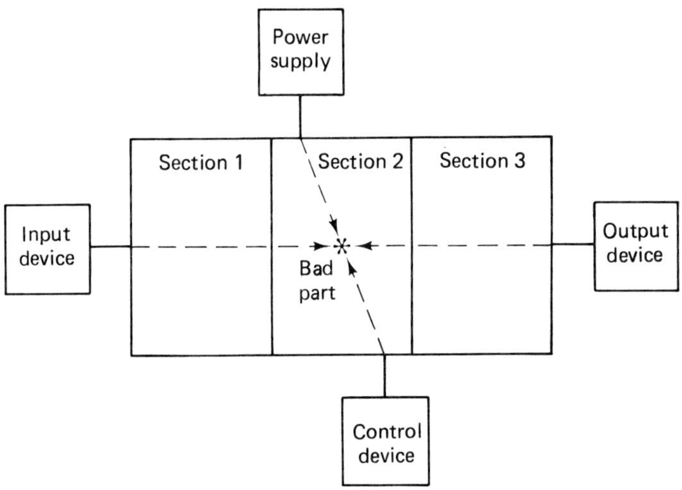

Figure 1-1 Dividing the electronic system into sections makes it easier to find the bad part. Note the four pathways to the target.

Inside the circuits you will find the individual parts. When any of these parts go bad, the entire system may fail. Figure 1-1 shows four possible pathways that lead you to the bad part. They start from an input device, an output device, a power source, and a control device.

Nearly all circuits are on the signal paths or bypaths between the input and output devices. Nearly all circuits have direct or indirect electrical lines from the power supply. Many circuits are connected to control devices.

1-1-1 The Game Plan

One of the main purposes of this chapter is to show you how to develop a game plan based on the way the system is set up. It will lead you down the pathway to the bad part. Along the way we preview some of the information found in the later chapters and sections. You can always come back here to see how they fit into the overall game plan.

To troubleshoot a problem, you will start from the outside and work your way in. This will be a step-by-step process. Each step will get you farther into the circuits. Each step will get you into the next step. This is your game plan. The major steps or phases of this plan will be as follows:

1. The outside: making the preliminary external checks, noting the symptoms, and deciding whether the trouble is in the equipment or in the user environment (including the user).
2. The gateways: checking the symptoms shown by the input, output and control devices, as well as the power supply; and making electrical checks on their lines and connections.
3. Working your way in: tracking down the bad electrical path from the gateways to the bad section or circuit with test instruments and common sense.
4. Nailing down the bad part: tracking down the circuit paths with test instruments to find and replace the bad part.
5. Cooking: operating the repaired equipment for 20 minutes or longer, sometimes for several days, depending on the type of fault.

We say that there are five major steps. More often than not, the first step or two can lead you directly to the defective circuit, sometimes right down to the bad part itself.

1-1-2 Divide and Conquer

At each step, the main idea is to divide your job into smaller segments. There are several advantages to this.

1. The smaller segments will be easier to work on because they have less things to contend with.
2. The smaller segments will give you more specific clues. This is where your knowledge, studies, and experience come in.

3. You can concentrate on one segment and not let the others bother you. If you pick the wrong segment, you can drop it and concentrate on another. Either way you will not have to follow up on all the imaginable possibilities.
4. Perhaps the most important advantage would be psychological. The dividing process fosters a "can do" attitude and a feeling of confidence. This is important for beginners.

1-2 PHASE 1: THE OUTSIDE

The worst thing a beginner or any technician can do is to start digging into the equipment without any preliminary investigation. Impatiently digging in like this can be a total waste of time. Unless something is burned out or visibly damaged, we have an iron-clad rule: Start from the outside.

1-2-1 The User Environment (The Preliminaries)

If possible, the preliminary checks should be made in the actual user environment. If this cannot be done, you will have to obtain as much secondhand information as you can over the phone or from the service ticket. You can also consult the field technician or whoever deals with the user.

Divide the user environment into:

1. The user
2. External wires, cables, and connections
3. Temperature, humidity, dust, and contamination
4. Interference (unwanted signals)
5. The controls (knobs, pushbuttons, and levers)
6. Input and output devices
7. The equipment itself

You must decide which of these seven possibilities to investigate first. With experience you will be able to look them over quickly and mentally eliminate most of them. Two major considerations will determine your choice: doing what is most likely to give you results, or doing what is easiest. Often, you can do both.

> TIP 1-1
>
> Never, never so much as lift a screwdriver without first going through the preliminary steps. Check the user environment if possible, and interrogate the user. Check all the external wires and connections, and check the controls, switches, back-panel fuses, and circuit breakers.

The detailed procedures for investigating most of the seven possibilities are given in the later chapters. Many are from Chapter 16, which deals with intermittent troubles. Here we preview a few things from the other chapters to illustrate the steps in the general game plan.

1-2-2 Interrogation

The first step in troubleshooting will involve the user. Remember, this person may not understand technical words. Use plain language. On the other hand, you must translate the user's language into your own. For example, when the user of a television set says "no picture," the actual condition can be any of the following: no raster (dark screen), no sync (broken or unstable picture), no vertical (horizontal line across the screen), no horizontal synchronization (many slanted lines), or no video (screen lights up with no picture). Each of these symptoms calls for a different line of action. Ask more questions to determine exactly what is meant by "no picture." Be tactful, and have confidence (in yourself and the user). Here are some important questions you can ask.

1. Can you turn the equipment on? The user may not know how to turn on and adjust the equipment.
2. How long does it take before it shows up?
   a. Parts and connections may go bad only after they warm up to a certain temperature. Once the equipment is on the bench, apply heat to the suspected parts to hasten the appearance of the symptoms. You cannot fix something that works okay. You will learn more about this in Chapter 16.
   b. When a fuse or circuit breaker blows out immediately, the power circuits are the prime suspects.
   c. Shorts in lower-power circuits blow the fuses and circuit breakers later.
3. When the trouble does occur, does it happen suddenly or gradually?
   a. Again, you can apply heat to the suspected parts to make the fault show up. You will also learn to monitor the electrical changes and learn about the symptoms that show up when a particular part or circuit fails.
   b. Sudden changes can occur when heat or vibrations break poor connections or cause unwanted connections to occur inside or outside a part. You will learn how to wiggle and tap the parts to locate the bad one.
   c. Sudden changes can also occur when part values reach a critical point that causes certain devices to switch on or off. You will learn to monitor voltages to see how far they are from the critical values.
4. What changes take place while it is acting badly?
   a. Each circuit changes the output in a certain way when a particular part in it changes value.
5. What time of the day does it show up, and under what circumstances?
   a. Power lines may have higher voltages at certain times of the day.
   b. Street traffic and spark-plug, as well as industrial machinery interference, may have peak periods.
   c. Nights and mornings may be cooler than afternoons.

d. Cranes at construction sites may change positions and deflect the signals at certain hours.

6. How long have you had this trouble—how many days, weeks?
   a. New industrial installations can cause interference.
   b. New high-rise construction can reflect signals and cause poor reception.
   c. Weather conditions, such as storm damage to cables and antennas, could cause troubles. Hot weather can cause trouble, too.

Here is another important reason why you should carefully question the user and obtain definite information. Sometimes you will spend hours on a problem of which the user may not even be aware. You will have trouble justifying your time. Before taking the job in, let the user know of any trouble you see or foresee. Note down clearly on the job ticket just exactly what the user wants you to work on.

One last reason why questioning is important. You may be wondering why we refer to Chapter 16 so much with regard to the user. The trouble may not even show up when the equipment is brought in for repairs. In these cases you must rely heavily on the user's comments.

1-2-3 What the Controls Tell Us

Besides the user, the control knobs and buttons can also tell us a few things. Not only can they tell us whether the equipment is really bad, they can also give us clues as to just where the trouble is internally.

Front- and back-panel controls, such as intensity or frequency controls, are connected to the circuits at strategic locations. Their actions give clues as to where the trouble is located. There are two types of controls. In Fig. 1-2(a) the control is an

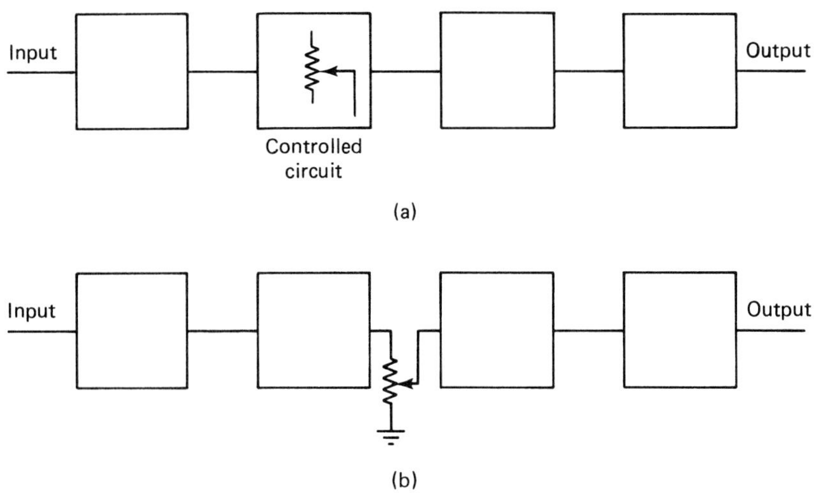

Figure 1-2 (a) Manipulating a circuit control may tell you whether or not the circuit is working. (b) A control in the signal path tells whether the trouble lies before it or after it.

integral part of the circuit. This type will give you clues by the changes that come about as you adjust it. A detailed knowledge of its role in the circuit will give you special clues. Also, the controlled circuit could lie in a signal path. This would give clues as to the location of the defect in the signal path.

The second type of control, shown in Fig. 1-2(b), is inserted in the path between circuits to modify the signal in some way, usually in strength. This control will also help you locate the origin of the defective signal.

In the following tips, "before" means toward the input device, and "after" means toward the output device.

> **TIP 1-2**
>
> If a control is in an extreme position to obtain a normal output, it may be compensating for a bad part.

> **TIP 1-3**
>
> If a signal symptom can be changed by adjusting a control that lies in a signal path, the trouble originates at the control or in the path before it. If the symptom cannot be changed, the trouble lies after it.
>
> If the signal cannot be changed, and it is excessive, the ground terminal of the control could be open. (In Chapter 2 we tell you about "grounds" and "opens.")

> **TIP 1-4**
>
> Dirty controls in audio systems generate scratchy sounds at the speaker when manipulated. In video systems "noise" often appears as streaks on the display tube as you manipulate the control. On an oscilloscope it appears as jumpy or jagged waveforms. Any such "noise" pick up indicates that everything between the control and the output or pickup point is probably okay.

TIP 1-5

If the output changes suddenly or erratically as you slowly turn a control, the control could be dirty or broken. Before you replace the control, spray some control cleaner into it and rapidly turn the control back and forth to clean it.

1-3 PHASE 2: THE GATEWAYS INTO THE CIRCUITS

All electronics is nothing more than the interfacing of input devices to output devices. A simple example is the audio amplifier that interfaces a microphone to a speaker. Through one path or another, all electronic circuit parts are directly or indirectly connected to input and output devices.

A more complex example of an electronic system with all types of output and input devices is the video cassette recorder (see Fig. 1-3). It has a tape head, a cassette loader, an antenna connector, an audio input, a cassette loader, motor controls, various kinds of sensing devices, and more. (The dew sensor in the illustration detects

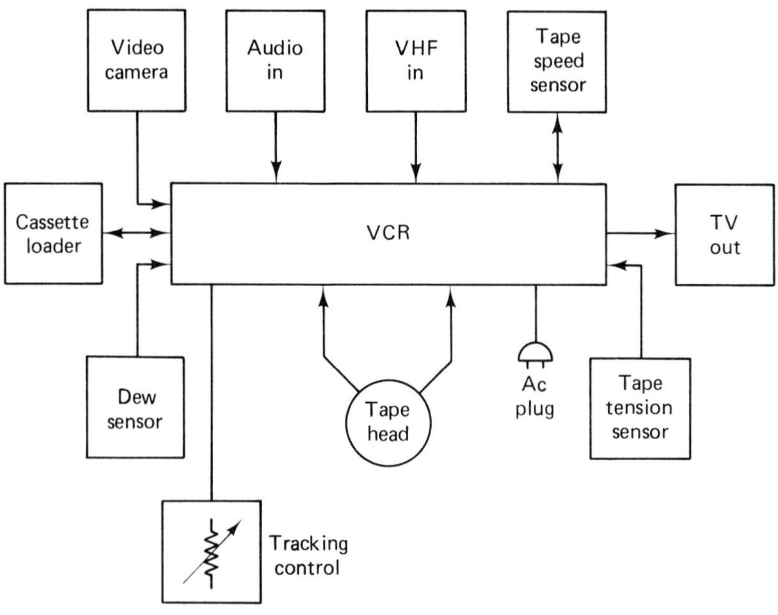

Figure 1-3 Electronic system with many input and output devices. Each can lead you to the defective circuit.

moisture on the tape.) Each one of the input and output devices in the illustration is a gateway into the circuits.

The power supply, as we noted before, is another gateway into the circuits. See the ac plug in the VCR diagram. We will learn in Chapter 6 that the power-distribution system reaches into just about every corner of the system.

The controls also constitute gateways into the circuits. Note the tracking control in the VCR system diagram.

1-3-1 The Input and Output Devices

There are two types of keys that open up the output gateways: clues that are picked up by your eyes, ears, nose, fingertips (symptoms), and clues that are picked up by your test instruments (electrical measurements). The symptoms will tell you which paths and circuits to check. The electrical clues at the output can be traced all the way back to the defective part.

Figure 1-4 is a simplified illustration of a defective color-monitor symptom and the dc signal path leading to the bad part. In this case the symptom is "no blue in the color display." We find a bad voltage on the blue output line. This line leads you to the blue color circuit. It leads you right to the troubled circuit and part. Note the break in the part that lowers the voltage below it. The up and down arrows indicate that the voltages have risen or fallen as a result of the break. In Chapter 7 you will learn why the arrow to the right of the amplifier points up instead of down.

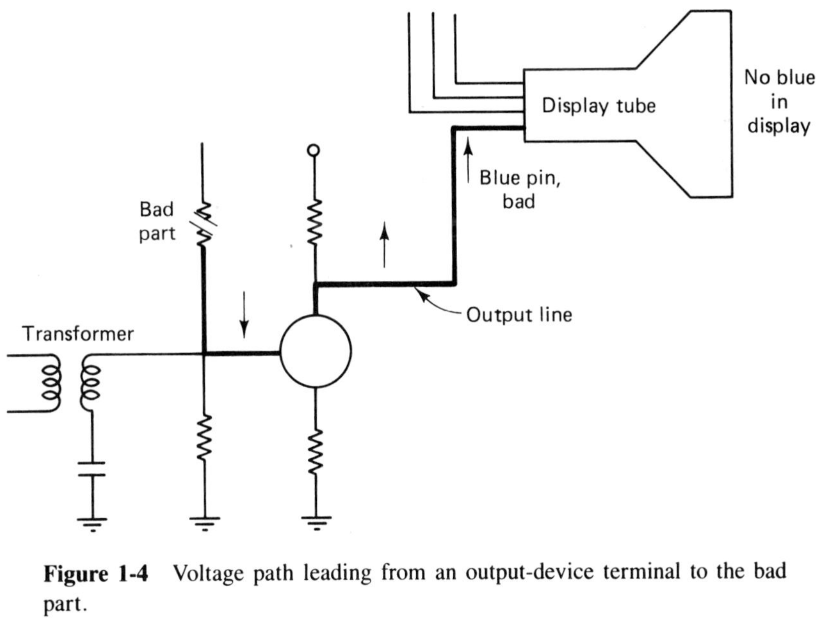

Figure 1-4 Voltage path leading from an output-device terminal to the bad part.

1-3-2 Symptoms

When electronic systems go bad, they show symptoms that we can see, hear, or feel. As noted before, the output devices provide us with the symptoms (except for burned, cold, hot, smelly, or visibly damaged parts). It pays to know what your output device can reveal.

Take, for instance, a radio speaker. It can reveal all the functions of a radio. Each function is associated with a certain section or combination of sections in the radio. For instance, a speaker should output only one station at a time. If it does not, something is wrong with the tuning section. Or perhaps the speaker volume cannot be turned up or down. If not, the trouble is in the volume control section. This may seem too simple, but you can upgrade it by giving it a fancy name: *symptom analysis*. Seriously, it deserves a special name.

Here are some more examples. If the disk drive in a computer system is not working right, the disk-driver circuit could be bad. If a VCR does not record, check the record circuits. If the display tube is dark, check the intensity (brightness) circuit, or other circuits that are responsible for brightness. Just these few examples show you the importance of symptoms. Of course, with more experience, you will be recognizing the more subtle ones.

1-3-3 Output Device Measurement

You will learn how to make electrical measurements on the output devices (you will learn about voltages). The speaker, for instance, requires zero dc voltage and low signal-voltage levels. The color display tube, on the other hand, has many different inputs for a wide range of signal and dc operating voltages. If you are going to do a lot of work on monitors and TV sets, it will pay for you to memorize the approximate voltages for every pin and connector on the more common display tubes. You might even memorize the pin numbers. As we noted before, defective voltages on these pins often lead directly to the defective part.

TIP 1-6

Learn the names and abbreviations of all the major input and output devices. Learn the names of all their terminals and their typical input and output voltages. Get to know the symptoms they reveal.

1-4 PHASE 3: WORKING YOUR WAY IN

Two types of paths are connected to the terminals of the output device: signal and dc voltage paths. Dc voltages are a type of voltage that is required by the system to process the signals. You will learn about them later. Both dc and signal paths get you into the circuits. You will pick up a lot of useful information about these paths and how to track them down in Chapters 7 and 9.

Not all dc voltage problems are easily detectable at the output, but visual output levels usually are detectable. Color monitor outputs, for example, will be excessively blue if the voltage on a certain pin is too high. Another example is the dc voltage on a speaker. A speaker should not have a steady dc voltage on it. It can be ''seen'' by turning the amplifier on and off. The cone will be pushed in or out. This may be more easily detectable than listening tests for some technicians. The sign of an experienced technician, however, is the ability to discern subtle differences in the symptoms just by looking and listening.

1-4-1 Working Through the Signal Paths

Figure 1-5 shows a typical system with several input and output devices. It shows that the output devices can be divided into the functions they perform. It also shows the sections responsible for each function. Each section is connected directly or indirectly to the output by means of signal paths. Suppose that function 1 of output device

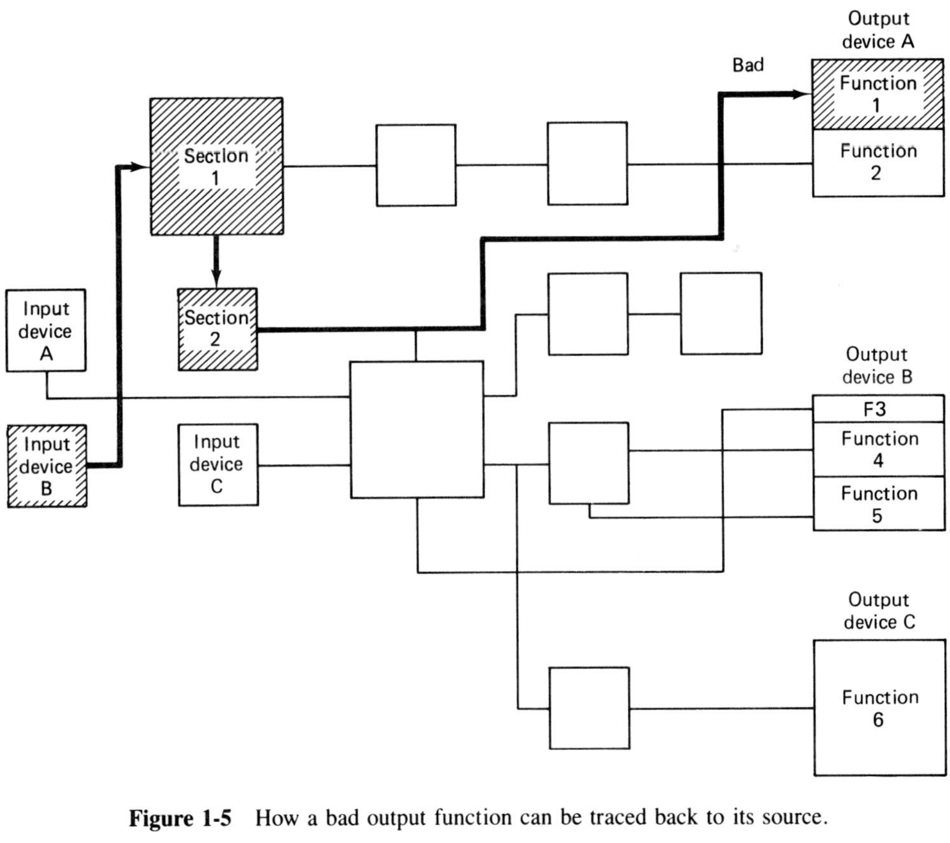

Figure 1-5 How a bad output function can be traced back to its source.

A is defective. You can see right away that input device *B*, and sections 1 and 2 are suspect.

You will learn to signal-trace basic amplifier stages in Chapter 9 when we study amplifiers. In general, it is simply a matter of following the signal paths from input to output, or from output to input, while taking signal measurements along the way. It is easy to understand, and beginners pick it up quickly. It is much harder to find the bad part.

1-4-2 Working Through the DC Paths

You will learn that to work properly, many output devices require dc operating and supply voltages. These dc voltages can be traced back to the internal circuits. In Chapter 9 you will work on dc signal paths that begin and end with certain types of devices. Back in Fig. 1-4, the path ended at a device called a transformer. You did not have to trace back any farther. You narrowed down the search area. This is good to know.

1-5 PHASE 4: NAILING DOWN THE BAD PART

Once you have found a circuit with bad voltages, the fundamental rules for locating bad parts in Chapter 2 will help you locate the bad part. These rules are important. We noted the arrow pointing down in Fig. 1-4. This indicates a voltage that is too low. The rules show what defective voltages can point to.

Chapters 2 and 4 provide general rules and tips for finding bad parts, and Chapter 5 gives you tips for finding specific types of defective parts.

1-6 ACTIVE TROUBLESHOOTING

Troubleshooting involves more than just passively reading test instruments. It also involves actively changing the circuits and observing the results. You will learn to connect trial parts into suspected circuits, disconnect leads, connect wires across part terminals, and introduce substitute voltages to study the results.

The feedback control circuit taken up in later chapters is an example that responds very well to active troubleshooting. Figure 1-6 shows a feedback control circuit. If something goes wrong in either the main system or the feedback control circuits, it may be impossible to locate the trouble simply by checking the voltages. If, for instance, the main system sends a bad voltage to the feedback circuit, this bad voltage will be fed back to the input of the main circuit. You'll find bad voltages all around the "loop." You'll have a vicious circle with no beginning or end.

By actively feeding a substitute dc voltage into the main circuit input, you can determine whether or not the trouble is in the main circuit. If you obtain a good output with a substitute control voltage at its input, the main circuit is okay, and the feedback circuit is bad. Otherwise, the opposite could be true.

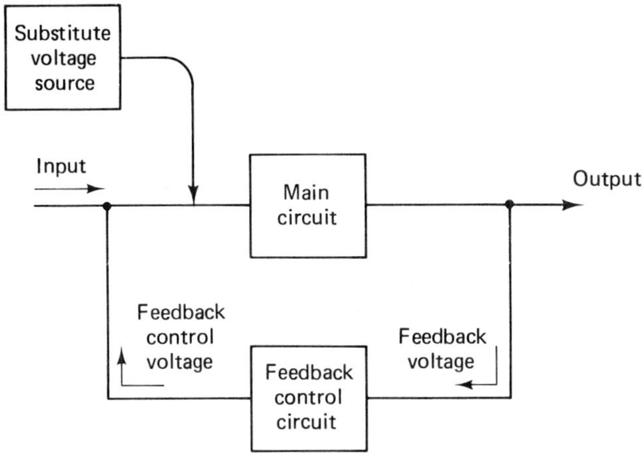

Figure 1-6 Introducing a substitute voltage to check a feedback system.

1-7 FIGURING THINGS OUT

Now that we have learned about the game plan, we can start thinking about the general ideas and approaches that will help us execute it. Experienced technicians will tell you that the most important test instrument is the brain. Some even disdain the use of test instruments other than the multimeter. That's okay if they specialize in a particular brand of equipment. Let them work on some new circuit that is on the bench for the first time. Then, you will catch them wiping the dust off their test equipment. In any case, it is true; your brain is the most important test instrument. Let's see how it can be used.

1-7-1 Trial and Error; Divide and Conquer

All thinking and all troubleshooting are basically trial-and-error processes. You make an educated guess and try something that might solve your problem. If it does not work, you try something else. This can be very inefficient, of course; there is more to thinking than just that.

It is also a comparison process. Every time you make a guess and try something out, you compare the result with what you think or remember is right. You do this almost unconsciously. This is the way the computer works, except that the kinds of things you remember and actually work with are more complex. You also have a lot of instinctive stuff buried in your brain system that no one really knows how to access. You are not a computer. You have sudden insights.

Dealing with complex problems is a very important part of thinking. This calls for a divide-and-conquer process. You divide your big problems into little ones, and attack them one at a time. You forget about all the other possibilities and concentrate on one.

This thinking process is also called the process of elimination. You follow up on one of your leads, and if your hunch is correct, you have eliminated all the other possibilities. If your hunch turns out to be wrong, at least you have eliminated it as a possibility.

The amazing thing about your brain is that you can do all this trial and error, all this dividing, comparing, and eliminating in your head. The trick is to keep your guesses to a minimum. The rules, tips and procedures in later chapters will speed up your thinking. Sometimes, you'll have to go over your own notes. Remember, it is a good idea to record your test instrument readings.

The main thing that keeps you on track is a procedure, a game plan. Once you have started out on one lead, it is usually not a good idea to get sidetracked on another, no matter how attractive it is. Following a set of procedures is very important; but don't get us wrong. Troubleshooting is not a cut-and-dried process. We are human beings, not computers. We can't always follow rigid procedures. By nature, we are constantly trying out new things and new hunches.

In summary, although troubleshooting is a trial-and-error process that may tend to get you off target, it is greatly facilitated and kept on target by a divide-and-conquer process of elimination. This procedure allows you to work your way in from the outside through the signal, power, and control lines.

1-7-2 Questioning and Thinking Things Over

Mental trial and error is done by asking yourself, "What would happen to the voltage at this point if part *A* is open?" then answering it. Do not plunge into the first hot lead you come across without stopping to think and question it. Every now and then you have to stop, step back, and ask questions. What have you done up to now? What have you not done? What other chain of actions can cause this voltage to go off? What other actions can this bad voltage cause? Did we overlook the power-supply voltages? Is device A plugged in? Is this really terminal *C*? In other words, you are like a computer that is polling all the possibilities that you, the programmer, can think of.

1-7-3 Playing the Percentages

All mental work makes use of two kinds of truths. The first kind has to do with laws or tips that are 100% true (on paper). For instance, take Ohm's law, which you'll study in the next chapter. It says that if you measure a higher-than-normal voltage on a good part, the chances are 100% (on paper) that more current is flowing through it. If it turns out not to be true, you know definitely that your measurements or something else was wrong, not the law. It is nice to have definite paper truths that you can rely on. Try to use them along with the basic tools in your hands. Paper truths enable you to think logically. That is, they enable you to say to yourself, "If this and this is true, then that must be true."

The second kind deals with tips or hunches that are less than 100% likely to be true. You take your chances. You try the most likely path. This is called "playing the percentages," or "playing the odds." Theoreticians say that you are using the laws of probability. They enable you to say, "If this and that are very likely to be true, then maybe if???"

Troubleshooters pay special attention to parts that are most likely to go bad. We cover the failure rates of various parts in Chapter 4. Generally speaking, except for parts in dangerous applications, the parts and sections associated with the greatest voltage, heat, and mechanical stress tend to have the highest failure rates. You will learn about these in the chapters on power supplies, power output amplifiers, and high-voltage circuits. Chapter 5 has a section on failure rates.

Connections cause a lot of trouble. They have high failure rates. You may see troubleshooters tapping and wiggling boards, parts, and connectors, looking for bad connections. The reason for this is obvious. If they can get things working by wiggling, they have found the bad connection. Generally speaking, there are at least twice as many connections as there are parts. In modern circuits, many parts have large numbers of connections that can go bad. The odds are pretty good for a bad connection to occur.

Bad connections are found not only by wiggling and tapping; they are also detected by the usual methods for finding bad parts. In Chapter 5 we show how to test connections with meters.

For the most part, for the sake of simplicity, we omit connections in this book. This does not mean that connections are not important. They are extremely important. They must always be considered.

The Divide-in-Half Procedure. Certain procedures are more likely to bring quicker results than others. For example, you can divide a big circuit in half, and either half can be tested. The bad half is then divided and tested. This dividing and testing into smaller and smaller halves can be repeated until only the bad part is left. A bad part in a circuit with 36 parts can be located with only four measurements. The odds are very high that this divide-in-half procedure is faster than testing the parts one by one.

TIP 1-7

Whenever possible, play the percentages; go after the parts with the highest failure rates first. Look for bad connections. Do the procedure that is most likely to be fastest.

1-7-4 Doing the Easy Things First

When the odds seem pretty equal—and sometimes when they're not—it is best to do the easiest thing first. Suppose that one part is 95% easier to work on than another part that is 90% more likely to be bad. It is better to work on the easier one first. At

> TIP 1-8
>
> Whenever possible, do the easy things first.

least you will quickly eliminate a possibility and feel better about taking on the more difficult and time-consuming alternative.

1-8 KEEPING NOTES

Keeping notes is good practice. Dealing with new circuits has always been a part of electronics. Keep a record of all the bad parts, symptoms, voltages, and resistances in the new circuits you work on. This will not only serve as an aid in the future, this will also help you to understand the troubles better.

Also record the model number and part. This will help you identify bad factory runs. Take note of the factory changes. Note the reasons why these changes were made.

With large-scale integrated-circuit chips (Chapters 14 and 15), it is even more important that you take notes. Record the symptoms, and record all the pin voltages. This is especially important, because it is not always possible to obtain good data from the manufacturer. Computer technicians in particular rely heavily on notes.

Many technicians record voltages right on the schematic. Some manufacturers' service centers will give you photocopies of their schematics. Study the notes that the technicians have scribbled on them.

1-9 HOW TO KEEP UP

To keep up with new developments, you have to read the literature. The odd thing about working with new devices and circuits is that you have to review the old. You have to recall long-forgotten subjects. There is no end to reading and learning about new topics and old.

1-9-1 What to Look for in the Literature

When troubleshooting a new circuit, we have to ask: What is it supposed to do? What are the symptoms when it goes bad? What kind of signal voltages does it input and output? What are its dc operating voltages? The answers to these questions are in the literature. Reading it can be a formidable task.

In most cases the language may be way over your head. In other cases, the new developments may come from people having their own language, their own terminology, even their own company jargon. The best you can do is to read the books, periodicals, and factory training manuals that are geared to your level of understanding. Look for interesting writing. Look for the things that are most important to you as a troubleshooter. You learn best what is most interesting and important to you at the given moment.

> TIP 1-9
>
> When looking up information about a circuit or a device, find out what it is supposed to do. Look for the possible trouble symptoms. Also look for typical input and output signal voltages, as well as dc operating voltages. Note the key words and abbreviations and the voltages associated with them.

When we talk about what a circuit does, we are talking about what voltages they produce for a particular output function. A synchronizing circuit, for instance, produces signal voltages that keep the display on a screen from breaking up. So the next time you see a broken-up display, you might trace the "sync" signal voltages back to the sync circuit.

One of the quickest ways to find out what a circuit or device does is to look at its name, or the name of the circuit or equipment in which it is used. A loudspeaker delivers sound. A display tube displays signals. An amplifier amplifies. An AGC circuit performs automatic gain control. Learning abbreviations and initials is important. Don't be intimidated by them. They will help you out.

Voltages have names, too. The names tell you where they come from and where they are headed. For example, the red screen voltage starts out from the red screen control resistor and ends up at the red screen pin of the color CRT. This may seem too simple, but even voltages with very strange names can be traced like this.

Illustrations. Another useful practice is to look through the illustrations. Sometimes, this is all you have to do. Most technical books are well illustrated. The illustrations are well organized and sequenced; they give you the key points of information. They give you the names of the input and output devices. They give you the names of the circuits. These names may tell you what the circuits do. Look through all the illustrations in this book if you have not done so already.

Going Over an Article or Chapter More Than Once. Finally, if you have trouble reading some article or book, you may have to read the whole thing over again, maybe more than two times. Chances are that when you read it the second time you will not even remember the trouble you had with it the first time. A lot of words have no meaning until you find out where the writer is coming from.

Oddly enough, you can only find out where the writers are coming from if you know where they are taking you. They have been there—at the end of the road. That is where they are coming from. You're just at the beginning. This is the greatest source of miscommunication between teacher and learner. It might be helpful to take a peek at the last paragraph first. You may have to read the whole chapter or article before you know where the writer is coming from. In other words, read the entire chapter quickly, then retrace it a second time. You may be surprised at how easy it is the second time.

Finally, don't expect to remember everything you read. You only remember what is useful or interesting to you at the moment. You must also understand that there are many levels of understanding. In fact, there is no end to what you can learn about a particular circuit. The best you can do is to see what's in it for you at your level of experience.

Beginners are confused by the atomic orbits and valence electrons they see at the beginning of a typical textbook (skipped over by most instructors). The authors seem to be coming from the fields of chemistry and physics. This is okay, for a so-called well-rounded education. The only problem is—we're trying to learn electronics in two years, not 10.

Actually, real circuit theory begins with Ohm's law, not with orbits. This theory, which underlies everything in the next chapter, comes from voltages and currents that are read with real meters in real circuits. Once the beginner learns to work in real circuits, things start to make sense. Your head stops spinning in orbits. The moral of all this is that you should get into actual troubleshooting as early as possible. You don't have to be a great theoretician to get started.

1-9-2 What to Look for in Mathematics

Mathematics has several purposes that are not always differentiated clearly. This confuses beginners. You should separate in your mind the following mathematical tasks. For one thing, mathematics tells you how much of something you can get for a given input. Second, mathematics helps you remember certain theoretical points, certain electronic actions. Third, mathematics often, but not always, provides equations that describe processes. Fourth, mathematics is used to prove the validity of an equation or statement.

You must clearly understand why mathematics has been introduced in the reading material. Too often the mathematical proof of an electronic action seems to be substituted for its explanation. They seem to be saying, "Here's the truth, don't question it." There is some validity to this approach, however. Knowing that something is valid will give you enough confidence to work with it. Later, you might even begin to see its real significance. Most important, it allows you to draw conclusions that you can work with. It enables you to think logically, even if you don't really know what you're doing.

Quite often, you'll wade through a long textbook presentation with pages of equations, only to find hidden in it the phrase "It can be shown that. . . . " Why not skip all that math and just say "It can be shown that the final equation is valid," and let it go at that? Actually, it is customary for academicians to establish the validity of a useful paper truth in this way. It's up to us to make use of it.

A math equation is like an electronic circuit; it has inputs and outputs. For the troubleshooter, the main thing is to know whether the output goes up or down when the inputs are changed. We do not have to "prove" it on paper. That is the mathematician's job. The "proof of the pudding" is in actual practice. The following tip may help you get around the problem.

> **TIP 1-10**
>
> When reading a heavily mathematical presentation, look at the final equation first. Read the beginning few paragraphs to find out what the input letters and symbols (variables) on the right-hand side stand for. Then see how they affect the output letter or symbol on the left-hand side. That is, see whether the output increases or decreases when each input is increased or decreased. (Plug small numbers into the equation.) Find out what these changes do in the actual circuit.

1-10 THE TROUBLESHOOTER

Troubleshooters are oriented toward finding a bad part, not the good part. For instance, you'll find out in Chapter 5 that a certain instrument may not be the best in the world for finding a *good* part in a circuit; but if this instrument shows a *bad* reading, the part is definitely *bad*. If you are looking for a good part, you would prefer another instrument. This does not mean, however, that it doesn't pay to check whether a part is good. If it's good, we can eliminate it as a "suspect," as detectives say. Of course, we still have to keep looking for the "culprit."

This point of view leads troubleshooters to look at circuit actions differently. They are always looking at circuits that are *not* performing normally. Throughout the later chapters you'll learn about circuits that are not working. In fact, if you are accustomed to reading textbooks that analyze working circuits, you may have difficulties. Try to remember that you are looking at a bad circuit action, not a good one.

Troubleshooters are manually oriented. Their thinking is often influenced by their cutters ("dykes" for diagonals), long-nose pliers, and soldering iron. These tools are always conveniently on hand. You always end up using them. Disconnecting one lead of a part to check a circuit, as you'll learn to do in Chapter 4, is a good example of the hands-on orientation.

The multimeter is always on hand, and regardless of what other test instruments are used, the multimeter is always involved in either the preliminary or the final checks. This influences your approach. Because of the limitations of the multimeter compared with other instruments, you rely heavily on symptoms and past experience.

Troubleshooters are efficiency oriented; they want to save time. Unless you are a specialist, you may prefer not to use an elaborate test instrument that's on the shelf. Its specialized leads and probes may or may not be stored (lost) in a special place. You don't want to lose time dusting things off (the instruction manual included) just to set them up for a single job. You develop ways to get around this problem.

Some excellent troubleshooters are very good in theory. You might see one of them in the library studying new circuits. Some otherwise good troubleshooters are not outstanding theoreticians. They learn just enough about the system to enable them to work their way into the circuits. They check the new circuits by applying the same old fundamentals. After all, these new circuits have dc voltages that are established in the same old ways. The parts and voltages go bad in the same old ways. This is why troubleshooters have a confident "can do" attitude.

Despite these differences, troubleshooters do have some traits in common. As noted above, they have confidence. They also have patience, but they know when it is time to try something different. They are not averse to doing the easy things first.

They don't go off the deep end when things don't go right. If you see them banging on a chassis, it is not out of anger and frustration. In Chapter 16 you learn the proper way to bang on a chassis.

WHAT YOU LEARNED TO DO

We went over the game plan in this chapter. We previewed some of the things you'll learn to do to execute the game plan. You learned that you start from the outside and work your way into the circuits. You learned that as a troubleshooter you will have to:

1. Make a preliminary investigation of the environment, the external connections, the user, and the equipment controls.
2. Check the input and output devices, the power supply, and controls that will lead you into the circuits.
3. Use a divide-and-conquer procedure.
4. *Think,* by using the process of elimination, playing the odds, asking questions, and doing the easy things first.
5. Keep records.
6. Read the literature to find out what a new device or circuit is supposed to do, and to find its input and output voltages as well as its circuit voltages.
7. Distinguish the different uses of mathematics in the literature and to utilize equations by looking at their inputs and outputs.
8. Gain confidence by doing the easy things first and by breaking big jobs down into little ones. You know that practical experience will clarify difficult concepts.

QUESTIONS

1. What are the five steps of the game plan?
2. What is the main tactic you will use in each step of the game plan?
3. Identify the seven parts into which the user environment is divided during a preliminary investigation.
4. What four questions should you ask the user if the trouble does not show up immediately?
5. What four things can the controls tell you?
6. What are the keys that can get you into the circuits?
7. By checking the output device, can a technician tell which section of a system is defective?
8. What two types of paths can lead you from the output device to the defective circuit?
9. Is it possible to go through the trial-and-error process in your head?
10. Cite four active troubleshooting procedures.
11. What four things should you look for when reading the literature?
12. Is the proof of the validity of the math equations associated with a given electronic action the same as an explanation of the action?

2

DC Voltage Checks for Locating Defective Parts

EXPERIENCE REQUIRED

In this chapter we show you how to locate defective parts in some very basic circuits. To understand this information, you should know how to:

1. Use a digital or analog multimeter. This requires some knowledge of electricity and metric numbers.
2. Insert batteries into radios or cassette players and connect jumper cables to start an automobile. In other words, you know there is a difference between positive and negative voltages. You know about conductors and insulators, and you are aware of shorts and bad connections.
3. Plug in ac adapters for dc equipment—you know there is a difference between ac and dc.
4. Identify the components of a simple lamp circuit: the plug, cord, switch, socket, and lamp.
5. Read simple circuit schematics.

A familiarity with the basic electrical equations would be very helpful but is not absolutely necessary. For the benefit of the readers with previous experience, Ohm's law, Kirchhoff's law, and voltage division concepts will be tied in with the tips and rules in this chapter.

2-1 WHY TROUBLESHOOTING IS DIFFERENT

2-1-1 Different Aims

Chapter 1 gave us an idea of how experienced troubleshooters worked their way into circuits. In this chapter we begin work on some of the more basic of these circuits. Our job is to locate the defective parts, conductors, or connections that cause such circuits to fail. To do this efficiently, troubleshooters have developed their own approaches.

Electronic designers and students usually work with correct values. They find correct-value parts for correct-value outputs. Our job is the exact opposite. We look for incorrect-value parts from incorrect-value outputs. The circuits have already been designed and the correct values have already been specified.

While good troubleshooters do record their voltage measurements, they keep their paperwork to a minimum. Rather than crunching numbers on paper or the calculator, they let the bad circuit and test instruments provide the answers. It is quicker that way. Let the designers and assemblers give us the correct voltages. When the voltages go bad, we take over.

2-1-2 Importance of DC Voltages

The voltmeter is our basic test instrument. Most electronic circuits are powered by dc voltages, the steady type that comes out of batteries and power supplies. We learn about fluctuating ac voltages later. As for now, for the sake of simplicity, we can say that practically all electronic parts are designed to maintain dc voltages at various points in the circuit. When a part goes bad, the dc voltages associated with it go bad.

> TIP 2-1
>
> An incorrect dc voltage reading indicates the presence of a defective component, connection, or conductor. A good dc voltage usually means that the circuit is good, so look for bad voltages first. (For information on apparently good dc voltages in defective circuits, see Tips 5-18, 5-21, and 5-26; and Section 7-3.)

Experienced troubleshooters are particularly good at tracking down bad voltages. It comes mostly from experience, but it will not hurt to review the fundamentals from the troubleshooter's point of view.

2-1-3 A Sense of Direction

In electronics there are two ways of looking at the direction of current flow. Many troubleshooters like to think of current as flowing "down" from the positive supply line to the negative supply line. (see Fig. 2-1). The negative supply line, which is often the metal chassis, is commonly called *ground.* Most supply lines have positive voltages, so we naturally think of current as flowing out of the supply point and "down" to ground. Schematics usually show the positive voltages on top and the ground (or the negative) voltages at the bottom. Current viewed as flowing from positive to negative is called *conventional current.*

A sense of direction helps us see how the power-supply voltage forces the current around the circuit and through the parts. This sense of direction helps us to see how defective parts can affect the flow.

If you have previous experience in electronics, you may be accustomed to *electron flow* that goes from negative to positive. This should present no problems because the direction of flow may be ignored in most cases. The exceptions will be noted later, when we take up semiconductors. Incidentally, even readers who are used to electron flow will find it easier to use conventional flow when studying power-supply rectifier circuits (Chapter 6). Theoretically, from our standpoint, we can usually ignore the direction of flow because for every electron that moves in a path, a positive charge can be thought of as moving in the opposite direction.

2-2 OPENS AND SHORTS

There are two fundamental faults that beginners can work on—opens and shorts. An open is a break in the circuit. A short is a bypassing of current around a part, or it is a complete loss of resistance in a part. There are also partial opens (or increased resistances) and partial shorts (or decreased resistances). These are harder to work on, but we will discover later that they have the same effect on voltages as do complete opens and shorts.

2-2-1 Voltage Is Electrical Pressure

Figure 2-1(a) shows a lamp connected to a 12-V supply line. The +12 V on the line exerts electrical pressure that forces electric current down through the lamp. Because the lamp resists the current, electrical pressure is built up on its high side. The pressure and current work against the resistance to energize the lamp. From the troubleshooting standpoint, we should note three things about this circuit.

1. The power supply exerts electrical pressure (voltage) on the high side of the lamp through a supply line. The low-side line is called the *ground return* or just plain *ground.* Look for both of these lines when reading a schematic.
2. In troubleshooting, the electrical pressure or voltage at a point is usually measured from ground [see Fig. 2-1(a)]. By this we mean that the common (black)

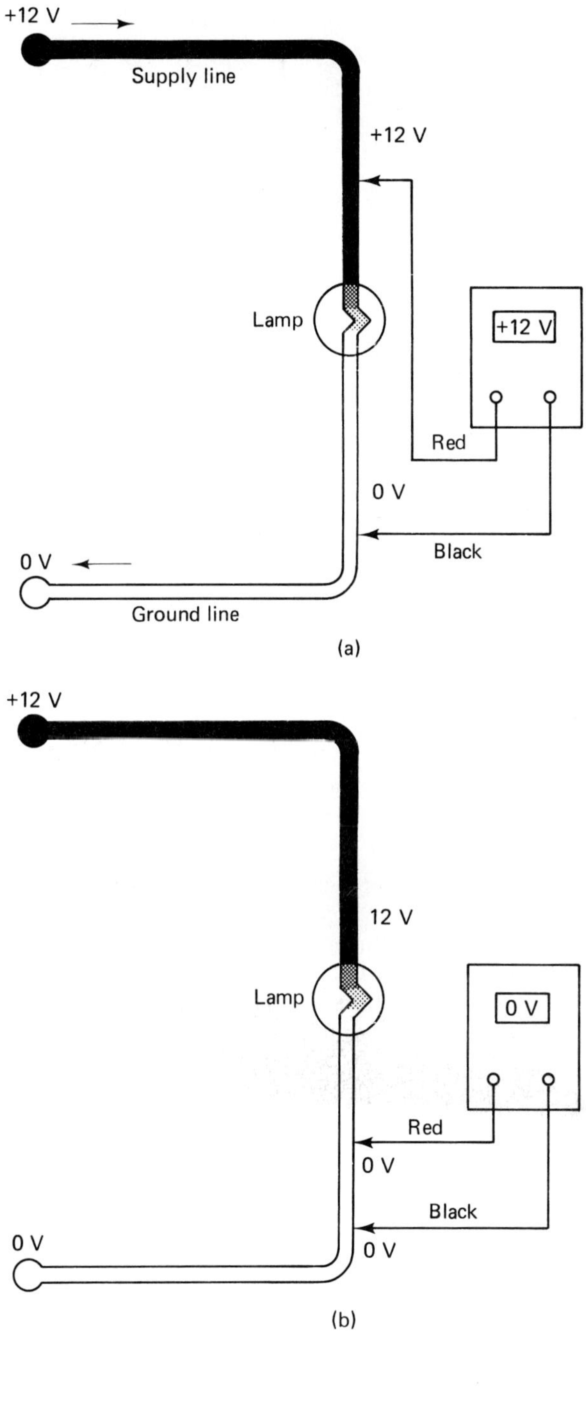

Figure 2-1 Measuring electrical pressure or voltage. (a) The dark line above the lamp has 12-V electrical pressure. (b) The bottom line has 0-V pressure when measured as shown.

lead of the voltmeter is connected to ground and the red probe is touched to the voltage point. The advantage of this is that the red probe can be moved quickly from point to point in the circuit while the black lead remains clipped to ground. Also, the schematics give voltages in this way.

3. We say that the low side has 0 V or zero pressure on it. This is because we are measuring all voltages with respect to ground. When measuring this point, the pressures on both leads of the meter will be the same. The meter will read 0 V, as shown in Fig. 2-1(b). (In some cases we say that the low side has a negative voltage on it. We take this up in Section 2-5.)

Note: If you are familiar with Ohm's law, you may want to review the fundamentals here. This law states that the current, I, through a part is directly proportional to the voltage, V, across it and indirectly (inversely) proportional to its resistance, R. That is, $I = V/R$ where I stands for current, V stands for voltage, and R stands for resistance.

If you want to know what happens to the voltage when the current or the resistance is varied, the equation is converted to $V = IR$. This equation is confirmed by the fact that the resistance of the lamp builds up pressure on its high side. Mathematically, it is easy to see by looking at $V = IR$ that V will increase if R increases.

Power is used up by the lamp when the current and pressure work against its resistance. This conforms with the power equations, $P = VI$, $P = V^2/R$, and $P = I^2$.

2-2-2 A Completely Open Part in a Simple Circuit

Figure 2-2(a) shows a simple circuit with only one part, a lamp. When a break occurs anywhere in this circuit we call it an *open,* and the current will cease to flow. With zero current, the lamp will not light.

As for the pressure or voltage, it falls to zero below the break. In other words, the lamp has no voltage on it—it does not light. Notice that the (0 V) in Fig. 2-2(a) is enclosed in parentheses to indicate a bad voltage.

Let's take the case where the lamp itself opens. In Fig. 2-2(b), the pressure falls to zero below the lamp and builds up on top of it. Thus the lamp has 12 V on top and 0 V below. However, these are exactly the same voltages that you have when the lamp is good. So how can you tell whether the lamp is open or not? Because it's not lit, of course.

The foregoing observation shows two important troubleshooting steps. First, note the symptom: The lamp is not lit. Second, read the voltages: 12 V and 0 V. With these voltages, the circuit must be okay, so the lamp should light, but it does not. Therefore, it must be bad. To simplify things, we omitted the lamp socket and its connections. They could be open.

Let's consider another case [see Fig. 2-2(c)]. What happens when an open occurs on the ground side of the lamp? In this case, the 12-V pressure drives the current down to the open point below the lamp. There, it is stopped in its track, and a maximum pressure of 12 V is built up at that point. This pressure backs up all the way

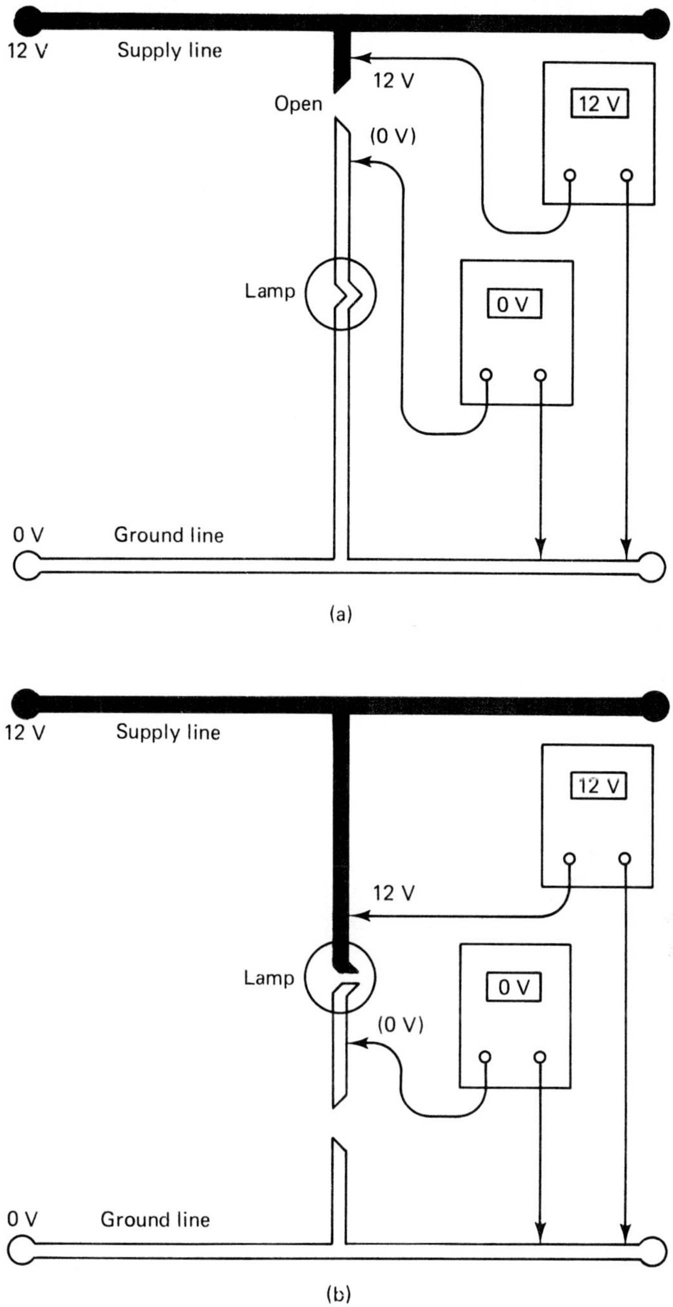

Figure 2-2 In all three cases, maximum pressure is exerted above the break, and minimum pressure is exerted below it. (Henceforth, defective voltages are shown in parentheses.)

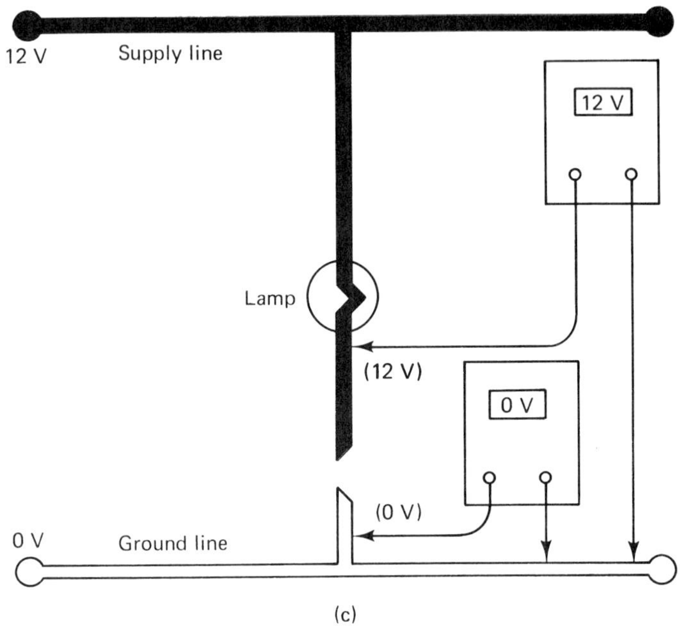

Figure 2-2 *Continued*

to the supply line. In other words, all points above the open have maximum voltage, the bottom of the lamp included.

So now, in Fig. 2-2(c), both the top and bottom of the lamp have 12-V pressure. Current can flow only from a higher-pressure point to a lower-pressure point. Therefore, you have no current through the lamp. It does not light up.

As for the voltage "on" the lamp, it is not 12 V; it is 0 V. The 12 V represents the pressure with respect to ground. The electrically effective pressure on the lamp itself is the difference between the pressure on top and the pressure below. Since both pressures are 12 V, the difference is 0 V. For more clarity, we should say that the lamp has 0 V "across" it. We can also say that the lamp "drops" 0 V across it, or that it has zero "voltage drop."

Summary: Open and Shorts. A look at the voltages in Fig. 2-2 leads to the following important points:

1. A complete open in a simple circuit has zero volts at all points "below" it (toward the ground side).
2. A complete open in a simple circuit has the maximum supply voltage at all points "above" it (toward the supply side). This happens even if the open is below a part (the lamp in this case).

2-2-3 A Completely Shorted Part in a Simple Circuit

In Fig. 2-3(a) we have a lamp that is "shorted" or "short circuited" by a piece of wire that is accidentally touching the two leads to the lamp. The lamp is bypassed or

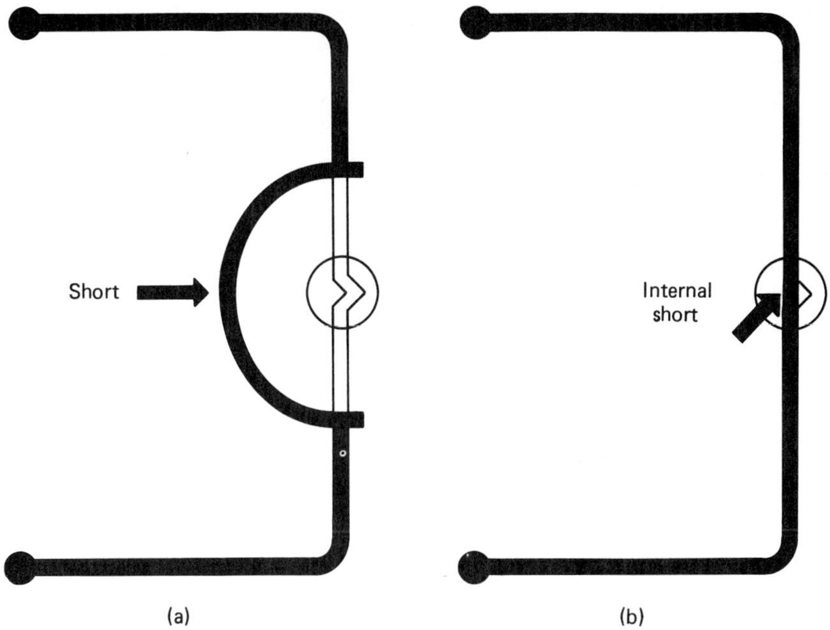

Figure 2-3 The effect of a short can be disastrous. (a) The lamp is shorted by a wire. Maximum pressure and current bypasses the lamp. (b) The lamp is shorted internally. Maximum pressure forces maximum current through the lamp and around the circuit.

"shunted." The wire or short has no resistance, and a tremendous amount of current rushes through it. The result can be disastrous. This is why we have protective devices (fuses and circuit breakers—Chapter 3) that cut off the current. Without these devices the wires or the power supply will burn out. If you have a strong power supply, a short like this can cause a blinding explosion.

Note that at the instant of the short, practically none of the current flows through the bypassed lamp. This means that there is practically no pressure or voltage across it. Actually, the circuit or power supply will burn out before we can read any voltages. *You cannot read the voltages when you have a complete short in a simple circuit.*

In Fig. 2-3(b) a short develops inside the lamp itself. Its internal leads have touched together (shorted), and the current rushes through. The result can be disastrous.

It should be noted here that those parts of the circuit that have the most resistance generate the most heat. If the external wires have more resistance than the shorted lamp, they will generate more heat. (If the lamp has 0 Ω, the power expended by it would be $P = I^2R$ or $P = I^2 \times 0 = 0$ W.)

Important: A short may generate excessive heat in a circuit and destroy it unless it is protected by a fuse or circuit breaker. (Fuses and-circuit breakers are taken up in Section 3-3.) Avoid shorting things out when you are working on a circuit. Be particularly careful about where and how you position your meter probes. Don't let them slip and short across two conductors.

Power. A normal lamp circuit generates normal amounts of power as heat and light. It has normal voltage and current. An open lamp has plenty of pressure on it, but it allows no current to flow—it expends no power. On the other hand, a shorted lamp allows excessive current to flow. This produces excessive heat in the power source and wires connected to the lamp. If, however, the power supply is weak and the pressure goes down, the heat or power will not be as great. Power requires both voltage and current ($P = I \times V$).

We have other types of energy besides heat and light. Speakers, for instance, develop sound energy. They are powered by audio amplifiers. We also have amplifiers that drive motors in computer equipment and videocassette recorders. They too require power. They require plenty of current and voltage.

2-2-4 A Complete Open or Short in a Series Circuit with Two Parts

Figure 2-4(a) shows two parts connected in series. The shading from dark to light represents the gradual loss of pressure (voltage) from top to bottom. The resistance of

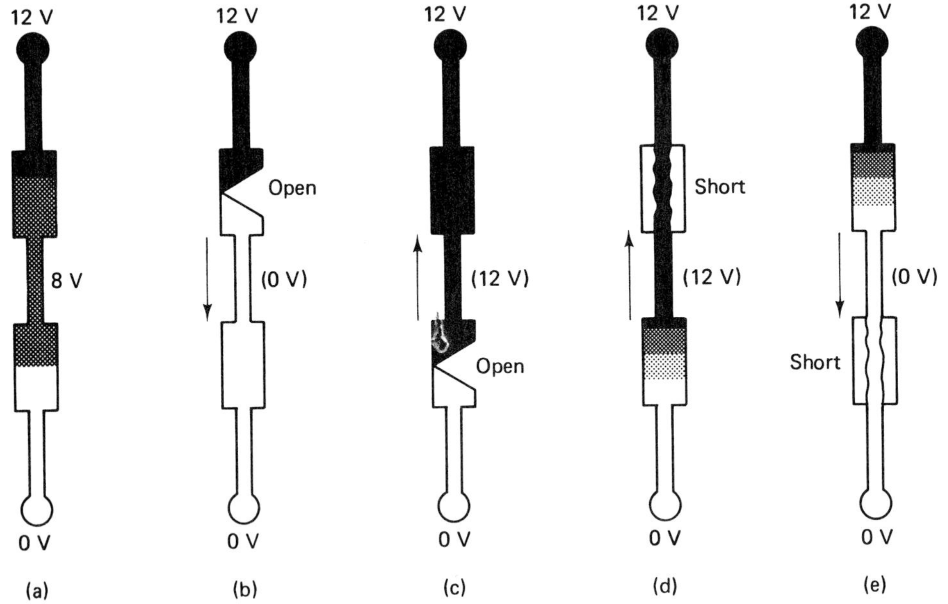

Figure 2-4 Effect of complete opens and shorts in a series circuit with two parts. (a) A typical voltage distribution. (b) The voltage below the open part drops to zero. (c) The voltage above the open part rises to the maximum. (d) The voltage below the short rises to the maximum. (e) The voltage above the short falls to zero.

the top part serves to reduce the pressure "below" it. The resistance of the bottom part serves to build up the pressure "above" it. The combined action produces 8 V at the junction of the two resistors. Let's say that this is the normal voltage as shown in a schematic.

> **Note:** If you are interested in seeing how the voltages in the circuit above conform with two very important equations, Kirchhoff's law and voltage division, we have the following:
>
> **Kirchhoff's Law.** According to Kirchhoff's law, the total voltage across a series circuit equals the sum of the voltages across the individual parts. In the case of two series parts, we have $V_T = V_1 + V_2$. Assuming that V_T, the supply voltage, does not change, if one of the voltages equals 0 V, the other must equal V_T.
>
> **Voltage Division.** According to the voltage division equation, the voltages across the parts in a series circuit divide in proportion to resistances of the parts. For two series parts we have $V_1/V_2 = R_1/R_2$. This converts mathematically to $V_1 = V_T(R_1/R_T)$ and $V_2 = V_T(R_2/R_T)$, where V_T and R_T are, respectively, equal to the total voltages and resistance across the two parts. Hence if a part has 0 Ω resistance, it will have 0 V across it.

Figure 2-4(b) shows that the top part is open. This cuts off the pressure to the bottom side, so we have (0 V) here. Note that we have the bad voltage in parentheses. We also have an arrow to show that the voltage has fallen. That's what we are primarily interested in—that it has fallen. Figure 2-4(c) shows that the bottom part is open, and you have full pressure on its top side.

Unfortunately, we still do not know which part is bad, because shorted parts can also give us the same readings. If the top part shorts, we have the same readings as in Fig. 2-4(c). The top part in Fig. 2-4(d) has no resistance, so it does not drop any voltage. We say that a complete short like this is a "dead short." The bottom part now has to take the full pressure (12 V)—the same as in Fig. 2-4(c). We cannot tell which part is bad just by looking at the voltages.

In Fig. 2-4(e) the bottom part shorts. It no longer has the resistance to build up the pressure, so it drops to 0 V. The top part has to drop the full voltage. The result is the same as in Fig. 2-4(b). We cannot tell which part is bad.

At least the voltages indicate that something is wrong. This is a good first step. Now we must find out which part is bad. There are several ways of tackling the problem.

2-2-5 Finding the Complete Open or Short in a Series Circuit with Two Parts

The quickest way to troubleshoot a series circuit with two parts may be to use an ohmmeter [see Fig. 2-5(a)]. A dead short should have 0 V and 0 Ω across it. Turn off

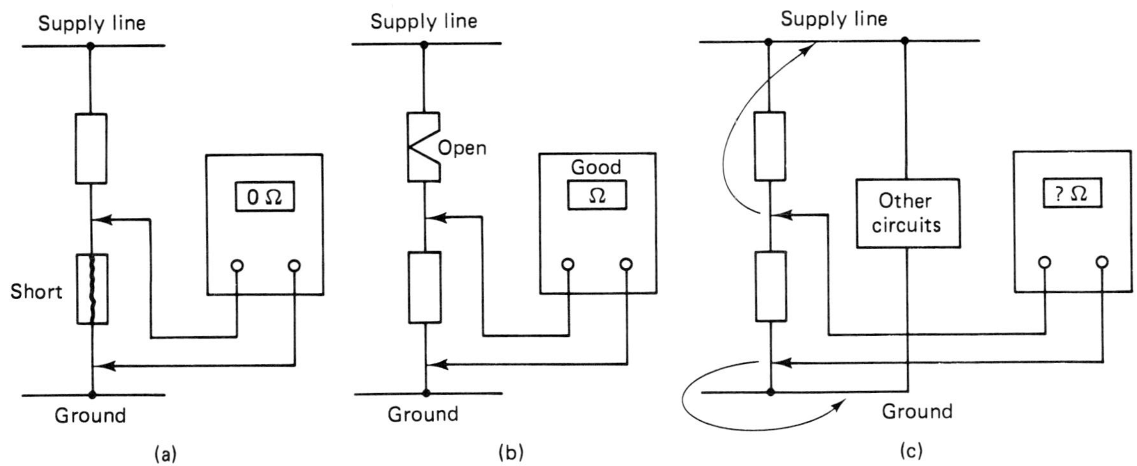

Figure 2-5 Checking for shorts and opens in a two-part series circuit with an ohmmeter. (a) Definitely shorted. (b) A strong indication that the upper part is open. (c) In most cases, resistance readings are inconclusive because other circuits affect the readings.

the power before using the ohmmeter. If you read 0 Ω across a part that normally should have more, it is definitely shorted.

If you read the normal out-of-circuit resistance across this part while it is still in the circuit as in Fig. 2-5(b), the other part could be open; but you cannot be sure of this. Figure 2-5(c) shows that other circuits can influence your readings.

If the ohmmeter shows a higher-than-normal out-of-circuit resistance across one of the parts as shown in Fig. 2-5(c), it is definitely bad. The influence of other circuit parts can only decrease the resistance reading. (We will learn more about this when we study the influence of other circuits in Chapter 3.)

If you read a lower-than-normal out-of-circuit resistance across a part while it is still in the circuit as in Fig. 2-5(c), you cannot tell whether the part is bad or not. Hence the ohmmeter cannot always give you an immediate answer. This makes things difficult. Furthermore, some defects show up only when power is applied to the circuit. With these reservations, we can only summarize the above as a possible first step in the following tip.

TIP 2-2

If one of the two parts of a series circuit drops 0 V when it should drop more, turn off the power and measure its in-circuit resistance. If it has 0 Ω and its out-of-circuit or rated resistance is not 0 Ω, it is shorted. If you read its rated resistance, it is probably good and the other part is probably open. If it has higher-than-rated resistance, it is definitely bad. If it has lower-than-rated resistance, you cannot tell immediately whether it is bad.

You can further confirm your suspicions by noting the temperature of the parts. The tip below applies to series circuits of two or more parts.

TIP 2-3

When a part shorts in a series circuit, the other parts may run warmer or hotter than normal. If a part opens, all parts will run colder than normal. Turn the power off before feeling the parts. Touch gently at first. Some parts normally run hot enough to sear your skin.

If the parts are warmer or hotter than normal, the part with an abnormally low voltage drop or resistance is shorted.

If the parts are colder than normal, a part with an abnormally high voltage drop is open.

For more confirmation, the quickest method may be to turn off the power and jump (connect) a good part across the part that dropped the full supply voltage. If normal circuit operation or voltage is restored, the jumped part is open.

If a substitute part is not available, use a resistor that has about twice the resistance of the jumped part, as in Fig. 2-6(a). If the voltage rises, the bottom part is not shorted and the top part is probably open. If the voltage remains stuck at 0 V, as in Fig. 2-6(b), the bottom part is shorted.

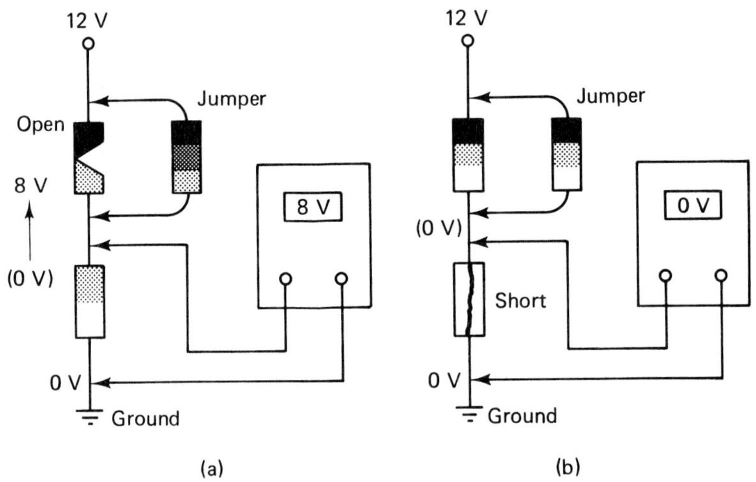

Figure 2-6 Jump-shorting a part to find an open. (a) If the voltage changes, the jumped part is open. (b) If the voltage does not change, the measured part is shorted.

If the top part is shorted or the bottom part is open, we would have the full supply voltage in between the two parts. In this case, jumping the bottom part will pull down the voltage if it is open; otherwise, the top part is shorted. In either case a jumped part that is open will change the voltage; otherwise, the unjumped part is shorted. (We assume only one part to be bad. See Tip 2-7.)

TIP 2-4

If one of the two parts in a defective series circuit drops the full supply voltage, turn off the power and jump it with a good part. Turn the power back on. If normal operation or voltages are restored, the jumped part is bad.

A resistor having about twice the resistance of the jumped part can also be used. If the voltage changes after the part is jumped, it is open. Otherwise, the unjumped part is shorted.

2-2-6 Partial Shorts and Opens in Series Circuits with Two Parts

Complete shorts and opens give you one of two readings: 0 V or full supply voltage. You do not have such definite readings with partial shorts and opens, but the voltages may be far enough from normal to make you suspicious. Since there are only two parts to contend with, the best method may be simply to remove one of them and substitute it with a good part. You know that at least one of the parts is definitely bad, and you will have to remove it anyway (see Fig. 2-7). Usually, this is a safe procedure. To be even safer, substitute the part that has the lower-than-normal voltage drop across it. If it is shorted, it may allow too much current to flow through the substitute part.

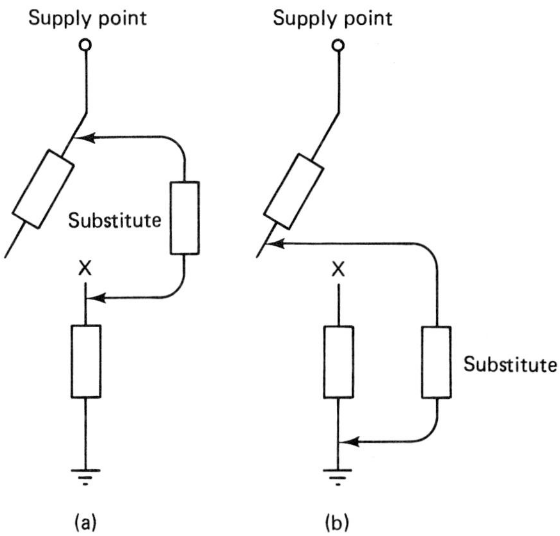

Figure 2-7 Partially defective parts in series circuits with only two parts are best found by substitution. You need only to disconnect one lead to take both parts out of the circuit.

Note that in Fig. 2-7(a) and (b), you do not have to disconnect both leads of a part to remove it electrically from a circuit. One lead disconnected is enough. This is one of the neat, timesaving tricks that you pick up in actual practice. Also note that only one disconnection was needed to substitute either part.

If a substitute part is not available, you can check the circuit part with an ohmmeter or other test instrument. Many test instruments can be used only after the part has been removed from the circuit. Even if they are capable of checking parts in-circuit, they work better out-of-circuit.

2-3 COMPLETE OPENS AND SHORTS IN SERIES CIRCUITS WITH THREE PARTS

In this section we learn the rules and tips for series circuits with three or more parts. We focus on three parts, but the same rules and tips can be applied to more than three. With just two parts in a series circuit, the single voltage between them could

TIP 2-5

If you have a partially defective voltage in a series circuit with two parts, disconnect the part with the lower-than-normal voltage drop and substitute it with a good one. If this restores the voltages or operation, fine. If not, the other part is bad.

You might disconnect the lead that is most convenient as long as you have the normal circuit when the substitute part is connected.

You can also check either part with a test instrument after disconnecting only one lead. (In Chapter 5 we learn to check parts with test instruments.)

not give us enough information. We had to remove or substitute a part to find out which was bad. Now, with more parts, we have more voltages to work with. The voltages alone can tell us which part is bad. Let's see how this can be done.

2-3-1 A Fundamental Rule

What you learn about these series paths can be applied to many electronic circuits with few changes. Figure 2-8 shows many series paths tied to a single supply line. This is similar to the way that many amplifier output circuits are connected. There is

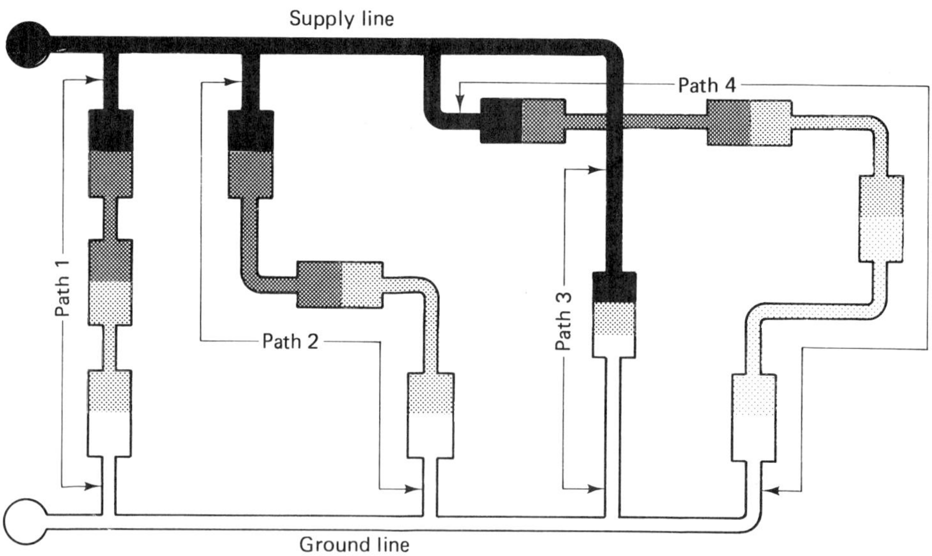

Figure 2-8 If the supply-line voltage remains unchanged in this system, a fault in one path will have no effect on the other paths.

one nice thing about these paths. The supply-line voltage usually does not change when something goes bad in one of the paths. Only the defective path will have the wrong voltages. The good paths have normal voltages. If the supply voltage changes, it usually changes drastically or blows a fuse. (This is taken care of in Chapter 6. The cases in which the supply-line voltage does change are discussed in Section 4-2-2.)

RULE 2-1

If you find a bad voltage in a series path between two good ones, the trouble lies in this path.

The rule above will help us find a bad part in a path between good supply and ground voltage points. To make use of this rule, we start by assuming only one part to be bad. Otherwise, things get too complicated.

TIP 2-6

Assume at first that only one part is defective. Take them on one at a time.

2-3-2 Complete Opens in Series Paths with Three Parts

Although we are focusing on a series path with only three parts, the rules and tips in this section can be applied to paths with more than three. At this point, you may be wondering what is the difference between a circuit and a path. A circuit includes the power source and all the parts in it. We will consider a path to be that part of a circuit that lies between two voltage points. The terms *circuit* and *path* are interchangeable, because any two voltage points can be considered as being a voltage source.

In Fig. 2-9(a) the top part is completely open, so it cuts off the pressure to everything below it. Hence all the voltages below it drop to zero. The arrows show that the voltages have fallen. Notice that they all point away from the open.

Similarly, if the middle part opens completely, the voltage below it drops to zero [see Fig. 2-9(b)]. The top-side pressure builds up to the maximum, so the voltage there rises to the supply voltage. Note that the arrows point away from the open. In Fig. 2-9(c) we see that if the bottom part opens completely, all the voltages above it will rise to equal the supply voltage. Note that in all three illustrations the arrows are pointing away from the open. We might say that an open tends to push the voltages away from it.

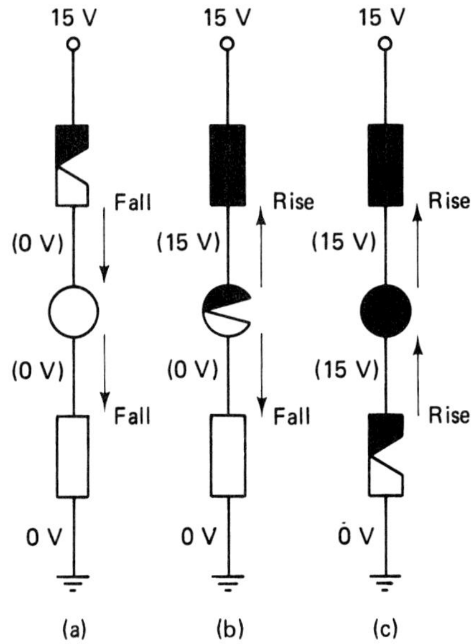

Figure 2-9 In all three cases, completely open parts have the maximum voltage above them and 0 V below them. Note that the arrows always point away from the open.

The following tip applies not only to completely open parts; it also applies to open connections and conductors.

> TIP 2-7
>
> Any point above a completely open part, connection, or conductor in a series circuit has the full supply voltage on it. Every point below the open has 0 V.

Remember that as a first step we assume only one part to be bad. Tip 2-7 is not valid in the unlikely event that all parts except one short out. Only the good part will then have the full supply voltage across it.

2-3-3 Complete Shorts in Series Paths with Three Parts

If any part shorts completely, it has no resistance. It cannot build up pressure. Its high-side voltage can be no higher than its low-side voltage. In other words, there is no voltage drop across a dead short. (Recall that $V = IR$, so that if $R = 0$, then $V = 0$.)

TIP 2-8

In a series path with three or more parts, if only one part drops zero volts, that part is shorted [see Fig. 2-10(a)].

If a part tied to the supply line shorts completely, its low-side voltage is pulled up to equal the supply voltage [see Fig. 2-10(b)].

If a part tied to ground shorts completely, its high-side voltage is pulled down to zero [see Fig. 2-10(c)].

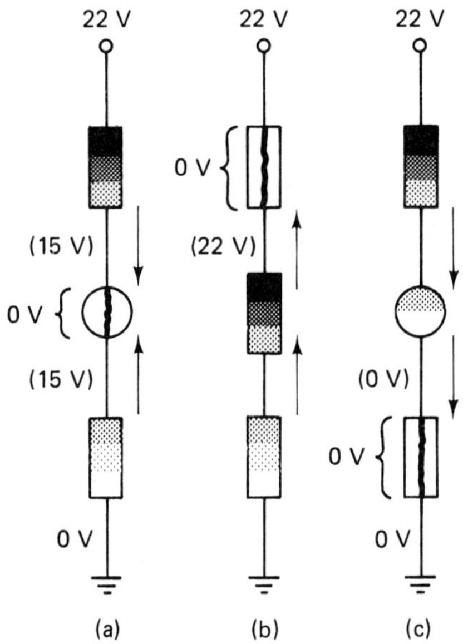

Figure 2-10 Completely shorted parts tend to pull down the voltages above them and pull up the voltages below. Note that they drop 0 V.

2-4 PARTIALLY DEFECTIVE PARTS IN SERIES CIRCUITS WITH THREE PARTS

Things were easy when we dealt with completely open and shorted parts in a series circuit. We simply looked for two kinds of voltages—full supply voltage or 0 V. At first sight, things seem to be different when the parts are only partially defective. It seems as though we have to work with many voltage levels, not just two. Fortunately, it turns out that we can still deal with only two voltage levels—*too high* or *too low.* Let us find out how partially defective parts can cause a voltage to be too high or too low.

Note: There are times when the difference between good and bad is blurred. Some circuits seemingly work fine with parts and voltages that are very far from normal. We discuss this at length in Section 9-7.

2-4-1 Two Basic Troubleshooting Rules

We now introduce two more fundamental rules for finding defective parts. We already encountered the first one. It said that if you find bad voltage points in a series path between two good voltage points, a bad part must lie somewhere in that path. Now we must find out exactly where that bad part is. As we know, the part must have either increased or decreased resistance, indicating that it is opened or shorted.

Let's say that the part is partially open or has increased resistance. We already know that a normal resistance drops a certain amount of voltage. Then it makes sense that if the resistance increases, the voltage below it will drop even more. (Recall that $V = IR$.)

We also know that a normal resistance builds up the voltage above it. Hence it makes sense that if the resistance increases, the voltage above it will rise ($V = IR$). This enables us to use the following rule:

> RULE 2-2
>
> An open or increased-resistance part tends to:
>
> **a.** Lower the voltages below it.
> **b.** Raise the voltages above it.

We say "tends to" because good supply and ground voltages often cannot be changed significantly. Other voltages can and will be changed by opens or increased resistances.

Now let us look at partial shorts or decreased resistances. If a part has decreased resistance, it will drop less pressure, so the voltage below it will rise. It will build up less pressure, so the voltage above it, toward the supply side, will fall. This leads to Rule 2-3.

Notice that a good part tends to hold the good voltages in place. When the part opens, it lets go of these voltages; it seems to "push" the voltages away from it. When a part shorts, it seems to "pull" in these voltages.

Although Rules 2-1, 2-2, and 2-3 will help you narrow down the location of a bad part in a series circuit, they will not pinpoint it immediately. The next section has two tips that will let you finish off the job.

RULE 2-3

A shorted or decreased-resistance part tends to:

a. Pull up the voltages below it.
b. Pull down the voltages above it.

2-4-2 Partial Opens (Increased Resistances) in Series Circuits

Let's see what happens in a series circuit when one of its parts opens partially. If the top part in Fig. 2-11(a) opens partially, its resistance increases. This increased resistance will decrease the pressure or voltage below it. This makes the voltage drop across the partially open part increase from 20 V to (40 V) (as earlier, defective voltages are shown in parentheses).

Furthermore, the voltage drop across the remaining good parts have decreased from 30 V to (15 V), and from 10 V to (5V). This is because all the pressures below the open have decreased.

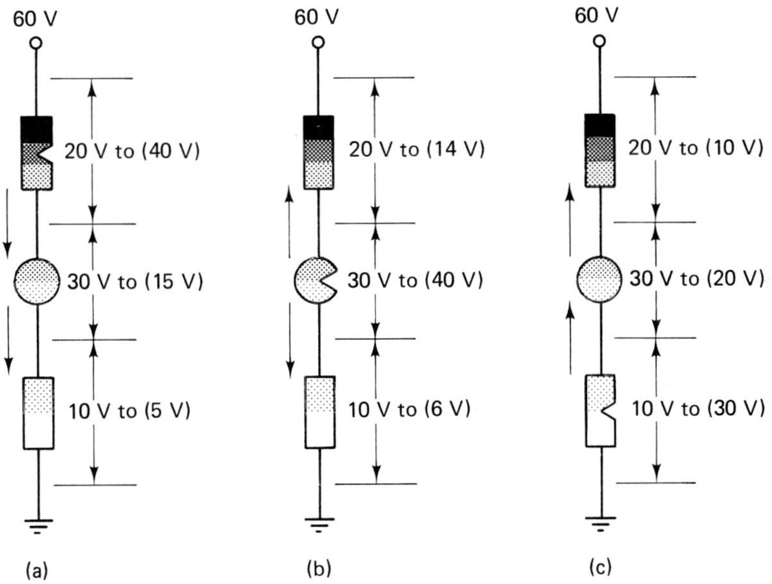

Figure 2-11 Partially open or increased resistance parts tend to drop the voltages below and push up the voltages above. The remaining good parts drop the voltages in proportion to their resistances.

In Fig. 2-11(b) the middle part partially opens and increases resistance. The pressure above it rises and the pressure below it falls. Thus, as the arrows show, the voltage drop on the open part increases and the drop on the other parts decrease. (We assume that the supply voltage does not change.)

If the bottom part in Fig. 2-11(c) partially opens, its resistance increases and the pressure above it will rise. Its voltage drop will increase, and the voltage drop on the other parts will decrease.

> *Note:* The above explanations are summarized beautifully with Kirchhoff's law and the voltage division equation. Kirchhoff's equation for three parts is $V_T = V_1 + V_2 + V_3$. If one voltage drop increases, the others must decrease (provided that V_T, the supply voltage, does not change).
>
> The voltage division equation states that the voltage drops are distributed in proportion to their resistances, or $V_X = V_T(R_X/R_T)$, where R_X and V_X respectively, stand for the resistance and voltage across any one of the series parts.

The information above on partial opens also applies to complete opens. All this is summarized in the following tip:

> TIP 2-9
>
> A partially open or increased-resistance part in a series circuit drops above normal voltage, while the good parts in its path drop less.

2-4-3 Partial Shorts (Decreased Resistances) in Series Circuits

We found out how the voltages can point to a partially open part in a series circuit. Let us now turn to partial shorts. In Fig. 2-12(a) the top part is partially shorted and has less resistance, so it drops less pressure. As a consequence, the partially shorted part will drop less voltage, and the parts below it must drop more. (We assume that the supply voltage does not change.)

If the middle part in Fig. 2-12(b) partially shorts and has less resistance, the pressure above it will fall and the pressure below it will rise. Consequently, it will drop less voltage while the others must drop more.

If the bottom part in Fig. 2-12(c) partially shorts, its voltage drop will decrease. The voltages above it will decrease; consequently, voltage drop on the other parts must increase.

In all the cases above, the voltage drop on the shorted part decreases while the drops on the other parts increase. These cases are summarized in the following tip,

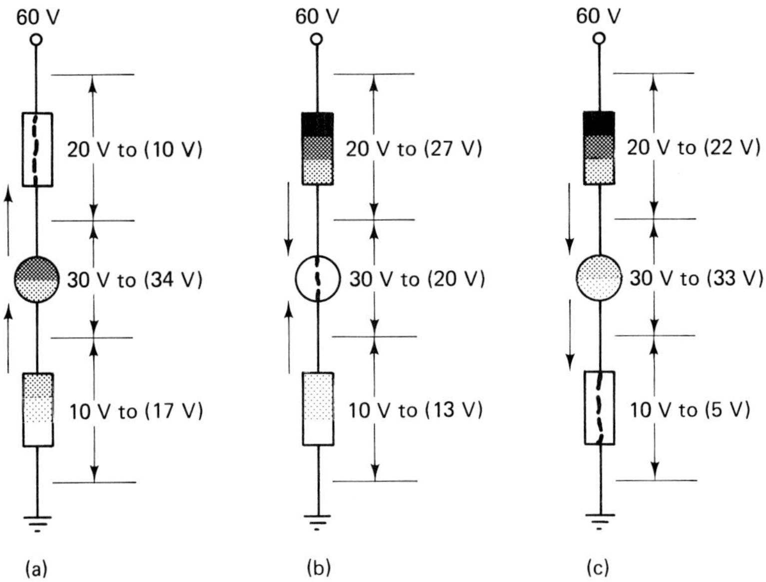

Figure 2-12 Partially shorted or decreased resistance parts tend to pull down the high-side voltages and pull up the low-side voltages. The remaining good parts tend to drop the voltages in proportion to their resistances.

which applies to complete shorts as well. (Note that the Kirchhoff and voltage division equations can be applied here and to all the cases above.)

> TIP 2-10
>
> A partially shorted or decreased-resistance part in a series circuit drops less than normal voltage, while the good parts in its path drop more.

Notice that the arrows in Figures 2-11 and 2-12 point away from a partial open and toward a partial short. As we noted before, a partial short "pulls" all the voltages toward itself, and a partial open appears to "push" them away.

Measuring a Voltage Drop. Since the tips for finding partially defective parts involve voltage drops, we should review two methods for reading them. In Fig. 2-13(a), two readings are taken. You subtract one reading from the other to obtain the voltage drop. In this case we subtract 10 V from 15 V to obtain a 5-V drop. Figure 2-13(b) shows another way. One voltmeter lead is connected above the part and the

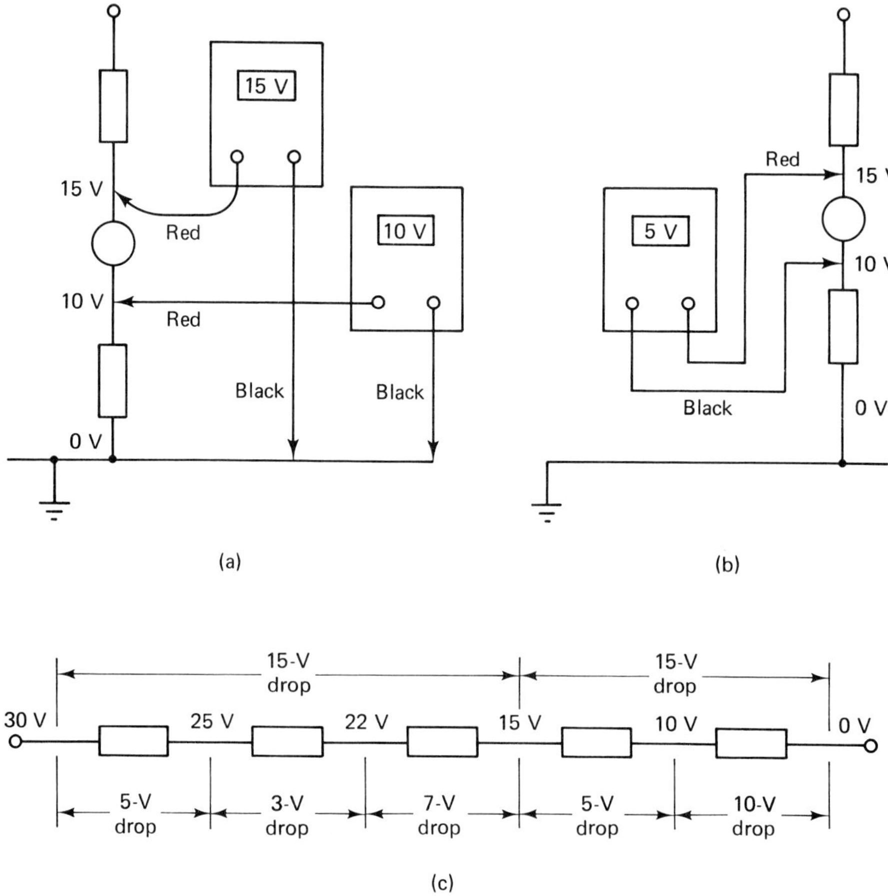

Figure 2-13 Measuring voltage drops. (a) Taking the difference between two voltages with respect to ground. (b) Measuring across the part. (c) The voltage drops across several parts.

other is connected below. The meter shows the 5-V drop directly. This second method is used when very small drops are measured. Many transistor circuits have voltage drops that typically measure about 0.7 V.

Figure 2-13(c) shows voltage drops across more than one part. Note how the smaller drops add up to equal the big drops. (*Note:* This is in conformance with Kirchhoff's law.)

The last two tips accord with perhaps the most basic of our rules, given on p. 45.

Intuitively, this important rule makes sense. If a part increases resistance, more pressure will build up on top, and less pressure gets through below. Just the opposite happens when a part decreases resistance. (All this is in accordance with Ohm's, Kirchhoff's, and the voltage division equations.)

RULE 2-4

In a series path, an increased-resistance part has a higher-than-normal voltage drop. A decreased-resistance part has a lower-than-normal voltage drop.

2-5 NEGATIVE VOLTAGES

Up to now, we have been working with circuits having positive voltages only. Sooner or later, however, you will be working with negative voltages [see Fig. 2-14(a)]. The battery represents the circuit in the power supply. The negative terminal is on the high side, the supply side. The positive terminal is now connected to ground. We say that ground has 0 V because that is where we measure from.

Figure 2-14(b) shows that the current is pushed up from ground. Ground or 0 V actually has a positive voltage with respect to the other points in this circuit. It has the highest positive pressure: the −20-V supply point has the least.

Figure 2-14(c) shows what happens when one of the parts in this circuit opens. In this particular situation, where the negative voltages are shown above the 0-V point, we can still get away with it if we consider 0 V to be less than −20 V. Then we might still say that opens tend to decrease the voltages "below" them (toward ground) and increase the voltages "above" them (toward supply). Similarly, Fig. 2-14(c) shows that shorts tend to pull the voltages "up" or "down." However, this is not consistent with the idea that a negative voltage is less than zero volts. Also, the "up" and "down" notion relates to the fact that the negative supply voltage is shown on top. Positive voltages are more often shown on top than negative voltages.

Furthermore, in circuits where the 0-V points are not connected directly to ground, or where ground is not immediately identifiable, you will run into trouble if you consider 0 V to be less than a negative voltage. We'll have to look into this.

2-5-1 Series Circuits with Both Positive and Negative Voltages

Let us start out by looking at a power supply that delivers both positive and negative voltages. Figure 2-15(a) shows one that delivers +20 V and −20 V. We say that the top terminal has a *positive polarity* and the bottom terminal has a *negative polarity*. We have two batteries representing the circuit inside the power supply.

The ground lead is connected to the negative terminal of one battery and the positive terminal of the other. Just what is the polarity of the ground terminal, positive or negative? We'll let you decide after you have read this section. The answer may be just what positive and negative is all about.

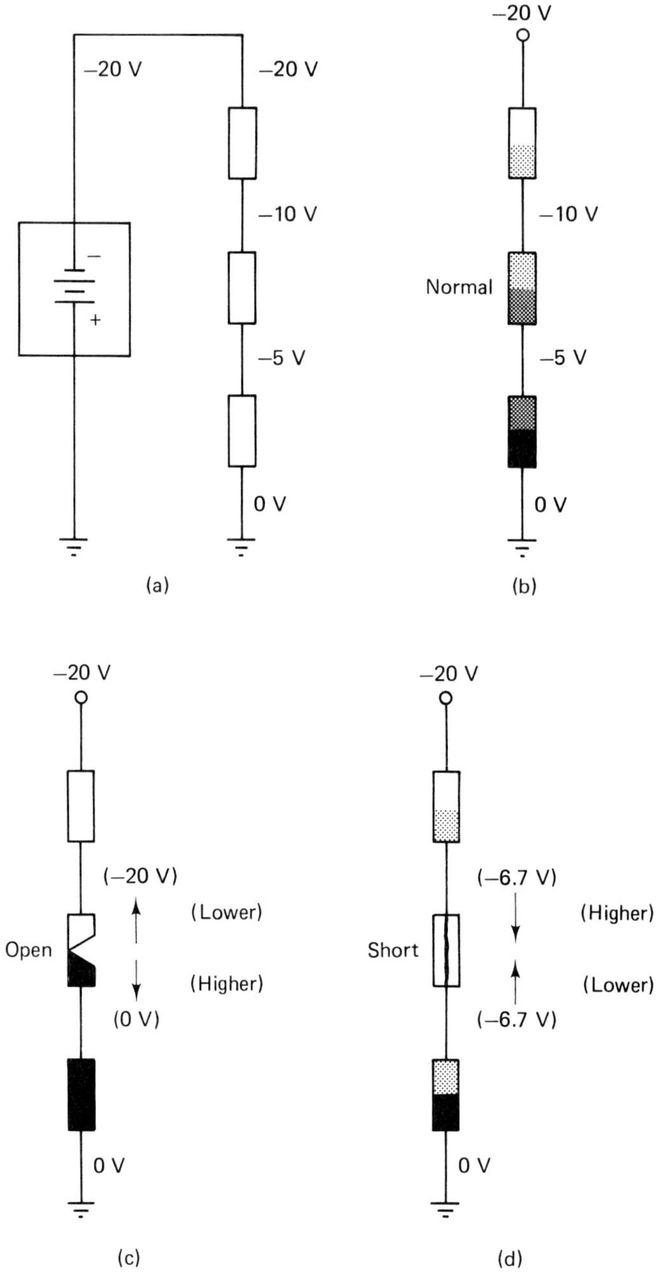

Figure 2-14 Negative power supply. To be consistent, the negative side is considered to have the lower pressure. (a) The positive supply terminal is grounded and may be said to have 0 V. (b) The voltage or pressure distribution. (c) An open pushes up the voltage above it and lowers the voltage below it. This accords with Rule 2-2. (d) A shorted part pulls down the voltage above it and pulls up the voltage below it. This accords with Rule 2-3.

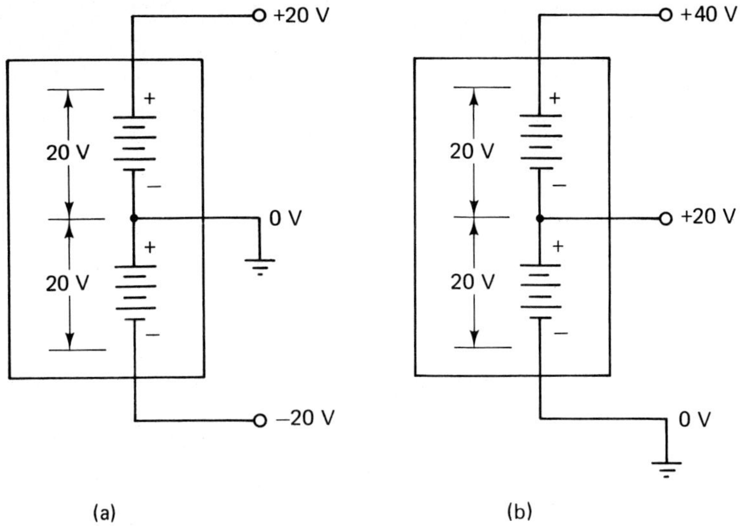

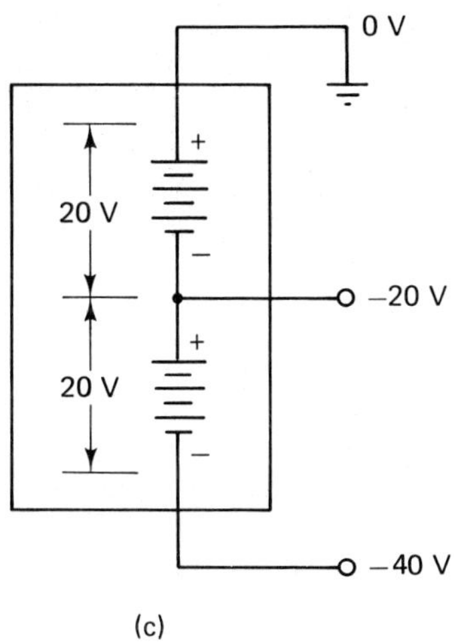

Figure 2.15 (a) Power supply with positive and negative voltages. (b) The same power supply with only positive voltages. Note that the ground point has been shifted to the negative terminal. (c) The same power supply with only negative voltages. Note that the ground point has been shifted to the positive terminal.

Figure 2-15(b) shows the *same* supply, but this time the ground lead is connected to the negative terminal of the bottom battery. Now we measure 40 V on top and 20 V at the middle.

Figure 2-15(c) shows the *same* supply again. This time, however, the ground lead has been shifted up to the positive terminal of the top battery. Now everything is reversed. We have 0 V on top and −40 V at the bottom.

What does all this tell us? For one thing, we can see that all these readings depend on where we happen to choose the ground or 0-V point. We also see that a negative voltage could be a positive voltage, depending on where we measure from. Another thing we learn is that the 0-V point is just another voltage point. As before, we see that 0 V is "greater" than −20 V. We also see that −20 V is greater than −40 V.

As the theoreticians say, polarity is "relative." A voltage is plus or minus only in relation to another voltage. Engineers and technicians talk about the "reference voltage" when referring to the point from where the other voltages are measured.

> *Note:* If we consider the numbers, 0, −1, −2, −3, −4, and so on, as decreasing in value from left to right, all the rules and tips in this chapter apply to series paths having negative as well as positive voltages provided that we leave out the terms *ground* and *supply.*

Let's look into this. Figure 2-16(a) shows a series path connected to the +20-V and −20-V supply terminals of our power supply. We show two sets of voltages in one and the same circuit. Note that the path is not connected directly to the power supply ground terminal. One set of readings, the unbracketed ones, +20 V and −20 V, are taken with respect to the power-supply ground terminal. The bracketed voltages between [+40 V] and [0 V] are taken with respect to the −20-V supply terminal. The shading gives you an idea of how the pressure is distributed. Since we are dealing with the same circuit, the pressure distribution is the same whether the 0-V point is at the bottom or in the middle. The voltage drops are the same. *They are independent of ground.* You will find out that this can be an important thing to know in some situations. For one thing, we shall see that the same old rules and tips apply, no matter where we put the 0-V point.

Figure 2-16(b) shows the same circuit with a partial open in it. The voltage above the defect has increased by 2 V from +10 V to +12 V. The voltage below the defect has "fallen" by 4 V from 0 V to −4 V. This accords with the rules.

Figure 2-16(c) shows what happens to the voltages when we have a partial short. We see that the same rules apply. The voltage above the partial short falls by 1 V, from +10 V to +9 V. Below the partial short, it rises by 3 V, from 0 V to +3 V. So far, so good.

However, you should be aware of what happens to the *voltage drops* across the parts with negative voltages. The voltage drop between (−4 V) and (+12 V) in Fig. 2-16(b) is 16 V. Look at it like this: There is a (0-V) point somewhere between the (+12-V) point and the (−4-V) point. A 12-V drop exists between the (+12-V) point and this (0-V) point. There is an additional 4-V drop from this (0-V) point to the (−4 V) point. This makes a total drop of 16 V.

Now let us look at the drop from the (−4-V) point to the −20-V point. We have a drop between two negative voltages. If you count down the volts from (−4 V) to (−20 V), you will find that there is a 16-V drop between these two points. It is as if you subtracted 4 V from 20 V.

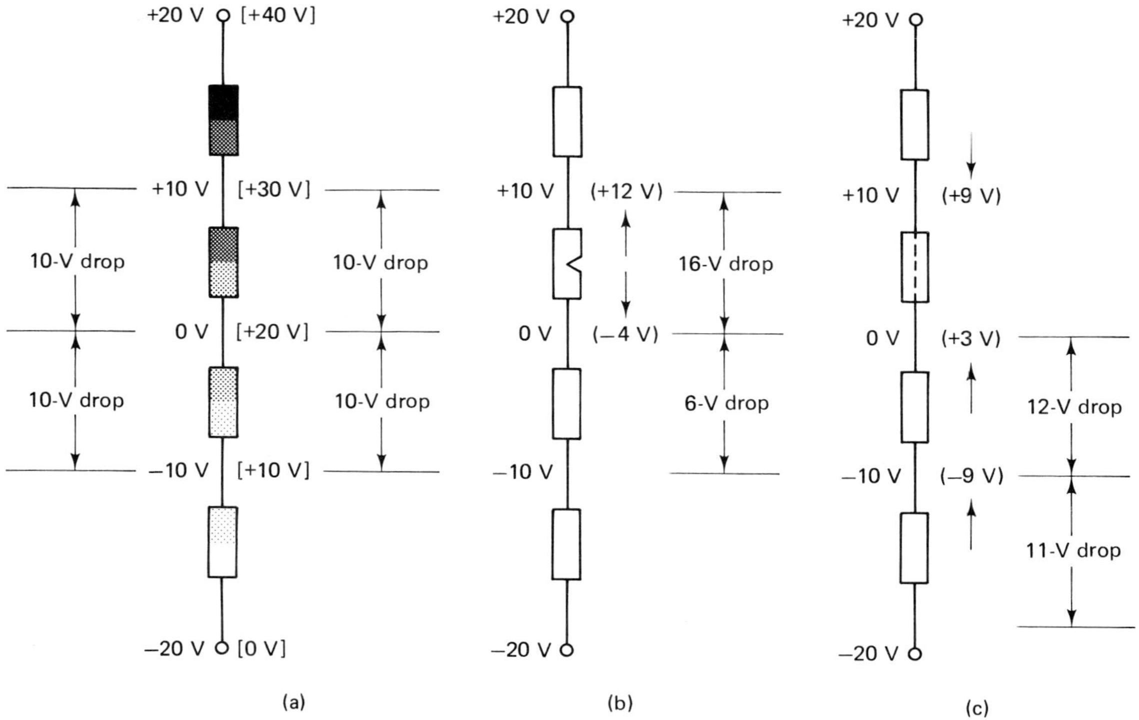

Figure 2-16 Measuring voltage drops across normal and defective parts in a series circuit with positive and negative voltages. (a) Normal voltage distribution. The voltage drops are independent of the point from which the voltages are measured. (b, c) The rules apply for negative and positive voltage circuits. (Note the voltage drops in all three cases.)

Note: The best way to find the voltage drop is simply to measure across the drop.

Voltage Drops between Plus and Minus Voltages. If you are not familiar with algebraic addition and subtraction, consider the following. The rule for finding the drop between a positive voltage and a negative voltage is to add the drops that are above and below the 0-V point. To find the drop between two negative voltages, remove the negative signs and subtract the smaller number from the larger.

Let us apply the rules above to the circuit in Fig. 2-16(c). The drop from the (+3-V) point to the imagined (0-V) point is 3 V, and the drop below it is 9 V. The total drop is 12 V. The drop from the (−9-V) point to the −20-V point is 11 V.

If you prefer, you can use the more general rules for algebra. They not only give you the voltage drop between two voltages, they also tell you whether you have a voltage drop or a voltage "rise." A voltage drop is indicated by a positive number, and a voltage rise is indicated by a negative number.

Rules for Algebraic Addition. If the signs are equal, add the numbers and take the sign of either of them. If the signs are unequal, subtract the smaller number from the larger and take the sign of the larger number. If the smaller number is negative, remove the parentheses if the term has them.

All possible examples involving the numbers 3 and 2:

$3 + 2 = 5$	(The signs are equal; they are both positive.)
$-3 - 2 = -5$	(The signs are equal; they are both negative.)
$3 + (-2) = 1$	(3 is larger and it is positive.)
$-3 + 2 = -1$	(3 is larger and it is negative.)
$-2 + 3 = 1$	(3 is larger and it is positive.)

Rules for Algebraic Subtraction. Change the sign of the subtrahend and add the numbers algebraically. If you have parentheses, remove them; and if you have two minus symbols next to each other, change them to a single positive symbol. If you have a negative and a positive together, eliminate the positive.

All possible examples involving the numbers 3 and 2:

3 minus 2	becomes	$3 + (-2) =$	$3 - 2 = 1$
−3 minus 2	becomes	$-3 + (-2) =$	$-3 - 2 = -5$
3 minus −2	becomes	$3 + (+2) =$	$3 + 2 = 5$
−3 minus −2	becomes	$-3 + (+2) =$	$-3 + 2 = -1$
2 minus 3	becomes	$2 + (-3) =$	$2 - 3 = -1$
−2 minus 3	becomes	$-2 + (-3) =$	$-2 - 3 = -5$

$$2 \text{ minus } -3 \quad \text{becomes} \quad 2 + (+3) = 2 + 3 = 5$$

$$-2 \text{ minus } -3 \quad \text{becomes} \quad -2 + (+3) = -2 + 3 = 1$$

Finally, the answer to the question of the actual polarity of the ground voltage depends on from where you are measuring. In fact, the polarity of any voltage depends on from where you are measuring. It depends, as we mentioned before, on the *reference point.* All voltages are referenced. There is no such thing as an independent voltage point. A voltage point in a schematic represents the drop between that point and ground.

2-6 A PRELIMINARY VOLTAGE-CHECKING PROCEDURE

It is possible, though not always smart, to locate bad parts solely with a volt-ohm-milliammeter. First you measure the voltages in the suspected circuits, then the resistance readings, and finally make a trial substitution of the suspected part. This works quite often. It can be fast and easy. It is recommended as a quick preliminary procedure that can be followed up by a more comprehensive one if necessary.

You start by checking voltages in suspected circuits; eventually, you will find a wrong voltage. Next you check the path in which you found the bad voltage. Then make use of Rules 2-1 to 2-4, the tips, and resistance readings.

This method is often abused by beginners. It can have disastrous results if you lack the patience to track the voltages carefully to the bad part. The catch is to check the voltages on *suspected* points.

The quickest way to find out what to suspect is to ask others who have worked on the same circuit. Keep your eyes and ears open. Look and listen to what the experienced troubleshooters are saying and doing. In the long run, though, you should develop your own methods. Starting with Chapter 6 we develop a sound approach for locating key test points. Recall that Chapter 1 puts it all together and gives you the overall troubleshooting procedure.

Another hangup comes with transistors. Good transistors can transfer bad voltages into circuits having good parts. This means that a bad voltage can be found in a good circuit. We handle this in Chapter 8. You will not be wasting your time, however, if you learn to make quick voltage readings and use the rules.

SUMMARY OF FUNDAMENTAL RULES FOR LOCATING BAD PARTS

Rules 2-2 and 2-3 apply to circuits having positive supply voltages if "above" is toward the positive supply terminal and "below" is toward ground. The same rules apply to circuits having negative voltages if "above" is toward ground, and "below" is toward the negative supply terminal. More generally, "above" is less negative, and "below" is more negative.

RULE 2-1

If you find a bad voltage in a series path between two good ones, the trouble lies in this path.

RULE 2-2

An open or increased-resistance part tends to:

a. Lower the voltages below it.
b. Raise the voltages above it.

RULE 2-3

A shorted or decreased-resistance part tends to:

a. Pull up the voltages below it.
b. Pull down the voltages above it.

RULE 2-4

In a series path, an increased-resistance part has a higher-than-normal voltage drop. A decreased-resistance part has a lower-than-normal voltage drop.

SUMMARY OF BASIC TIPS

You need not memorize these tips. We have them here for review and quick reference. A review of these tips should give you an idea of how things are done in troubleshooting. Recall that in circuits with positive voltages only, ''above'' refers to the supply side and ''below'' refers to the ground side.

You can also study these tips to enhance your understanding of the basics: Ohm's law, Kirchhoff's voltage law, and voltage division.

Tip 2-1. An incorrect dc voltage indicates the presence of a defective component, connection, or conductor. A good dc voltage usually means that the circuit is good, so look for bad voltages first. (For information on apparently good dc voltages in defective circuits, see Tips 5-18, 5-21, and 5-26; and Section 7-3.)

Tip 2-2. If one of the two parts of a series circuit drops 0 V when it should drop more, turn off the power and measure its in-circuit resistance. If it has 0 Ω and its out-of-circuit or rated resistance is not 0 Ω it is shorted. If you read its rated resistance, it is probably good and the other part is probably open. If it has higher-than-rated resistance, it is definitely bad. If it has lower-than-rated resistance, you cannot tell immediately whether it is bad.

Tip 2-3. When a part shorts in a series circuit, the other parts may run warmer or hotter than normal. If a part opens, all parts will run colder than normal. Turn the power off before feeling the parts. Touch gently at first. Some parts normally run hot enough to sear your skin.

If the parts are warmer or hotter than normal, the part with an abnormally low voltage drop or resistance is shorted.

If the parts are colder than normal, a part with an abnormally high voltage drop is open.

Tip 2-4. If one of the two parts in a defective series circuit drops the full supply voltage, turn off the power and jump it with a good part. Turn the power back on. If normal operation or voltages are restored, the jumped part is bad.

A resistor having about twice the resistance of the jumped part can also be used. If the voltage changes after the part is jumped, it is open. Otherwise, the unjumped part is shorted.

Tip 2-5. If you have a partially defective voltage in a series circuit with two parts, disconnect the part with the lower-than-normal voltage drop and substitute it with a good one. If this restores the voltages or operation, fine. If not, the other part is bad.

You might disconnect the lead that is most convenient as long as you have the normal circuit when the substitute part is connected.

You can also check either part with a test instrument after disconnecting only one lead. (In Chapter 5 we learn to check parts with test instruments.)

Tip 2-6. Assume at first that only one part is defective. Take them on one at a time.

Tip 2-7. Any point above a completely open part, connection, or conductor in a series circuit has the full supply voltage on it. Any point below it has 0 V.

Tip 2-8. In a series path with three or more parts, if only one part drops zero volts, that part is shorted. If a part tied to the supply line shorts completely, its low-side voltage is pulled up to equal the supply voltage. If a part tied to ground shorts completely, its high-side voltage is pulled down to zero.

Tip 2-9. A partially open or increased-resistance part in a series circuit drops above normal voltage, while the good parts in its path drop less.

Tip 2-10. A partially shorted or decreased-resistance part in a series circuit drops less than normal voltage, while the good parts in its path drop more.

WHAT YOU LEARNED TO DO

In this chapter you learned how to:

1. Check whether series parts are dropping voltages in proportion to their resistances.

2. Look for defective parts by noting whether the voltages are either too high or too low.
3. Use the fundamental rules for locating bad parts in series circuits.
4. Use some tips for interpreting incorrect voltages to locate a defective part.
5. Use the VOM to find bad parts in series paths.
6. Locate defective parts in circuits having negative as well as positive voltages.

QUESTIONS

The following questions apply only to parts in a series path that lie between a good supply voltage and ground.

1. If all the voltages below a part equal 0 V, what is likely to be wrong with it? (Assume that the part is not tied to ground.)
2. If all the voltages above a part equal the supply voltage, what is likely to be wrong with it? (Assume that the part is not tied to the supply line.)
3. If a part has an above-normal voltage drop and all the other parts have below-normal voltage drops, what is wrong with it?
4. If a part has a below-normal voltage drop and all the other parts have above-normal voltage drops, what is wrong with it?
5. In a series path with only two parts, the midpoint voltage is 0 V, and the resistance measured from that point to ground is much greater than 0 Ω. Which part is bad, and what is wrong with it?
6. We have a series path with three parts. The voltage above the middle part has fallen from 10 V to 6 V, and the voltage below it has fallen from 5 V to 3 V. Which part is bad, and has it increased or decreased resistance?
7. We have another three-part series path. The voltage above the part increases from 8 V to 10 V, and the voltage below it decreases from 5 V to 3 V. Which part is bad, and has it increased or decreased resistance?
8. The voltage above a defective part has changed from a normal −4 V to a defective -3 V. The voltage below it has changed from a normal −8 V to a defective -9 V. Has the resistance of the part increased or decreased?
9. What is the voltage across a part that has +2 V on one side and −3 V on the other?

PRACTICAL TROUBLESHOOTING EXPERIENCE

Our first job calls for tack-soldering parts together temporarily without splicing or twisting them together. It is done by forming a bead of solder on one lead and then applying a thin coating of solder (tinning) another lead and soldering the two together. Make sure that the leads are clean. You might let your bench act as a third hand by placing the roll of solder on it with the solder wire pointing up at a convenient angle. This frees both hands.

Tack-soldering is done when parts are substituted into circuits temporarily. It is also done for experimental purposes. For the latter purpose, however, it is better to use a protoboard. You connect the parts together into circuits by inserting their leads into pin holes on the board. If you do not need to learn how to solder, you can use one of these boards.

1. Connect three 6-V lamps with screw-type sockets in series with a 6.0-V battery or power supply. The negative terminal of the power supply is considered to be ground.
   a. Unscrew one of the lamps. How many volts are on each terminal of the open socket when measured to ground? How many volts are on each terminal of the two nonopen sockets?
   b. Obtain a jumper wire—a length of wire with alligator clips on both ends. Jump (shunt) it across one of the lamps to simulate a short. You can tack-solder a jumper wire if you prefer.
      How many volts are on the shorted lamp? How many volts are on the nonshorted sockets? Are the lamps brighter or dimmer than before? How many volts are on the lighted sockets?

2. Obtain three resistors of widely varying values. The lowest value should be at least 100 Ω, and the highest should be no more than 1000 Ω. Tack-solder them together in series and connect them across a 6-V battery or power supply.
   a. Which resistor has the highest voltage drop? The least?
   b. Short one of the resistors with a jumper wire. What happens to the voltage drop across the shorted resistor? What happens to the voltage across non-shorted resistors?
   c. Replace one of the resistors with a resistance of less value to simulate a partial short. Is the voltage across the simulated partial short higher or lower than before the change? What happens to the other voltage drops: Do they increase or decrease?
   d. Replace the reduced-value resistor with one of higher value than the original. What happens to the voltage across the increased value part: Does it increase or decrease? What happens to the voltage drops across the unchanged parts: Do they increase or decrease?

(Optional for Steps c and d: Note the before and after voltage drops across the unchanged parts. Do these voltage drops maintain the same proportions before and after?)

3

Checking Conductors, Connectors, Switches, and Power-Input Circuits

EXPERIENCE REQUIRED

To make use of this chapter, you should know how to:

1. Use the multimeter to find partial or complete opens and shorts in series circuits.
2. Identify 120-V ac wall outlets, fuses, circuit breakers, switches, wires, conduits, and cables.
3. Identify electrical insulators.
4. Identify coaxial or shielded cables and ribbon lines.
5. Identify printed-circuit boards and connections.
6. Identify power transformers, electrolytic capacitors, power chokes, and wire-wound resistors.
7. Use the basic electrical hand tools.

3-1 SAFETY PRECAUTIONS: A REVIEW OF BASIC ELECTRICITY

A basic understanding of the wiring behind the 120-V ac wall outlets will help you to work safely and confidently. The following precautions and explanations are actually a review of basic electricity.

1. When working on a live circuit, avoid touching metal pipes of any kind. Stay away from all types of structural beams, bars, and metal walls.

Explanation. Electrical power is connected to your 120-V ac outlet through two wires. One is called the "hot wire," the other is the "ground wire." As you should know, all electrical devices have to be connected to at least two wires. To get a shock, you must touch two conductors, one hot and the other grounded (connected to ground).

Ground or earth is used as a conductor of electricity. Touching ground and a hot wire can be dangerous. The ground wire that enters a building is connected directly to the water pipes in the building. All the water pipes thus become a part of the ac circuit. The lamp in Fig. 3-1 is connected to the hot side of your outlet and to a water pipe. It is lit up even though only one lead is connected to the socket.

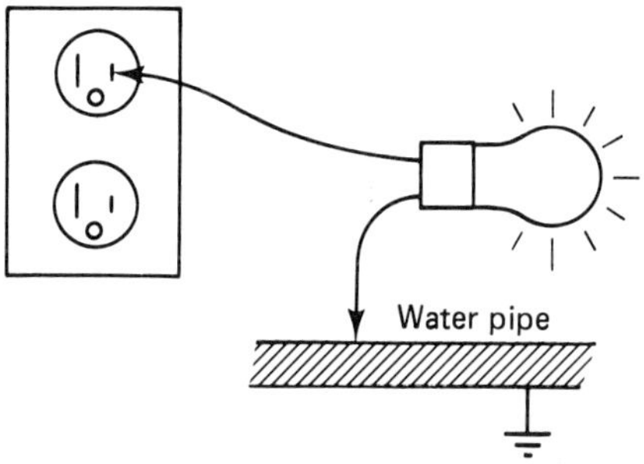

Figure 3-1 The lamp indicates that the water pipe carries a potentially dangerous electric current.

Not only water pipes—but all metal pipes, bars, and beams that are deeply embedded in earth—are a part of the ground circuit. Sometimes metal structures are purposefully grounded. This is a safety measure. If lightning strikes a metal structure, or if a live wire shorts to it, the voltage on it will be held to ground level, or zero volts. As long as you touch only ground or grounded metal, you are safe. However, if you accidentally touch grounded metal and a hot wire, you are in for a shock.

2. Beware of metal conduits for electrical wires. They are connected directly to the ground line.
3. Beware of the metal cabinets of surrounding equipment while working on a circuit. Do not touch them.

Explanation. Metal cabinets are often grounded. This is true of 120-V ac equipment having three-wire electric cords and plugs. The third wire is connected to ground. This keeps the metal cabinet from having a "hot" voltage if a hot wire accidentally shorts to it. This is a safety measure for the users but not for the technician.

For the technician, a grounded cabinet can be dangerous. When you work on it, you are within a few inches of shorting yourself between a hot wire and the metal cabinet. Don't touch the cabinet.

4. Don't touch the exposed chassis of surrounding equipment. They may be hot.

Explanation. In some equipment with two-wire ac cords, one of the wires is connected directly to the internal metal chassis. For safety to the user (not for the technician), the cabinet is insulated from the chassis. An additional safety measure is the use of *polarized plugs*. They can be inserted in only one way because one prong is wider than the other. This ensures that the chassis is always grounded; it is always connected directly to the wider prong.

However, some older equipment have nonpolarized plugs. Use an isolation transformer on this type of equipment. Furthermore, many older installations do not have polarized outlets. The polarized plug is made so that it can be clipped to fit the nonpolarized slots. This can be dangerous for the technician who works on a bare chassis. Always use an isolation transformer.

Isolation Transformers. There are no direct connections between the input and output of an isolation transformer. It is a device that transfers power from its input to its output by means of a process called electromagnetic induction. There can be no complete circuit from the ac hot line to the ac ground line through the transformer. A lamp connected from any output of an isolation transformer to a ground pipe will not light.

5. Signal cables also have grounded shields and connectors. Avoid them when working on hot circuits.
6. Use insulated tools. Don't touch the metal parts of the tool even when you are not working on a live circuit. You want to develop "fail-safe" habits.
7. Use only one hand when working on a circuit.

Explanation. Electricity will flow through a complete circuit. If you accidentally touch the hot line with one hand, and touch grounded metal with the other hand, you will become one part of a complete 120-V ac circuit. You want no part of that. Use one hand only. Many technicians are trained to keep one hand in their pocket when working on a chassis.

8. If you have to use your other hand to hold the equipment while working on it, hold on to an insulated part. Safer yet, use props or other devices to hold your work securely. Turn off the power before repositioning it.
9. Do not stand on concrete or damp floors of any kind. In fact for safety, all concrete floors should be covered with insulating mats. Wear rubber-soled shoes.

Explanation. Wet wood or concrete can conduct electricity. Even if the concrete is not in direct contact with the wet soil, it can have grounded metal pipes and reinforcement embedded in it.

10. Unplug all unused equipment. Place the plug where you can see it.
11. Whenever possible, use the ohmmeter function of your multimeter. It is the safest. You must unplug the chassis before using it. Place the plug where you can see it. Later, you'll feel that a chassis is dangerous if you don't see its plug.
12. Before handling any part of the circuit, make a visual check for bare wires. Do not touch any exposed wires.

13. Inspect ac power cords for wear and tear. Replace worn power cords.
14. A beginner should not work on chasses or circuits that are directly connected to the ac line unless under direct supervision.
15. Do not work on the other side of an ac outlet. That's the electrician's job. An electronic technician habitually thinks that black wires are grounded. In electrical work, black wires are hot.

Explanation. In electronics, ground wires are often black. In electrical work they are hot. This fact shows that the two fields of work are widely separated. There must be good reasons for this. Play it safe. You do your work and electricians do theirs.

16. We have covered some of the more important precautions. Undoubtedly, there are situations we have not covered. A knowledge of the fundamental safety precautions should be constantly applied in all circumstances.

3-2 CHECKING CONDUCTORS AND CONNECTIONS

A good way to get started in electronics is to work with wires and connections. A high percentage of troubles involve line cords and connectors. They are not always easy to check, and the symptoms can be misleading. For the most part, however, the theory is easy. The conductors either conduct or they do not. The insulators either insulate or they do not. The contacts either contact or they do not.

Conductors and contacts should have zero resistance. Insulators should have infinite resistance. That's the theory. In real life, however, you have poor connections, and invisible breaks in the conductors that may or may not lose contact.

3-2-1 Inspecting Line Cords and Cables

Line cords are subject to countless cycles of bending and pulling. Eventually, metal fatigue sets in and they fail. The insulation also fails eventually. It wears out or deteriorates with age. Ac power cords for portable equipment fail quite often. Then, of course, you have microphone cables that are constantly being whipped around. You also have cable connectors that are repeatedly pulled out and pushed in.

> TIP 3-1
>
> Line cords usually break at their ends. Since you cannot see the broken wire, look for signs of wear on the insulation at these points.

Always check the line cords, cables, and connectors first. We all know of the beginner who spends several frustrating hours on a big chassis, only to find that a line cord is bad. Do the easy things first!

> TIP 3-2
>
> Make the easy checks, such as cords, wires, cables, connectors, switches, and fuses or circuit breakers first.

Wiggling and Shaking Wires. Microphone, earphone, and speaker lines are usually prechecked without instruments. They often break at the connectors. With the equipment turned on, listen to the output. Wiggle and shake the cord at one end; make sure that you do not disturb the other end. You can do so by pressing down on the line with one hand or foot to isolate the disturbance to one side. If you now get a response, you know which side is bad.

You can check the connector in the same way. Hold it tightly against the chassis and move the cord. If the equipment output responds to the disturbance, the cord, not the connector, is bad.

If you get no response, push and squeeze the cord so that the inner, hidden break makes contact. If you get a response, you have an open wire. For a further check, pull the wire instead of pushing it and see if the response goes away. If it does, you are even more certain of an open at that point.

If the checks above fail, you can try something more drastic. Pull and stretch the cord. If the cord stretches excessively and too easily, the inner wire is broken. Note how the cord insulation thins out. Bending the cord at this point may indicate the presence of the broken wire more definitely. If you break an already weakened inner wire, that is okay; it should be replaced anyway. Better to do it now than later.

The entire procedure should not take more than a minute or so. It takes longer to explain it than to do it. Do not check ac power cords as above unless the wire covering and all other insulation is in good condition. Look for exposed strands of wire. Use one hand only.

Shielded Lines. A shielded line has an inner wire surrounded by an outer conductor of braided wire or foil. The outer and inner conductors are separated by insulating material. Look for loose strands of braided wire that may be shorting to the inner wire.

Sometimes a poor soldering job at the connector will melt the inner insulation and short the outer braid to the inner wire. Inspect the inner insulation for evidence of poor solder work.

3-2-2 Ohmmeter Continuity Checks

The ohmmeter is used when the voltage is too dangerous or when there is no voltage at all. It is often used to check the continuity of wires. A few feet of copper wire measures practically zero ohms. An open conductor reads infinite ohms. A long line, about 100 ft, may have a fraction of an ohm. If you are going to do a lot of work with them, it may be a good idea to look up their ohms-per-foot values. In most cases, however, all you need to know is whether or not the line is broken.

Here is an interesting method for checking a long multiwire color-coded cable (see Fig. 3-2). At the far end, connect all the wires together. At the near end, check the continuity between any two of the wires, say *A* and *B*. If you have no continuity, one of them is open. To find out which, take the continuity between one of them, say wire *B* and a third wire *C*. If you have continuity now, wire *B* is good. Wire *A* is open.

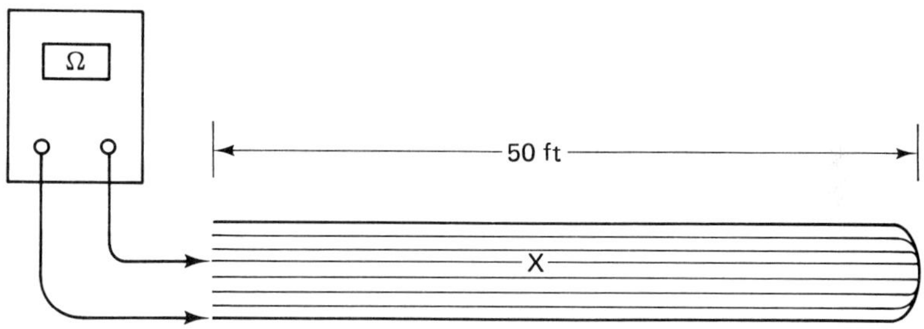

Figure 3-2 Quick way to check the continuity of long cables. Tie all the wires together at the other end and check the continuity between them—two at a time.

If not, wire *B* is almost sure to be open. If you want to be even more certain, make a third check with a fourth wire. Why not? It only takes a second. Coaxial lines may be checked in a similar manner.

Many cables end with connectors. You might rig up a connector that shorts all the lines together when you connect it to the cable.

TIP 3-3

Multiwire cables and coaxial lines may be checked for continuity by shorting all the conductors together at the far end and then checking the near end for continuity between the wires with an ohmmeter. Find two wires that have no continuity between them. Then, if necessary, check one of them against a third wire.

3-2-3 Voltmeter Continuity Checks

The voltmeter check for conductor continuity is simple. All points on a conductor should have the same voltage. There should be no voltage drop across any two points of a conductor. Otherwise, the conductor is open.

There are two ways for checking continuity with a voltmeter. The first way is to check several points on the line with reference to ground; all checks should show the same voltage. The other way is to measure the voltage between the ends of the conductor; you should read 0 V between the two ends. Theoretically, you could be reading equal voltages on two separate lines. This is not likely, if your dealing with a wire of the same size and color. If necessary, turn off the power and check the line with an ohmmeter.

3-2-4 Inspecting Connections

Check connections with care. Bad connections are often associated with oxidation, which makes a hard coating that develops around exposed metal. It is commonly known as corrosion. If you thoughtlessly disturb the connection and break the coating, the circuit may renew operation, and you may never know why. You must be aware of every move you make when you first start out on a job. Be very deliberate.

TIP 3-4

Your very first measurements in the troubleshooting process should be done with care and attentiveness. If you unknowingly disturb an open connection, the circuit might start operating, and you may never know why.

After considering the user complaint, your first move is often to turn on the power and wiggle the connectors. Do it in such a manner as to restore contact and operation.

> **Warning:** Do not perform these wiggle checks on ac power connectors until you have learned the safety rules in the beginning of this chapter. Do not make these checks in high-voltage circuits (see Section 11-6). A beginner should not work on dangerous circuits unless under direct supervision.

You can also clean dirty connectors with a knife or sandpaper. Commercial cleaning fluids are also used for this purpose.

3-2-5 Inspecting Solder Connections

Soldered terminals often have to be checked. The solder should be shiny and show evidence of a smooth flow. The edges of the solder should show a thin coating, as in Fig. 3-3(a). A beady solder joint is suspect [see Fig. 3-3(b)]. There should also be enough solder on the joint to ensure good mechanical strength. This is true in cases where the wire is loosely attached to the terminal. However, there should not be too much solder; this will hide the thin coating that lets you and others know that the electrical contact is good.

For mobile equipment, the wire is tightly wrapped on the terminal. Here again a smooth shiny flow and a thin coating on the edges of the joint should be seen on both the terminal lug and the wires connected to it. Look for shorts caused by excess solder, hairline solder splatter, and solder beads that rolled across conductors.

Open solder connections can often be found by carefully wiggling the parts with a plastic stick. If operation is restored temporarily, one of the connections of the wiggled part could be open.

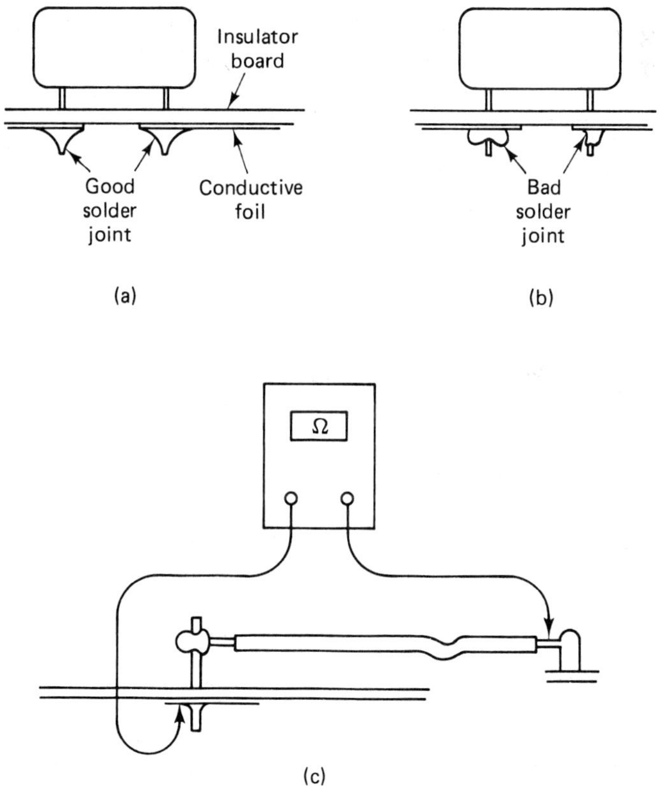

Figure 3-3 (a) Good solder connections shows evidence of good flow and adherence. (b) Bad solder connections are beady and show signs of poor adherence. (c) 0 Ω between both sides of a connection indicates good electrical contact.

3-2-6 Checking Connections with an Ohmmeter

Suspected connections can be checked with an ohmmeter. Don't forget to turn off the power. Figure 3-3(c) shows a questionable solder connection. Gently touch one ohmmeter test prod to an unsoldered part of the foil near the terminal lug, and touch the other prod to an unsoldered part of the connecting wire. Zero ohms indicates good continuity.

> TIP 3-5
>
> Do not overlook the solder connections. Check for continuity between the connecting leads or conductors.

3-2-7 Checking Connections with a Voltmeter

You can also check a connection with a voltmeter. The voltage to ground on both sides of a connection should be equal. Clip one terminal of your voltmeter to ground and touch the red prod to the unsoldered parts of the terminal and connecting wires.

Coaxial Shields. The outer shield of a coaxial line is grounded. There should usually be zero volts on the shield when measured to ground. Use either the ac or dc function of your meter, whichever is appropriate.

3-3 CHECKING SWITCHES, FUSES, AND CIRCUIT BREAKERS

Theoretically, the operation of switches, fuses, and circuit breakers are often omitted when studying the circuitry. This would be a big mistake if you were actually working on a circuit. The experienced troubleshooter usually checks these out first. These are often easy to access and easy to check.

3-3-1 Checking Switches

A faulty switch will not turn on the equipment or turn it off. It may have a spongy feel or not have the right snap. It may feel too loose or as if something in it is broken.

To check a switch for continuity with an ohmmeter, unplug the equipment and flip the switch on and off. You should have zero ohms across its terminals when it is on and a higher resistance when it is off. The resistance should equal the circuit resistance, but don't worry about it if you are just checking the switch.

Switches can also be tested with a voltmeter. The low-voltage side of the switch should have zero or low volts when the switch is off and a higher voltage when it is on. You can also measure the voltage across its terminals. It should have zero volts across its terminals when it is on and a higher voltage when it is off.

3-3-2 Checking Fuses and Circuit Breakers

Fuses and circuit breakers should have zero resistance and zero voltage across their terminals. Otherwise, they are open. You can use an ohmmeter with the power off or a voltmeter with the power on. For circuit breakers, check for an open before you press the reset button, and check for continuity after you press it.

You must be familiar with the particular type of circuit breaker before you can check it by feel. Some spring-loaded types have a certain identifiable feel when they are being reset and another feel after reset.

3-4 TROUBLESHOOTING PC BOARDS

Many electronic circuits are mounted on printed-circuit (pc) boards. Figure 3-4 shows a small one. The flat copper foils or *traces* on the surface of the board serve as conductors. The large number of traces and solder joints on the bigger boards can cause a lot of trouble, as you might guess.

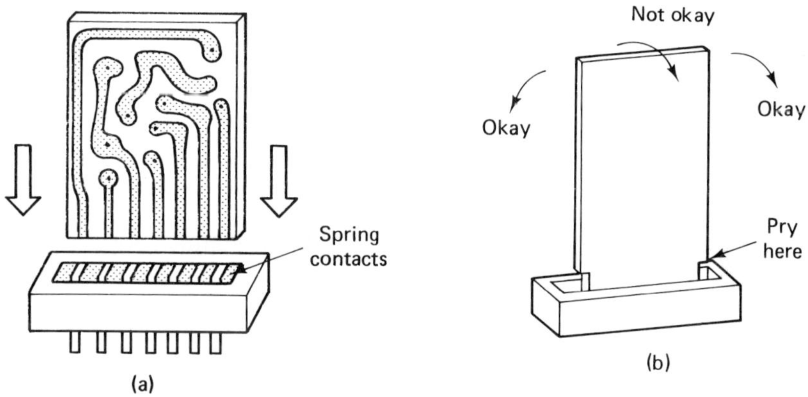

Figure 3-4 Printed-circuit-board module or card and the proper and improper ways to remove it.

3-4-1 Inspecting and Checking PC Boards

A PC board can be visually inspected. Look for bad soldering connections. There should be smooth flow between the part leads and the traces. Look for loose solder beads or splashes that could short across the traces. Hairline splashes can be hard to see.

Also look for hairline cracks in the board and traces. A magnifying glass and light can be helpful. A light shining through the board will also show up cracks.

Sometimes foreign matter such as a loose hair or lint may result in an open trace during the manufacturing process. These may be difficult to spot. Sometimes they can only be found by advanced troubleshooting methods, or by a painstaking ohmmeter search, one line at a time.

Look for traces that seem as if the copper was deposited very thinly, as if it were sprayed on. Sometimes these develop into opens.

Don't forget to check the mounting screws at the four corners of the boards. If you suspect grounding problems, loosen them a little and then tighten them again. Do not tighten too much; you can break the board or strip the copper or the chassis threads.

The components can be carefully pushed to one side or another to see if operation can be restored. Limit yourself to the suspected locations. You don't want to create new problems by wiggling indiscriminately.

Bending the entire board carefully may also restore contact and operation. Note the angle and location where a board is bent to restore contact. This may help you locate the broken trace or bad connection. Tapping the board very lightly over various locations with a plastic probe will also help isolate the open trace or connection.

As a last resort, you can rap the entire chassis sharply with a screwdriver handle to see if the problem is mechanical. Make sure, of course, that there are no delicate parts or mechanisms on the chassis. The problem with this is that once the circuits are working, you will not be able to locate the fault. However, at least you know the trouble is an open or shorting contact.

An ohmmeter is almost indispensible for checking long traces that seem to travel all over the board. With the help of a schematic, identify the ends of the traces, then check them for continuity.

After you find an open trace, you must find where on the trace the break is located. First check the continuity between one end and the midpoint of the trace. Then find the midpoint of the bad half and check for continuity to the corresponding endpoint. Repeat this until you find the break.

Sometimes you'll have to clean off the insulating coating on the trace so that you can contact it with an ohmmeter. A sharp steel point on your test probe is ideal for poking or scratching through the insulation. The insulation is called solder resist. It keeps solder off the traces during the manufacturing process. Note that there is no solder resist at the ends of the traces. Don't get any of that stuff on the solder connections or you'll have a tough time resoldering them.

If the break is hard to see, scrape the trace clean and slide the ohmmeter test prod over the trace until you see an abrubt change in the meter display. Analog meters are superior in this kind of work. You can catch the flick of the needle with the corner of your eyes as you focus on the trace. Usually, the scraping itself will reveal the break, either to the eye or to the touch, as you scrape across it with a sharp tool.

Voltmeter tests on pc board connections and traces are made by clipping the black lead to ground and moving the red probe from point to point. All points on a trace and around a connection should have the same voltage. A different voltage between any two points indicates a lack of continuity. Slide the test prod over the trace. An abrupt change in the meter display will reveal a break that is hard to see. Scrape the trace clean to see the break and to solder it.

TIP 3-6

A PC board trace can be checked for a hairline break by cleaning it and sliding one of your ohmmeter probe tips over it while the other end is touched to one end of the trace. Look for a sudden change in the meter reading.

3-4-2 PC Modules or Cards

PC boards often serve as modules or cards that fit into slotted receptacles called edge connectors. The traces on a card come out to the edge of the board like many fingers, as shown in Fig. 3-4(a). These connect to spring contacts in the edge connector. These contacts are subject to corrosion and may have to be cleaned occasionally. The spring contacts can be checked and carefully bent and tightened. The card can be carefully wiggled to check for bad edge connections. The faults occur at two places: at the spring contact and at the solder terminals to the main board.

The ohmmeter or voltmeter can be used to check the connections. There should be zero resistance and zero voltage drop between the trace and the soldered terminal.

When the edge connector receptacles are mounted on a mother PC board, care should be taken when pulling out a card. Keep the flat sides vertical to the motherboard at all times.

Often the card is shaped, as shown in the illustration, so that you can pry one end up a little with a screwdriver. This will put zero force on motherboard solder connections. Alternately lift or pry each end a little at a time until it comes out easily [see Fig. 3-4(b)].

With more experience and knowledge, the type of the response might tell you which edge connection is bad.

TIP 3-7

When pulling out a vertically mounted card, make sure that its flat sides are always vertical as you lift. For zero-force leverage, you can lift or pry one side up at a time. The force is exerted to free the slide connection and not the solder connections.

TIP 3-8

Edge connectors should be routinely under suspicion. In particular, suspect their solder connections. You may gently wiggle the cards with a plastic stick while the power is on. If you get a response at the system output, tighten the contacts, and resolder the connections.

3-5 A BATTERY AND LAMP CIRCUIT: TROUBLESHOOTING BASICS

We have various ways of checking voltages and resistances in lamp circuits. Basically, we apply what we learned in Chapter 2 about opens and shorts. We also have some commonsense visual checks. Each check calls for a decision as to what must be done next. The type of thinking that underlies these simple checks and decisions is fundamental to all troubleshooting.

3-5-1 The Basic Battery and Lamp Circuit

The battery-powered lamp circuit in Fig. 3-5(a) is much like the power-input circuit of more elaborate equipment. The troubleshooting approach is basically the same. A high percentage of troubleshooting jobs on the bench begin with a check of the power-input circuit. The circuit here consists of a lamp, battery, and switch. Points *A* to *F* represent the terminals of the parts and conductors. There are three conductors, lines *AB*, *CD*, and *EF*. Line *EF* can be a metal chassis used as the ground conductor.

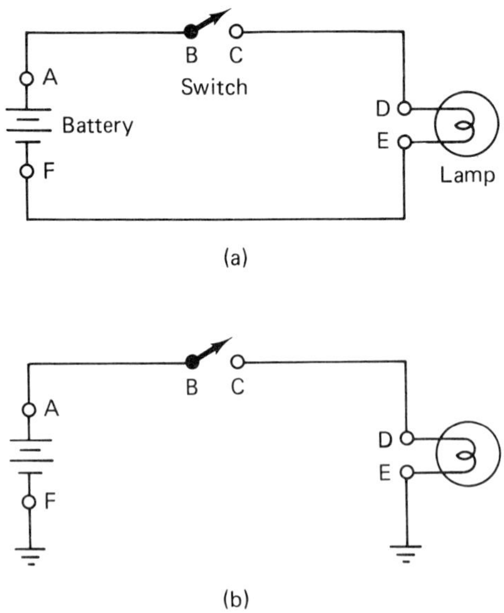

Figure 3-5 (a) Simple lamp circuit with a switch. (b) The same circuit with ground symbols to represent the line from *F* to *E*.

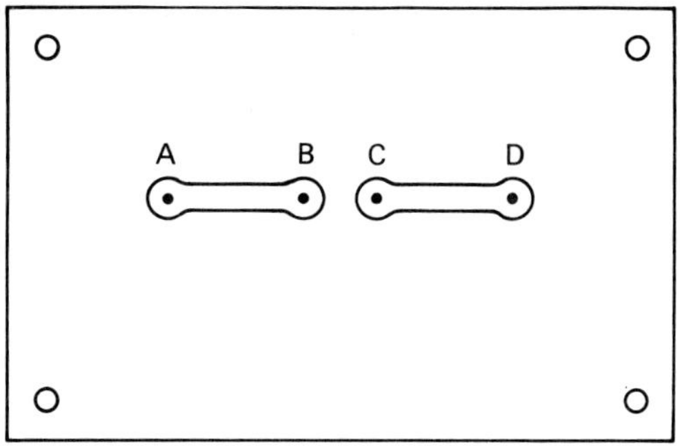

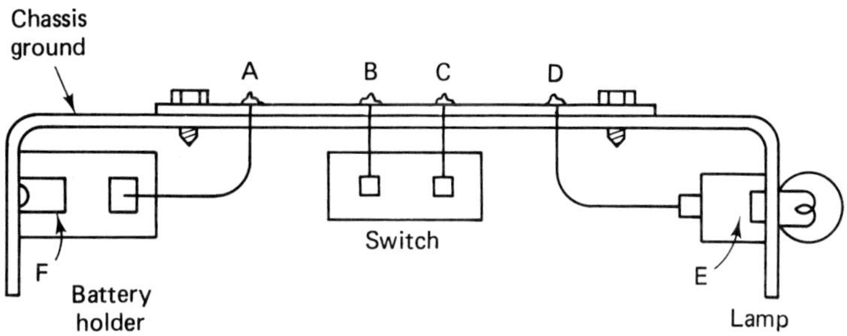

Figure 3-6 (a) Printed circuit for the simple lamp circuit. (b) Actual wiring and chassis for the same lamp circuit.

Note the two ground symbols in Fig. 3-5(b). They indicate that *E* and *F* are connected to a common ground line, in this case a chassis.

Figure 3-6 shows the PC board and chassis for the circuit above. The board has traces *AB* and *CD*. The side view of the unit, Fig. 3-6(b), shows the battery holder terminals connected to *A* and *F*. The lamp socket terminals are connected to *D* and *E*. *E* and *F* are connected to the chassis. The chassis provides the "ground-return" line.

3-5-2 The Preliminary Steps

Some of the preliminary steps for troubleshooting this lamp circuit have already been taken. The symptom has been noted—the lamp does not light. The chassis and circuit board have been removed from the outer case. The lamp and battery have been checked out by substitution.

There is a slight possibility that a short might overheat and damage the new battery; but it is usually not worthwhile to check it. You can usually tell it is the batteries that are bad because they steadily weaken in a normal way. This type of commonsense thinking and weighing of chances is a fundamental part of troubleshooting. In the process of installing the new lamp or battery, you might look for visual damage in the socket or the battery holder.

If by chance the new battery overheats, you will have to turn the lamp off and use an ohmmeter to check for a short. If the battery does not overheat, you can use a voltmeter to check for an open. These two alternatives are fundamental for troubleshooting power circuits. We pursue the line of attack for opens first.

The preliminary checks show that the battery did not burn out, so you know there is an open. The chassis has already been removed from the cover. Now it is time to visually inspect the PC board. Look for cracks in board and traces. Look for defectively etched traces. Look for bad soldering connections. After visually checking out the board, you can check the voltages.

3-5-3 Using the Voltmeter to Find an Open

As you know, it is common practice to measure the schematic test-point voltages to ground. You clip the black (common) lead of your voltmeter to the chassis (ground) and move the red prod from one test point to the next.

In the procedures below, we concentrate on the sequence of checks. The actual details for checking the connections and traces were given in the preceding section. The switch, lamp, and battery holder are usually replaced rather than repaired.

This is another basic troubleshooting decision. Should a part, a circuit, or even the entire equipment be repaired or replaced? Troubleshooting is not only a technical activity; it is also an economic activity. We opt to repair. The time and expense of repairing it is less than that for chasing after a new lamp unit. Besides, the user has a sentimental attachment to this lamp. The whole procedure should not take more than 2 or 3 minutes anyway—15 minutes to an hour for a beginner.

Two Procedures for Checking Voltages. There are two procedures for checking voltages. Each has its advantages and disadvantages. The first is to proceed around the circuit, checking each point in order, starting at the positive battery terminal and ending at the negative terminal. For more complicated circuits, this can be slow and tedious. However, you always know where you are at. It is easy to remember what you did and what you didn't do. This can be an advantage in some situations.

In the second procedure, you first check the point that has a good chance of taking care of the most possibilities at one crack. You do this with every succeeding check until the defect is found. This procedure is faster over the long run. This illustrates another fundamental point. Play the percentages. Over the long run, the law of averages says that it is faster.

Both methods use Rules 1-1 and 1-3. The first rule, if you recall, states that the trouble lies between two good points. The second states that an open tends to lower the voltages below it and raise the voltages above it. In this case the voltages are either pushed up to equal the supply voltage, or pushed down to equal the ground voltage. Either you have the supply voltage or you do not. There are no extra resistances besides the lamp to drop the voltage to any other level.

The Faster Procedure. Figure 3-7 indicates the order in which the checks are made. Check 1 is first. Depending on the outcome of Check 1, either Check 2 or Check 2′ is next. Check 3 follows check 2, and Check 3′ follows Check 2′. All voltages are with respect to point *G*.

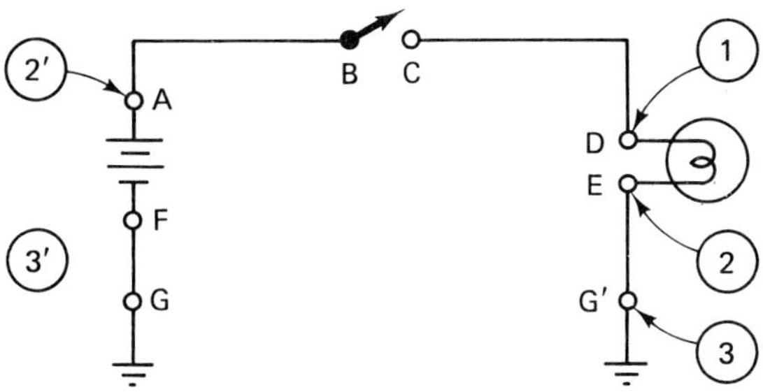

Figure 3-7 Quick procedure for checking voltages to find an open in a simple circuit. The circled numbers indicate the order in which the checks are made.

Check 1. Turn on the switch and check the voltage above the lamp at *D*. If point *D* has voltage, the battery and switch terminals are okay, and you do not have to check points *A, B, C, F,* and *G*. With one stroke, you eliminated five possibilities. The trouble must be at or between points *D, E, G′,* and *G*.

Do Check 2 next. If point *D* has no voltage, do Check 2′ next.

Check 2. Check the voltage at point *E*. The lamp should drop the full voltage, so you should have 0 V below it. If not, there is an open below. (The preliminary checks indicated a good lamp.)

Check 3. Check the voltage at point *G′*. You should have 0 V there. If not, there is an open below.

Check 2′. Going back to the first check, if the voltage at *D* is missing, check the voltage at *A*. If the voltage at *A* is good, there is an open below. Check the voltages at *B* and *C*. The switch could be open.

Check 3′. If the voltage at *A* is bad, check the voltage at point *F*. You should have 0 V there. An open between points *F* and *G* will give you a negative voltage reading at *F*. (The red probe of your meter is contacting the negative terminal of the battery.)

3-5-4 Using an Ohmmeter to Find an Open

To use an ohmmeter to check out the battery-and-lamp circuit for an open, refer to Fig. 3-8. Remove the battery first. Ohmmeter tests are always made with the power removed. Clip one ohmmeter lead to the battery terminal *A*. The other lead will be moved to various points in the circuit to measure their continuity to terminal *A*.

Check 1. Check for continuity between *A* and *F*. Turn the switch on and off. This will confirm whether you have an open in the circuit. You should be reading the lamp resistance if you have continuity. If not, do Check 2.

If you do not find an open, go back to the preliminaries. Now, if you still can't find anything wrong, you have what is known as an *intermittent*. Intermittent troubles are covered in Chapter 16.

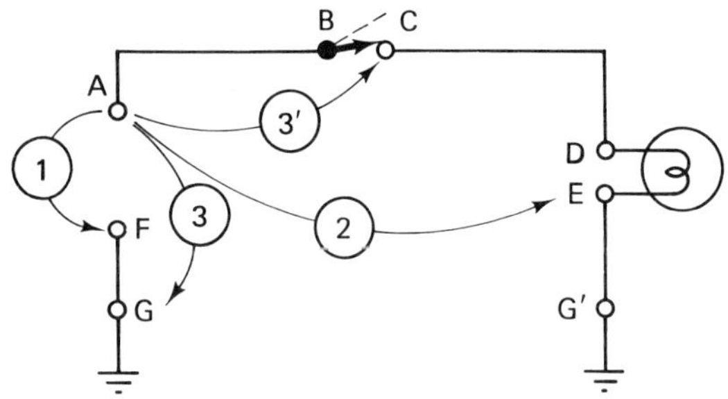

Figure 3-8 Procedure for finding an open in a simple circuit with an ohmmeter.

Check 2. Turn the switch on. Check for continuity between *A* and *E*. You are checking the switch and lamp socket among other things. (The lamp was already checked.) You should be reading the lamp resistance. If you do, the lamp socket and switch are good; do Check 3 next.

If you don't, the switch may be open; do Check 3′ next.

Check 3. Measure the continuity between *A* and *G*. You should be reading the lamp resistance. If you don't, a ground-side connection is most likely open. That is, any connection or conductor between points *F* and *E* can be open.

Check 3′. Measure the continuity between *A* and *C*. Flip the switch a couple of times. If you have continuity through the switch, it is good. The lamp socket or any conductor or connection at or between points *C* and *E* could be open.

If you have no continuity between *A* and *C*, the switch or anything at or between *A* and *C* could be open.

3-5-5 Using an Ohmmeter to Find a Short

A short will damage or weaken the battery, so you had better use an ohmmeter (refer to Fig. 3-9).

Check 1. Remove both the battery and the lamp. Turn *off* the switch and measure the resistance across the battery terminals *A* and *F*. Use the highest resistance range. If the meter reads infinite ohms, the supply side of the switch is okay. Do Check 2 next.

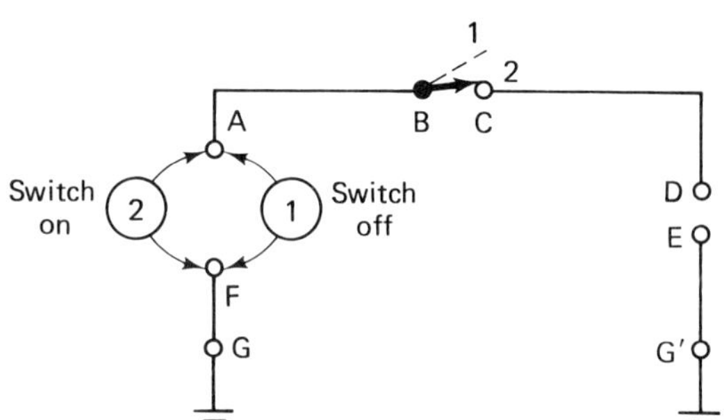

Figure 3-9 Quick procedure for finding a short in a simple circuit with an ohmmeter.

If you have a high-resistance reading in the high ohmmeter range, you have a partial short. The battery holder or board can be dirty. They should be cleaned out.

If you have near-zero ohms, you have a dead short. Look for hairline solder splatter and loose wire that may be shorting between trace *AB* and the chassis or the corner mounting screw.

Check 2. This time turn *on* the switch and take the same measurement as before across *A* and *F.* You should have infinite ohms. If not, the socket is the most likely part to be shorted. There may be a loose part in it.

As before, a high-resistance reading in the highest ohmmeter range indicates a dirty board.

> TIP 3-9
>
> When checking for an open or high resistance, use the higher ranges of your meter. When checking for a short or low resistance, use the low-resistance ranges.

3-6 TROUBLESHOOTING AN AC POWER-INPUT CIRCUIT

In Fig. 3-5 we saw a basic lamp circuit with a switch and battery. Figure 3-10 shows a similar circuit. It is a basic power-input circuit. Instead of a battery, we have an ac plug. Instead of a lamp, we have a block representing the main circuits. In addition, we have a fuse for protecting the circuit parts. When there is a short and excessive current flows, the fuse melts and cuts off the current.

The circuit block can be treated as a single part. That is, it can open and short or change resistance—just like a single part. Working on the input circuit will teach you a lot of things about troubleshooting in general.

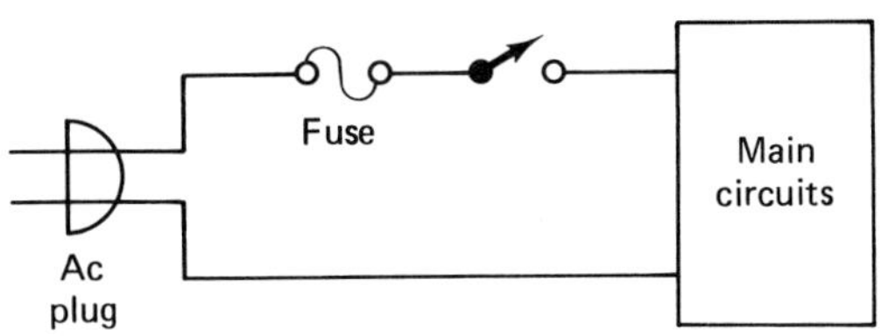

Figure 3-10 Power input circuit. Do not check this circuit with a voltmeter. It is safer and quicker to use an ohmmeter. The procedures are the same as for the simple lamp circuit.

3-6-1 The Game Plan

The first step is to determine whether there is any trouble at all. In this case we find that the symptom does indicate trouble. There is no sign of operation or output of any kind.

Next, we must determine whether the cause is in the equipment, in the ac outlet, or in the building circuit-breaker panel. There may be other possibilities lurking, but we'll take the most likely or easiest ones first. We cover these in the preliminary steps. If nothing is found in the preliminaries, we look into the equipment, possibly removing the chassis.

Since there is no sign of power, the possibilities in the equipment are: Either there is no power or signal input, or there is no power or signal output. Since there is no sign of any kind of action at all, we decide to check the power input first. Furthermore, the power-input circuit is easy to check, and the status of the fuse may also tell us something.

We next determine what kind of test instrument to use. We decide on the ohmmeter for safety. (**Caution:** Always disconnect the ac plug and always use an ohmmeter to check out a power-input circuit.) The input circuit is connected directly to the ac power line. It is always best to stay away from this danger source. Now that we have a game plan and the ohmmeter handy, let us look at the step-by-step procedures.

3-6-2 The Preliminaries

Let's see whether we have any trouble at all. Plug the equipment in and turn on the power and see what happens. If nothing happens, wiggle the ac line cord and plug and see if you can get a response. If nothing happens, check the fuse. The glass type can be checked visually. The glass could be discolored. Look for the melted or broken fuse wire. You can also check it out-of-circuit for continuity with an ohmmeter.

The following tip shows how the fuses or circuit breakers can help you determine the source of the trouble.

TIP 3-10

If the equipment is dead, a good fuse or circuit breaker indicates an open after it or a short before it.

If the building circuit breaker trips, and the equipment fuse or circuit breaker does not, the short is between the two fuses or breakers.

If the building circuit breaker does not trip, but the equipment fuse or circuit breaker does, the short is in the equipment.

If all the fuses above are okay, the equipment probably has an open.

If the equipment is small, you can take it over to another outlet and test it. Otherwise, visually check the plug and ac cord insulation. If it is okay, wiggle the cord again for a response from the equipment. If there is a response, the wall socket or power cord may be defective.

For a more positive test of the wall outlet, try a test lamp or any other electrical device such as a small radio that happens to be around. You can even run an extension cord from the equipment to another outlet.

If you are going to do a lot of work with big equipment, you should carry along one of those little three-prong checkers that plug right into the outlet. They'll check the outlet for you.

If a test lamp, socket tester, or any other electrical device is not handy, an experienced technician can show you how to check an outlet with an ac voltmeter. An inexperienced technician can cause sparks to fly. Tools and parts can disintegrate. Needless to say, this can be bad for your eyes. Do not attempt to learn electrical work unless you are under direct supervision.

The external circuit breaker can be checked by trying the other lights and sockets in the room or building. If more than one outlet or light is bad, check the external circuit breaker.

A prime suspect is the cord, particularly in portable equipment. For safety, unplug the cord. Recall that we learned how to check line cords in Section 3-2-1.

If everything checks okay up to this point, the trouble is in the equipment. It is time to remove the chassis.

3-6-3 Finding Power-Input Circuit Opens with an Ohmmeter

For safety, use the ohmmeter for checking an ac power-input circuit. (Don't forget to pull out the ac plug.) Basically, we are checking for opens as in the lamp circuit in the preceding section. Figure 3-11 shows a simplified ac power-input circuit with various check points. We will show the significance of each check and skip the details. This may give you a better feeling for the underlying thinking. Refer to the circled numbers for each check.

Check 1. Check for continuity across the plug prongs, A to A', while flipping the switch on and off. In most simple equipment, this confirms whether or not there really is an open.

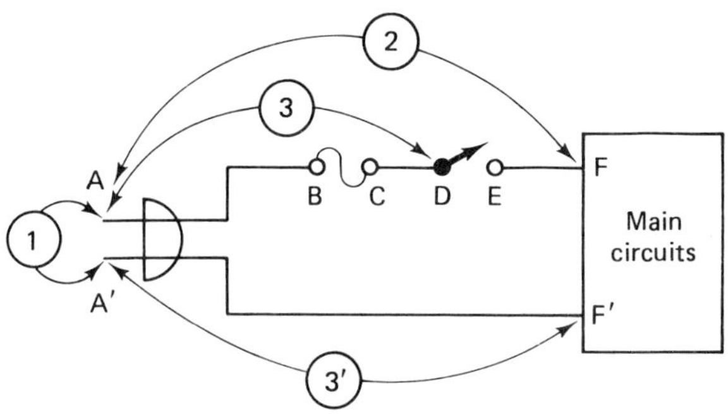

Figure 3-11 Procedure for checking a power-input circuit for an open with an ohmmeter.

Check 2. Check for continuity between the hot-side prong and the high side of the circuit block, *A* to *F*. This tells you whether an open is in the circuit block or the hot side of the input circuit.

Check 3. Check for continuity between prong *A* and the switch input, *A* to *C* or *D*. This tells you whether the switch or the fuse is open.

Check 3′. Check for continuity between the ground prong of the ac plug and the low side of the system block, *A′* to *F′*.

3-6-4 Finding Power-Input Circuit Shorts with an Ohmmeter

You can not use the voltmeter when the power-input circuit shorts out. The fuse blows and you have no voltages to measure. The following checks with the ohmmeter may prove instructive (refer to Fig. 3-11).

Check 1. With the *switch open,* check across the prongs, *A* to *A′*. This isolates line cord from the rest of the circuit.

If you get less than infinity now, the short or partial short is on the input side of the switch. If you get infinity, the short is on the output side of the switch.

Check 2. With the *switch closed,* check across *A—A′* again. If the input circuit is good, you should get some response with your ohmmeter. The amount of resistance you get might give you a clue as to where the trouble might be. You will learn more about this in Chapter 6 when we deal with power supplies.

3-6-5 Power-Line Interference Filters

Many ac power-input circuits have interference filters built into them (see Fig. 3-12). You will learn more about filters in Chapter 7. This one filters out interference voltages on the power line. For now we are mainly interested in whether the coils are open or not, and whether the capacitors are shorted or not. If the coils are open, they will have infinite resistance and so will the resistance across *A—A′*, even with the switch on. These coils are made of thick wire and you will probably never see an open or shorted one. You may find one that is poorly soldered. You will probably see burn

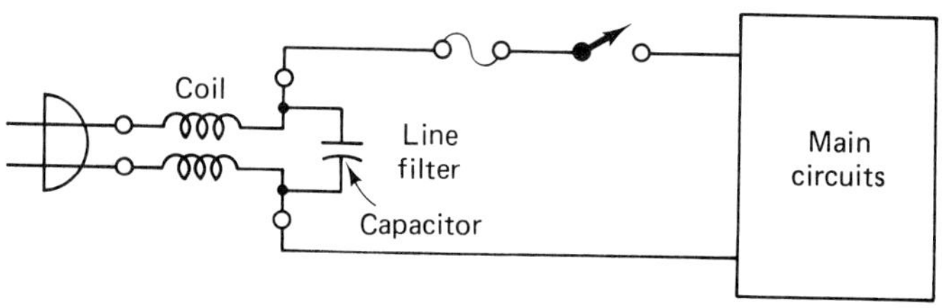

Figure 3-12 Power-input circuit with a line filter that reduces the interference voltages coming in or going out on the ac line. The coils have practically 0 Ω and the capacitor should have infinite resistance.

marks if they are poorly connected. They are thick and unbending, so their leads might break if subject to movement.

If the capacitor is shorted, the building circuit breaker will trip. If it partially shorts, the input resistance *A—A′* will have less than infinite resistance with the switch open. However, it is more likely to overheat and possibly explode or show visible signs of damage. These capacitors are rated for high working voltage, and you will probably never see one that's shorted. Most of the time, it is the solder connections that open.

WHAT YOU LEARNED TO DO

In this chapter you learned how to:

1. Work safely and confidently around equipment powered by 120-V ac. You learned that earth ground, and many things connected to it, including water pipes and damp concrete floors, conduct 120-V ac.
2. Inspect line cords, connectors, switches, fuses, circuit breakers, solder connections, PC boards, and PC card edge connections.
3. Take the preliminary steps for troubleshooting power-input circuits.
4. Identify power-input circuits and troubleshoot them with a game plan that covers all the possible causes, and to choose the easiest, fastest, and most-probable possibility first.
5. Use the ohmmeter for the fastest and safest way of checking the power-input circuit for opens and shorts.

QUESTIONS

1. Suppose that someone had clipped the ground prong of a polarized plug and inserted it so that the hot side was connected to an electronic chassis. What would happen if a test-instrument lead that is connected to ac ground is clipped to the chassis?
2. What would happen if a cable-TV line connector accidentally touched the chassis above?
3. Give three reasons why an isolation transformer should be used.
4. Why should you use only one hand when working on electronic circuits?
5. Should an ohmmeter be used to measure a resistance in a chassis for which the power has not been turned off?
6. In Fig. 3-7 suppose that points *D* and *E* have maximum voltage with respect to chassis. What parts or connections could be bad?
7. In Fig. 3-8 there is zero ohms between points *A* and *F* with the switch closed. What would you do next to isolate the trouble?

PRACTICAL TROUBLESHOOTING EXPERIENCE

1. Remove the batteries from a regular flashlight. Identify the positive conductor and connections. Identify the negative conductors and connections. Identify the switch contacts and the switch terminal connections to the conductors. How can you check the switch action with an ohmmeter?
2. Look at the schematics of various types of equipment. Find the power-input circuit in each of them. Identify the ac plug and cord, the power-input fuse or circuit breaker, and the ground and hot-side terminals to the main circuits in each schematic.
3. The following will acquaint you with a procedure that you will be doing quite often—checking the power-input circuit with an ohmmeter.
   a. Obtain an ac-powered electronic chassis.
   b. Check the continuity across the prongs with an ohmmeter. With the chassis still unplugged, flip the power switch on and off.
   c. Follow the hot wire of the line cord to its terminal connection.
   d. Find the hot-side wire or conductor leading to the fuse or circuit breaker. You may find parts that we have not mentioned yet. The ones in series with the fuse and switch should have zero resistance. The others should have infinite or very high resistances unless they are coils.
   e. Check the continuity between the hot prong and the switch with an ohmmeter. How can you tell which side of the switch leads to the hot prong?
   f. Follow the wire and/or conductor that leads from the fuse or circuit breaker to the switch.
   g. Follow the wire from the switch to the main circuit.
   h. Follow the ground wire. Where does it terminate?

4

Troubleshooting Branching Paths

EXPERIENCE REQUIRED

To use the information in this chapter, you should know that higher-than-normal resistances drop higher-than-normal voltages, and that lower-than-normal resistances drop lower-than-normal voltages. You should know that complete shorts drop 0 V and that complete opens drop the maximum voltage.

4-1 TROUBLESHOOTING PARALLEL CIRCUITS

In Chapter 2 we learned how to troubleshoot simple series paths. They had no paths splitting off from them, nor did they have any paths joining up with them. We now have to find out how these other paths affect the voltages.

4-1-1 Two Lamps Connected in Parallel

Figure 4-1(a) shows two lamps connected in series. The lamps do not have the same resistances; therefore, they do not have the same voltages. Figure 4-1(b) shows the same two lamps connected in parallel. Both lamps now have the same voltage; they are connected to the same power supply.

Let us now look at the current through these lamps. The series-connected lamps are dimmer. This is because neither of the series lamps has the full supply voltage. Each of the parallel lamps, on the other hand, has the full supply voltage on it. They glow brighter.

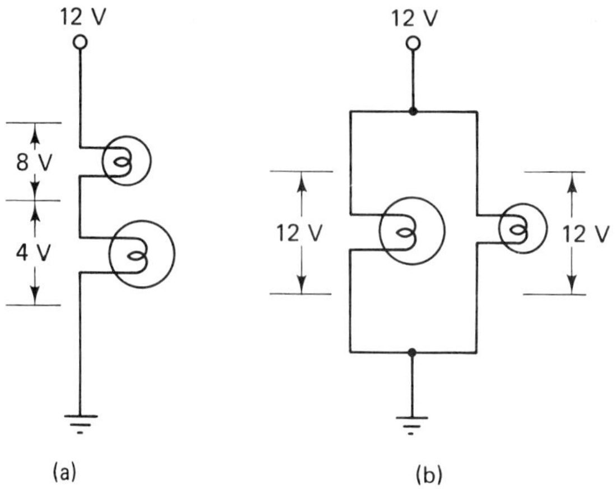

Figure 4-1 (a) The series lamps taken together have more resistance, hence less current and voltage on each lamp. (b) The parallel lamps taken together offer less resistance and together pass more current, and each lamp has the supply voltage.

Now let us look at the resistance offered by the two lamps. In the case of the series lamps, the supply voltage must squeeze the current through two parts, one after the other, so the resistance must be greater. Therefore, less current flows through. The dimness of the lamps shows this. In the case of parallel lamps, the two pathways allow more current to flow through than one pathway. Together, they have less resistance.

If the two lamps had equal resistances, we could give you a simple formula for figuring out their resistance in parallel: $R_P = R/2$. You simply take one-half of their individual resistance value.

In the case of unequal resistances, the formula is more complex. In actual practice it is very difficult to know just which and how many parts in the surrounding circuits are actually in parallel. Fortunately, we hardly ever perform any calculations. The most important thing to know when you measure the resistance across a part in parallel with other parts is this: The measured resistance across the part is less than the out-of-circuit resistance of the part. In fact, it is less than the out-of-circuit resistance of any of the parallel parts. Common sense would tell you that if you have more pathways for the current, you are going to read less resistance to it.

Equation for Parallel Resistances. Sometimes when you cannot find the right-value resistor for a trial substitution, you will have to tie two or more resistors in parallel to obtain the right value. You can then use the formula $1/R_T = 1/R_1 + 1/R_2$ for two resistors, where R_T stands for their combined resistance. If you have to use three resistors, the formula becomes $1/R_T = 1/R_1 + 1/R_2 + 1/R_3$. The formulas are difficult to work out with a pencil, because you have to find the reciprocal of each resistance. With a scientific calculator, it is easy. Just press the 1/X key to obtain the reciprocal for a resistance. After adding the reciprocals, you press the reciprocal button one last time to obtain the parallel resistance of all the parts tied together.

In troubleshooting, we first try to determine whether a resistance in a parallel circuit is okay without disconnecting it from the circuit. The following tip should be helpful.

> **TIP 4-1**
>
> If a part in a parallel circuit opens completely or partially, the resistance of the whole circuit will increase. (Excepting capacitors, Section 7-2.) The resistance across the circuit will decrease if a part in it shorts completely or partially. (Excepting coils, Section 7-2.)
>
> The resistance across a parallel circuit should be less than its lowest-resistance part.
>
> The resistance across a parallel circuit with two parts should not be more than half the resistance of its highest-value part.

Interesting Sidelight. It may be easier to understand parallel resistances if you see them as parallel pathways rather than as resistances. In electronics math, these pathways are called conductances. Mathematically, the current-carrying ability of a pathway is called *conductance* or G, where $G = 1/R$, or G is the reciprocal of R. In other words, we are saying the greater the resistance, the smaller the conductance. When the resistance of a part increases, its conductance decreases.

The equation for parallel conductances is easy: Just add up the two conductances. Tip 4-1 is easier, too, if we replace *resistance* with *conductance* as shown.

> **TIP 4-1A**
>
> If a part in a parallel circuit opens completely or partially, the conductance of the whole circuit will decrease. (Excepting capacitors.)
>
> The conductance across the circuit will increase if a part in it shorts completely or partially. (Excepting coils.)
>
> The conductance across a parallel circuit is greater than its highest-conductance part.
>
> The conductance across a parallel circuit with two parts is less than twice the conductance of either part. (Too bad that ohmmeters do not display conductances.)

Very often you are better off simply disconnecting a part and measuring its resistance, because it is harder to find the other resistances that are in parallel with it. However, if there are many parts, you may find yourself disconnecting part after part without finding a bad one.

> TIP 4-2
>
> If there are only a few parts in parallel, it may be best simply to disconnect the parts and measure their resistances.

If there are more than a few parts, we have problems. The following information may give us some help.

4-1-2 Opens or Partial Opens in Parallel Circuits

Troubleshooting a parallel lamp circuit is easy. There are three very important troubleshooting reasons for this. First, an open lamp will not light up. Second, a shorted lamp is rare. Third, most lamps are detachable. You can easily remove a lamp and check it. However, not all parts give you a visual indication when they go bad. They short about as often as they open; and worse yet, they are soldered in. Checking parallel parts in general is not easy. In addition to the above, they present another problem. In Fig. 4-2(a), P_1 is good and P_2 is bad. If you measure the voltages on each part, they will show the same voltage. The voltage will not tell you which is bad.

Figure 4-2(b) shows the measured resistance across each part; both readings are equal. That is because the ohmmeter is connected to both parts simultaneously. The only thing the resistance reading tells you is whether the circuit is bad—but not which part.

The situation is worse when more than a few parts are connected in parallel. They will all show the same voltages or resistances across them [see Fig. 4-2(c)]. If one of them is bad, you will not be able to locate it with a multimeter. If it were not for parallel paths, troubleshooting would be a snap.

This is where your ingenuity comes in. One or a combination of the following may enable you to find an open.

1. Check the temperature of the parts. An open or increased resistance part may be noticeably colder than normal in some circuits. Be careful—turn off the power. Touch lightly; some parts normally run hot enough to sear your fingers.
2. Microvoltmeters, micro-ohmmeters, current tracers, and short finders can be used in computer circuits to find opens.
3. As in the case of the lamp, the special functions of a part may give you a clue. We take this up in Chapters 5 and 8.

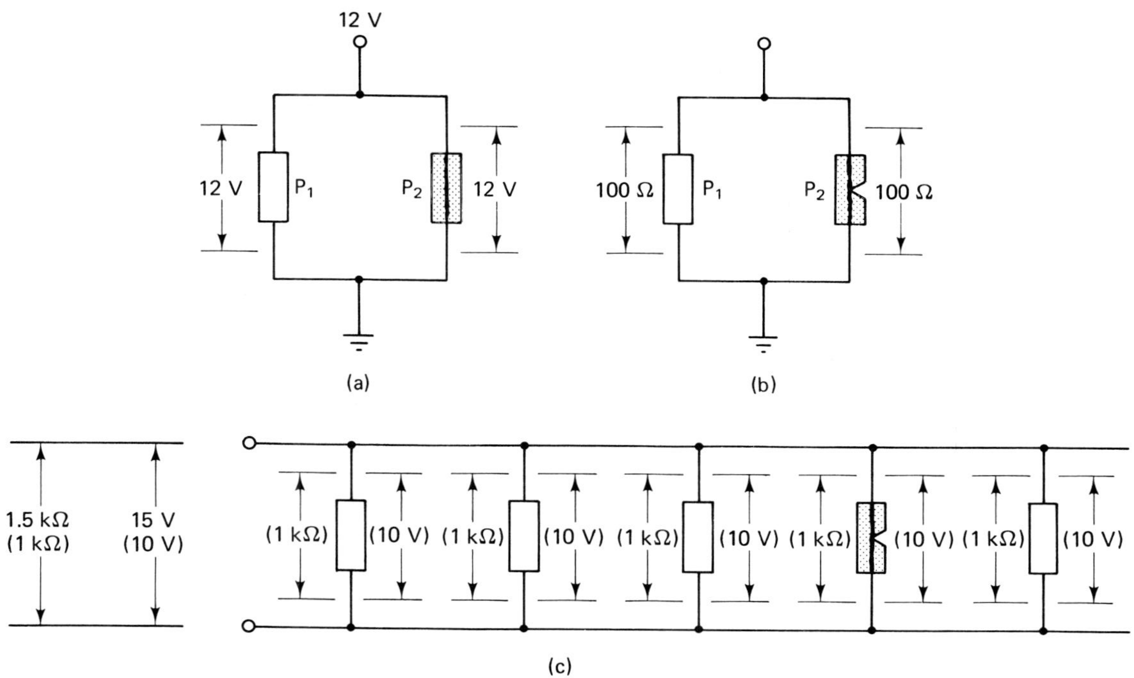

Figure 4-2 Parallel circuits have equal resistances and voltages across all their parts, including the bad part. This makes it impossible to tell which part is bad with measurements only.

4. Some parts are more likely to open or increase resistance than others. Disconnect the most likely one and check it. We take this up in the next chapter (Section 5-4).
5. A high percentage of opens are actually bad connections, either at the solder joints or inside the parts. In either case, wiggling the part leads may reveal the open. Wiggle gently; you don't want to create new problems.

Here is a tip that may help when it takes too long to find the other parts in a parallel circuit.

> TIP 4-3
>
> If the resistance across an in-circuit part is higher than its rated out-of-circuit resistance, it is bad.

This tip works very well on coils. The coil has very little resistance, so no matter what part is connected across it, its in-circuit resistance will rise if it opens. Actually, this tip works with any kind of part in any kind of circuit.

4-1-3 Finding Dead Shorts in Parallel Paths

If there are many parts connected together in parallel and one of them shorts, you will have 0 Ω across all the parts. There is no way you can tell which of the parts is bad. You might try and disconnect them one at a time. If you are lucky, you'll find the short with just one check. In Fig. 4-3 there are eight parts in parallel. Chances are 8:1 you won't find the bad part on the first trial. For better odds, use the divide-in-half method in Section 1-7-3 as follows:

1. Cut the supply line at point *A* so that half of the resistors can be eliminated. If you still read a short, the ohmmeter side of the cut has a short. If you no longer read a short, the short is on the other side.

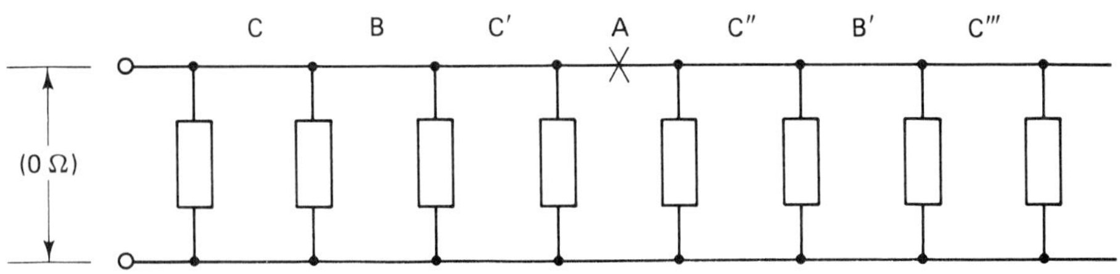

Figure 4-3 Finding a shorted part. The circuit is divided at *A* to find which side has the short. Then the bad half is divided at *B* or *B′*. Keep dividing the bad half and checking until only one possibility remains.

2. Next, reconnect the cut if necessary, and cut the supply line at B or B' so that only half of the remaining half is measured by the ohmmeter. Determine which half has the short, and check that side as you did previously.
3. Repeat the steps above at C, C', C'', or C''', cutting the circuit in half, until you have narrowed the number of parts being tested to just two. Four parts require two cuts. Eight requires three.

4-1-4 The Series Test Lamp Method for Finding Shorts

Figure 4-4 shows the time-honored electrical benchman's method for finding a short in 120-V ac equipment with parallel circuits. Use it when other methods fail. A lamp is placed in series with the supply line. All safety precautions are taken. Use one hand only. If any one of the parts is shorted, the lamp will glow brightly. If the shorted part is removed, the lamp will glow less brightly. The parallel parts are removed one by one until the lamp ceases to indicate a short. Unplug the lamp each time you remove a part. The last part removed is shorted.

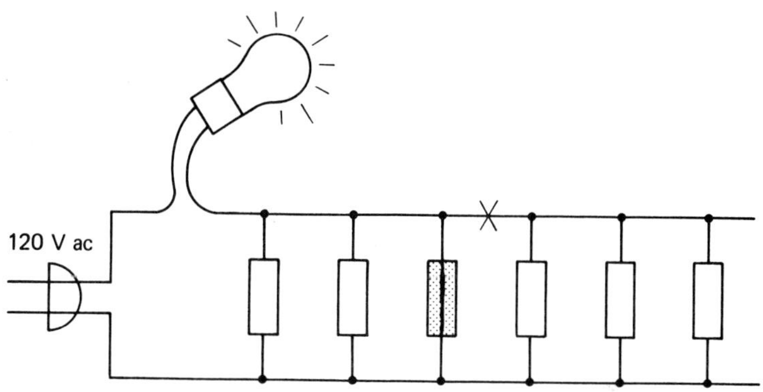

Figure 4-4 Finding a short with a series lamp. The short causes the lamp to glow brightly. Removal of the short will dim the lamp. Keep opening the line at the halfway points until the lamp ceases to glow brightly.

If there are many parallel parts, use the same procedure that was used for the ohmmeter test just previously. Unplug the lamp and cut the supply line at the halfway point as you did in Fig. 4-3. Turn on the power. If the lamp still glows brightly, the short is on the lamp side of the cut. Otherwise, it is on the other side of the cut. Proceed as you did with the ohmmeter test, cutting the circuit in half each time until only one part is left.

4-2 TROUBLESHOOTING SERIES–PARALLEL PATHS

The explanations in this section may seem a bit involved at times. This is mainly because we are dealing with theoretical part resistances and circuits. Real circuits have real parts that give us real clues—special clues. While a lot of practical hands-on information is found here, the main thing we will learn is how the basic rules for partial and complete opens and shorts can be applied. Do not try to memorize everything.

4-2-1 A Series–Parallel Circuit with Three Parts

Figure 4-5 shows a simple series–parallel circuit. This circuit does not give you much more information than the parallel circuit. The voltage at point *A* is the only one that can change when something goes wrong. We assume that the supply and ground voltages do not change.

The resistance of part P_1 serves to drop the voltage to the level at *A*. Parts P_2 and P_3 combine to hold up voltage *A*. If the voltage at *A* falls, it can be caused not only by an open P_1, but also by a short in the combined resistance of P_2 and P_3. Similarly, a rise in the voltage at *A* can also be caused by any one of the three parts. The voltage at point *A* cannot tell you which part is bad. Furthermore, in real circuits, resistance checks are unreliable because other circuits tied to the same supply and ground lines will affect your reading.

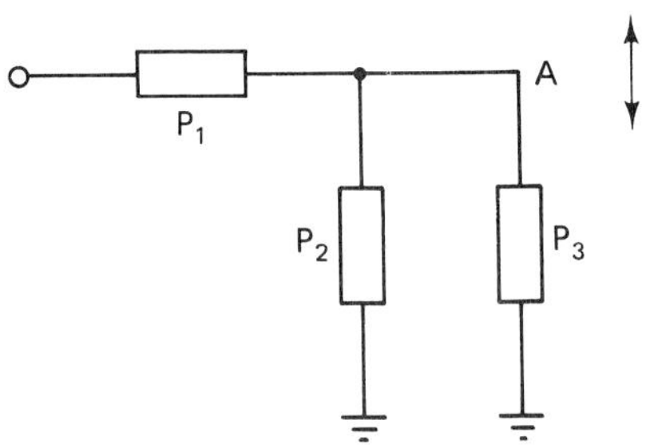

Figure 4-5 Simple series–parallel circuit. Removing a branch part will cause the remaining parts to divide the voltage in proportion to their resistances.

Fortunately, this circuit has a low parts count, so you might just as well go ahead and disconnect one of them. Otherwise, for the two parallel parts, use the methods described in the preceding section. As for the lone third part, P_1, it is connected in series with the two parallel parts taken as a unit, or block. If one of the two parts in the block shorts or partially shorts, the voltage above the block will fall. The whole block acts like a partial short: similarly with an open or partially open part; the entire block will act like a partial open.

In short, the simple series–parallel circuit is equivalent to a simple series circuit consisting of two parts. One part is P_1. The other part is represented by the block containing P_2 and P_3. Hence you can check out a simple series–parallel circuit using the methods developed for checking a series circuit with two parts in Section 2-5. You can also use a tip that is similar to the one developed for two-part series circuits.

TIP 4-4

In a simple series–parallel circuit, if the voltage below the single non-parallel part equals 0 V, and the resistance below it to ground does not equal 0 Ω, the part is open.

We are skipping connections, as usual. Don't forget them. Here are two more tips for the simple series–parallel circuit.

TIP 4-5

In a simple series–parallel circuit, if the single nonparallel part is the only one with zero voltage drop, it probably has an open connection, or it is shorted.

Tip 4-5 is almost 100% valid because it would take two open parts to cause a zero voltage drop across the series part. This is not likely to happen.

Here's another tip for series–parallel circuits. We cannot skip over this one. It makes use of the fact that in a series circuit, the voltages across each part are proportional to their resistances. It also makes use of the fact that when one of the branches opens, the remaining parts are in series.

TIP 4-6

In a simple series–parallel circuit, if one of the parallel parts opens completely, the voltage above it will rise and the voltage will divide the supply voltage in proportion to the remaining series resistances. Otherwise, the top part is partially shorted.

4-2-2 A Series–Parallel Circuit with Five Parts

Figure 4-6 shows a series–parallel circuit with five parts. Each of the parallel branches has two series parts. If we look at how the pressures are distributed in this circuit, we will find all the information we need to troubleshoot it with a voltmeter. Remember that electrical pressure is voltage.

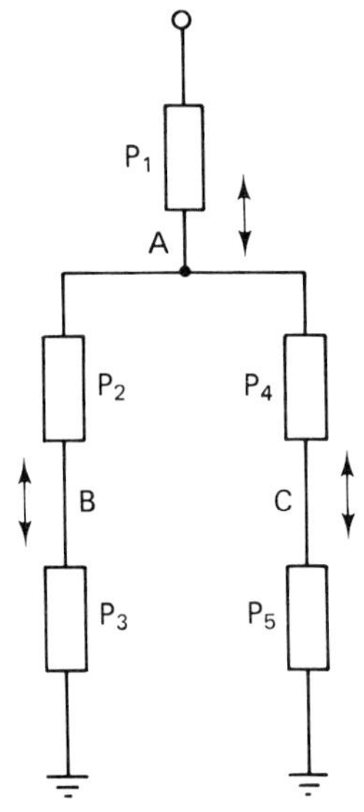

Figure 4-6 Series–parallel circuit with branches having series parts. The arrows indicate how voltages can change when a part goes bad.

If the resistance of P_1 increases, the pressure below at point A will drop. Voltage A falls. If the resistance of P_1 decreases, the pressure below at A will rise. Voltage A rises. This accords with the fundamental rules for partial and complete opens and shorts.

When the pressure at A changes, the pressures or voltages at B and C will change along with it by the same percentage. This is in accordance with the rule for voltage division, which says that the voltage drops in a series circuit are proportional to the resistances.

If one of the parts below changes resistance, the voltages above and below it will rise or fall in accordance with the rules we already learned for simple series circuits. In view of the foregoing observations, the following rule seems reasonable.

> RULE 4-1
>
> The rules for partial and complete opens and shorts can be applied to series–parallel circuits.

With this understanding, we should be able to see what happens as each part in the circuit goes bad. Let us take complete opens and shorts first.

Complete Opens and Shorts. If one of the very bottom parts opens completely, as in Fig. 4-7(a), its branch voltage will *rise,* and it will have the full branch voltage across it. It will also have the full branch voltage across it when the part immediately above it shorts, as in Fig. 4-7(b); however, the branch voltage will *fall* in this case.

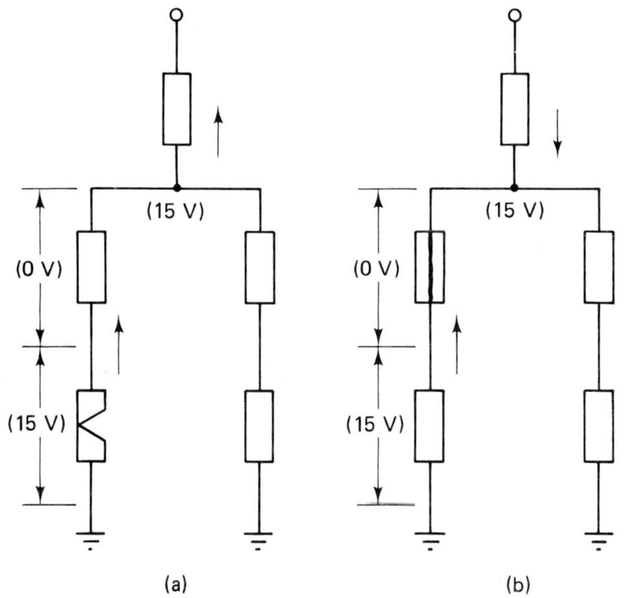

Figure 4-7 (a) An open branch part raises and equals the branch voltage. (b) A shorted branch part drops zero volts and lowers the branch voltage. It also tends to raise the voltage below it.

Partially Defective Parts. Now let us look at partially defective parts in a series–parallel circuit with five parts. We can use the basic rules and the voltage division laws to check the circuit. Let us skip the details and note the following:

1. If the top part partially opens or shorts, as in Fig. 4-8(a), all the voltages below it will change by the same percentage.

   The voltage drop across the bad part changes by the greatest percentage.
2. If one of the branch parts partially opens, as in Fig. 4- 8(b), the voltage below it will tend to fall and the voltage above it will rise. The branch voltage at *A* will rise. All the voltages in the good branch will rise by the same percentage.

   The voltage drop across the bad part will change by the greatest percentage.
3. If a branch part shorts, as in Fig. 4-8(c), the voltage above it will fall and the voltage below it will tend to rise. The branch voltage at *A* will fall. All the voltages in the good branch will fall by the same percentage.

   The voltage drop across the bad part changes by the greatest percentage.

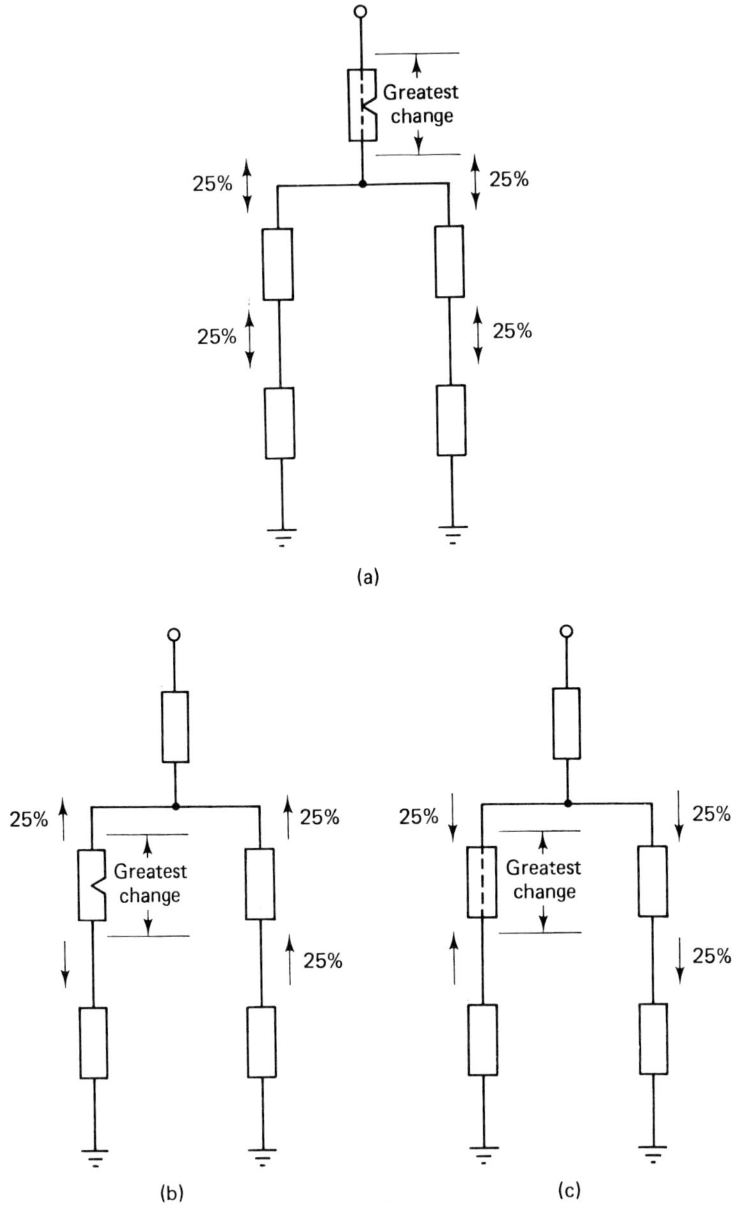

Figure 4-8 (a), (b), and (c) show that the voltage across the bad part changes by the greatest percentage. Also, the voltages in a good branch change by the same percentage.

In the following tip, the voltages in one branch are not necessarily considered to be "above" or "below" the voltages in another branch.

> **TIP 4-7**
>
> In a series–parallel circuit:
>
> The voltage drop across a bad series part will change by the greatest percentage. "Look for the voltage or voltage drop that has changed most drastically."
>
> All the voltages above a partial short will tend to fall and all the voltages below it will tend to rise. All the voltages above a partial open will tend to rise and all the voltages below it will tend to fall.
>
> All the voltages in a good branch will change by the same percentage. The voltages in the bad branch will not change by the same percentage.
>
> If all the voltages in the branch parts change by the same percentage, the single nonbranch part is bad.

Note: The tip above should be studied very carefully. Most power distribution systems (see Section 6-2-4) are wired in series–parallel. Very often there are more than two branching parts. The tip above applies to any number of branches. "Look for the voltage or the voltage drop that has changed most drastically."

The next chapter has more information for finding defects in series–parallel circuits. It deals with specific parts that give you specific clues that make it easier to locate the defective parts. For this reason we can skip a lot of detailed explanations for some of the tips given here. However, you should study this chapter to get some notion of how the rules are applied generally.

4-3 TROUBLESHOOTING PI CIRCUITS

The pi circuit is so called because it looks like the Greek capital letter for pi (Π; see Fig. 4-9). It has many uses. It is similar to a dc power-supply filter circuit we will see in Chapter 6. It is also used in signal circuits in Chapter 7, where you may have to use an oscilloscope. In this section we learn about simple dc voltage and resistance checks.

Let's see what happens to voltages *A* and *B* when each part goes bad.

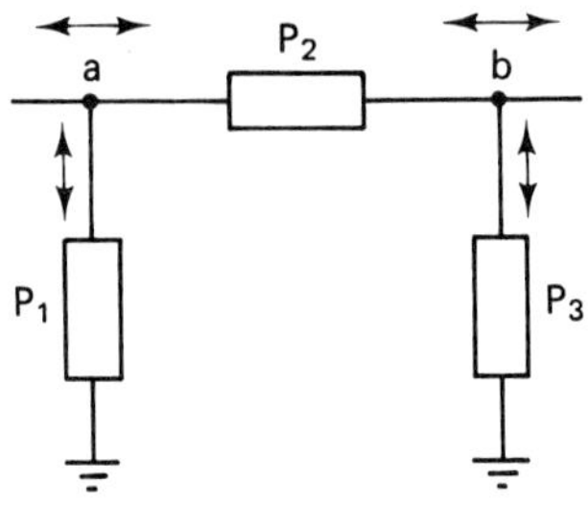

Figure 4-9 The voltages changes at points *a* and *b* in this pi circuit will tell you which part is bad.

4-3-1 Complete Opens and Shorts in Pi Circuits

Complete opens and shorts are often easy to spot. The following cases demonstrate this in pi circuits (refer to Fig. 4-9). Remember, we are not talking about coils and capacitors in this section. We will get to them in Chapters 5 and 7.

1. If the part P_1 shorts completely, the voltage at both A and B will fall to zero.
2. If part P_3 shorts completely, the voltage at B will fall to zero, but the voltage at A will not.
3. If part P_2 shorts completely, the voltage at A and B will be equal to each other.
4. If part P_2 opens completely, the voltage at A will rise and the voltage at B will fall to zero.

The remaining cases are not as easy. They are covered in the next section.

4-3-2 Checking a Pi Circuit with a Voltmeter

The pi circuit is a variation of the series–parallel circuit in the preceding section. The same tips can be used. The voltage drop across a bad series part (P_2 or P_3) changes by a greater percentage than do the good parts. If P_1 goes bad, all voltage drops will change by the same percentage.

1. When P_1 increases resistance, voltages A and B go up by the same percentage [see Fig. 4-10(a)].
2. When P_1 decreases resistance, voltages A and B go down by the same percentage [see Fig. 4-10(b)].
3. When P_2 increases resistances, voltage A goes up. Voltage B goes down [see Fig. 4-10(c)].
4. When P_2 decreases resistance, voltage A goes down. Voltage B goes up [see Fig. 4-10(d)].
5. When P_3 increases resistance, voltages B and A go up, but voltage B will rise by a greater percentage. [see Fig. 4-10(e)].
6. When P_3 decreases resistances, voltages A and B go down, but voltage B will fall by a greater percentage [see Fig. 4-10(f)].

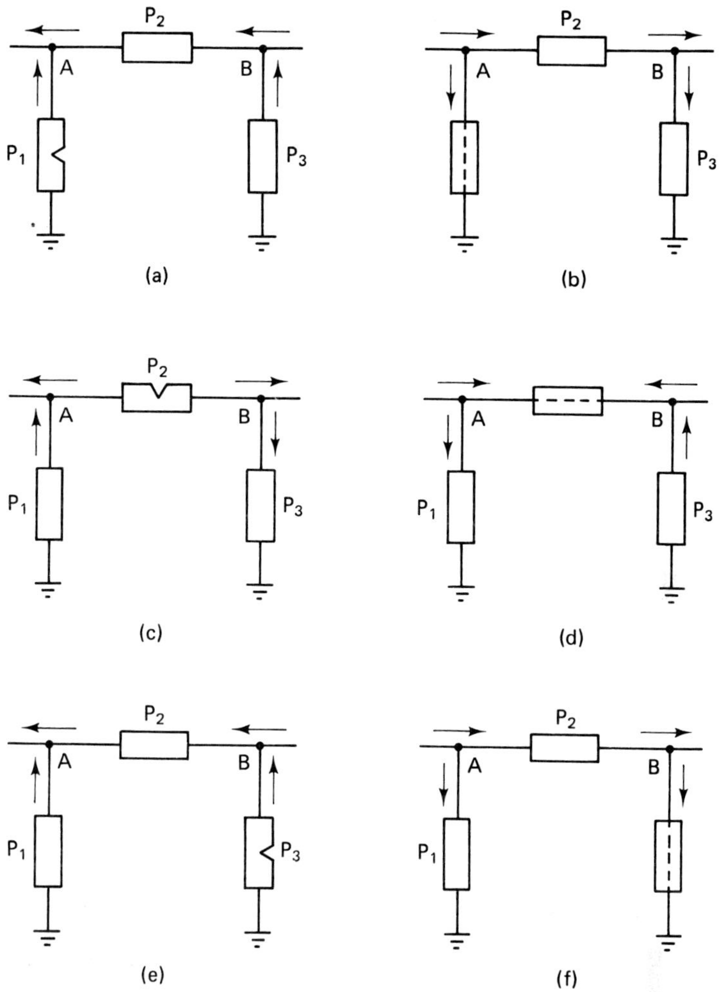

Figure 4-10 If P_1 goes bad, both voltages A and B will change by the same percentage. If P_2 or P_3 go bad, its voltage drop will change by the greatest percentage. Note that a partial open tends to push up all the voltages above it and push down all the voltages below it. A partial short tends to pull down all the voltages above it and pull up all the voltages below it.

4-3-3 Checking a Pi Circuit with an Ohmmeter

Pi circuits are often checked with an ohmmeter when voltages cannot be measured. We have a few checks that may prove useful.

1. If one leg, P_1 or P_3, partially opens or shorts, both resistances will change but the bad leg will change more in terms of percentage.

2. If one leg has a dead short, it will measure zero ohms; the other will not.
3. If the top part shorts completely, the legs will show equal resistance.

4-4 THE BRIDGE CIRCUIT, A MULTIPLE-SOURCE CIRCUIT

Figure 4-11(a) shows a bridge circuit with two sources. When the side parts divide the voltages so that voltages at *A* and *B* are equal to each other, the bridge is said to be in a balanced condition. In this condition, a change in the resistance of P_5 will have no effect on voltages *A* and *B*.

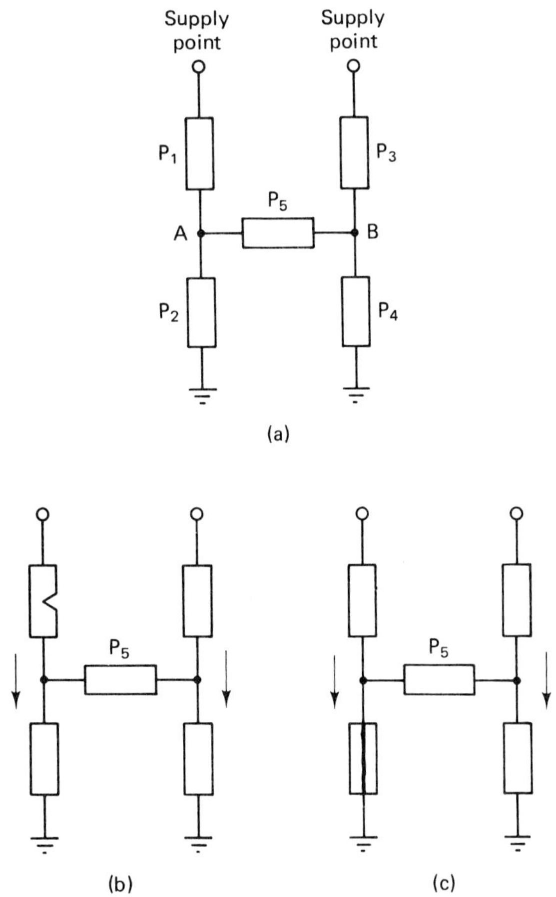

Figure 4-11 The direction of the voltage changes in a bridge circuit at points *a* and *b* are the same for partial opens and shorts. There are several tips in the text that show how to deal with this.

The balanced bridge circuit is one of the exceptions to the troubleshooter's assumption that a good voltage probably indicates a good circuit (Tip 1-1). A bad P_5 has no effect on the voltages at A and B when the bridge is balanced.

Another problem arises when troubleshooting this circuit. Figure 4-11(b) shows what happens when a top part opens partially, and Fig. 4-11(c) shows what happens when a bottom part shorts partially. In both cases voltages A and B go down. Simply noting whether a voltage goes up or down is not enough.

> *Note:* In a bridge circuit, the voltage drop across the bad part changes by the greatest percentage.
>
> The side having the bad part will have the greater percentage change of voltage.
>
> Zero dc voltage drop across any side part indicates that the part has a dead short.

If the circuit is balanced, we noted that P_5 will give no indication of changed resistance value (unless it has a bad connection). In this case, unbalance the bridge. Jump (shunt) one of the side parts with a comparable value resistance. For example, if P_1 equals 1-k Ω, jump it with a 1-k Ω part. The results of this test are given below.

> *Note:* In a balanced bridge circuit, if something is wrong but the dc voltages are okay, the bridge member could be bad. Jump a side part with a resistance of comparable value.
>
> If you can change the voltage but it remains balanced, the bridge part is completely shorted.
>
> If you cannot change the voltage on the other side, the bridge part is open.

Otherwise, disconnect one side of P_5 and measure its resistance. Do not waste your time trying to figure things out. To solve this on paper, you must use advanced network theorems.

4-5 CIRCUIT BLOCKS

Complex electronic circuits can be broken down into groups of smaller circuits called blocks. If we can treat a block as a single part, things will be a lot easier. For instance, look at Fig. 4-12. Each block can be treated as a single part. The arrows show whether the blocks are partially shorted or open, just as if they were single parts.

On the other hand, as we saw in the case of a balanced bridge, not all blocks can be treated as single parts. However, when a block does have an abnormal voltage across its terminals, we can treat it just as we would a single part. The following rule should help us locate bad blocks.

RULE 4-2

The same rules for finding defective parts can be used for finding defective circuit blocks.

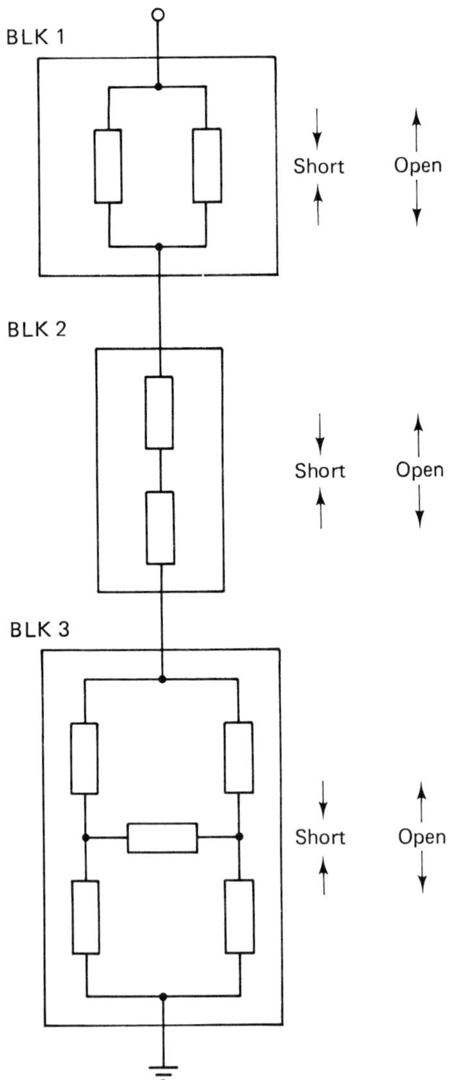

Figure 4-12 If a block has a bad voltage across it, a part in it is bad. However, if a block has a good voltage, it does not necessarily mean that it is good. Note that block 3 might show a good voltage if the bridge part in it goes bad. Similarly, a shorted coil or an open capacitor will not show up a bad dc drop.

The following tip assumes that we can treat blocks like parts.

TIP 4-8

A short in a circuit block may tend to pull down the voltage above it and pull up the voltage below the block.

An open in a circuit block may tend to push down the voltage below the block. (Coils, capacitors, and balanced bridges excepted.)

The rule and tip above permit us to find a fault in two stages:

1. Find the bad block.
2. Find the bad part in it.

Let's take a simple example. In Fig. 4-12, block BLK 1 is made up of two parallel parts. If either part opens, the voltage below it will fall. If either part shorts, the voltage below it will rise.

BLK 2 is composed of two parts in series. If either part opens or shorts, the voltages above and below it change accordingly.

BLK 3 is a bridge circuit with the top legs connected to the input line. If the circuit is normally unbalanced, any open in it will cause the voltage above the block to rise and any short in it will cause the voltage to fall.

If the bridge circuit in BLK 3 is normally balanced, only the bridge part will fail to indicate a bad block or part.

We have just seen what complete opens and shorts inside the block do to the external voltages outside the block. Partial opens and shorts would have the same results, partially. Analyses of more complex circuits is beyond our purposes. Actual practice will show that in most cases an open part will make the entire block act like an open part, and similarly with a shorted part. In all cases, a defective block voltage indicates a bad internal part. A shorted block points to a short inside it. The same goes for an open.

4-6 THE LOAD AND VOLTAGE TRANSFER CIRCUITS

The purpose of electronics is to transfer voltage or power from one circuit or device to another. A microphone must pass signal voltage to an amplifier. An audio amplifier transfers signal power to a speaker. A video amplifier must transfer signal voltages to a display tube. A tuner will pass signal voltage to an amplifier. These circuits are designed to transfer the maximum voltage or power from a source to a load. Let's see how shorts and opens in these transfer circuits affect the voltages.

4-6-1 Troubleshooting the Source and Load Circuits

Figure 4-13 shows a typical source circuit block that passes voltage to a load circuit block. The source circuit consists of a voltage source, a series part P_1, and an output part P_2 that helps to divide the source voltage to obtain the output voltage V. The load circuit, block L, is in parallel with P_2, so it has the output voltage on its terminals.

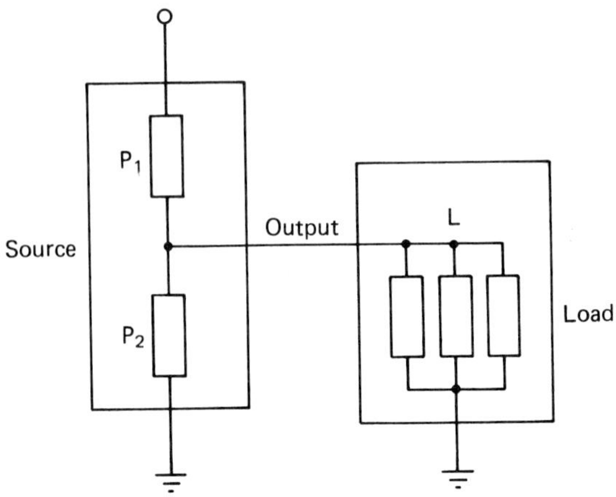

Figure 4-13 A defect in the load circuit block will pull down or raise the source voltage. This concept of load is very important in electronics.

An investigation of this voltage-transfer circuit will show that if the load L partially opens, the output voltage will rise; and if the load L partially shorts, the output voltage will fall. You will encounter this circuit very often in troubleshooting. Let's see what happens when key parts in this circuit go bad.

If the series part, P_1, in the source circuit opens, the output voltage will fall. If it opens, the output voltage will rise.

If the parallel part, P_2, in the source circuit opens, the output voltage will rise. If it shorts, the output voltage will fall.

We already mentioned that if the load opens, the output voltage will rise; and just the opposite will happen if the load is shorted.

The key voltage is at the output. If it is too high, look for an open in the load L, an open P_2, or a shorted P_1. If it is too low, look for the opposite conditions.

The following checks will help you further isolate the bad part or block. Check the load circuit for bad voltages and resistances. Review the sections on series–parallel circuits. You can disconnect the load and see whether P_1 and P_2 divide the voltage in proportion to their rated resistances. You can also investigate the voltages in the source circuit. Check the temperature of P_1; it will be cooler if there is an open below it and warmer if there is a short below it.

4-6-2 "Loading Down a Circuit"

We often come across the words "heavy load" and "light load." A heavy load has a low resistance and "draws" a heavy current. A light load has a high resistance and draws a light current. We also hear the expression, "A short is loading down the voltage." It means that the short is "pulling down the voltage" across it while drawing heavy current through it. A "loss of load" may cause a source voltage to rise. Understanding these phrases is very important in troubleshooting.

4-7 OPEN CONNECTIONS IN BRANCHING CIRCUITS

To keep things simple we did not always mention open connections in series–parallel circuits. It should be remembered that connections do go bad—they open. Troubleshoot them as you would parts, as noted below.

Notes on Open Connections

1. In a series–parallel circuit all the voltages below an open connection will fall below normal—all except ground voltage, of course.
2. If the part in series with a supply point has an open connection at its bottom terminal, all the voltages below it will fall to zero.
3. In a series–parallel circuit, all the voltages above an open connection will rise above normal—all except supply point voltage, that is.
4. All the voltages above an open connection will rise to the full supply voltage if all the parts above the open are in series with the supply point.
5. If a part connection opens, you will be able to measure the true resistance of the part. (Always turn off the power before using an ohmmeter.)
6. Recall that if a connection to a two-terminal part opens, the part will have zero voltage drop.

The foils or traces in a printed circuit come in many shapes, sizes, and lengths. Do not forget that the schematic does not generally show how the connections and conductors are actually made and located. Schematics are valid for electrical representation only. Electrically, a dot is the same as a line. A zigzag, curlicue line is the same as a straight line. A connection to the left of a line is the same as connection to the right. Seeing the relationship between a schematic line and the actual printed circuit conductor is not always easy. The trick is to check each path separately from one part terminal to another on a schematic, then trace out a path between the same two terminals on the board. Doing this one path at a time can be a tedious job, but it has to be done. Many times, you will have to keep turning the board or chassis over and over as you look for the terminals on one side and the paths on the other. Perhaps this is the most time-consuming part of troubleshooting.

Note: To speed things up, you can turn off the power and use an ohmmeter to check the continuity of the path between each pair of terminals.

4-7-1 Checking Simple Series–Parallel Connections

Figure 4-14 shows the PC connections of a simple series–parallel circuit. There are eight connections in all! Chances for an open are pretty high. Let's see what happens to all the voltages when each connection opens.

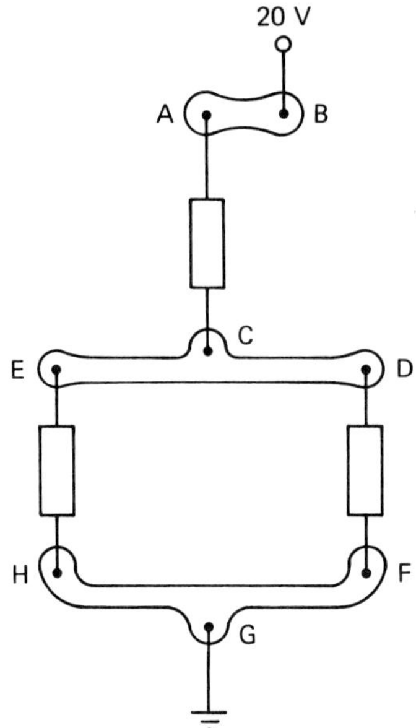

Figure 4-14 How a simple series–parallel circuit might look on a PC board. Note all the connections that can go bad.

1. If either connection *A* or *B* opens, all the voltages on the connections "below" point *A* will fall to 0 V.
2. If connection *C* opens, all the voltages on the connections and leads "above" it will rise to 20.0 V; all those below it will fall to 0 V.
3. If connection *D* opens, the voltages on *D*, *C*, and *E* will rise. The voltage below on the part lead connected to *D* will fall to zero. A similar thing will happen if *E* opens.
4. If connection *F* opens, all the voltages above it will rise, but not to 20 V. Point *C* will divide the 20 V in proportion to the resistances of the parts with good connections.
5. If connection *G* opens, all the voltages above it will rise to 20 V.

4-7-2 Checking Connections in a Bridge Circuit

There are 12 PC connections in the bridge circuit shown in Fig. 4-15. Let's take a look at some of the more interesting open-connection voltage readings.

If connection *B* opens, all the voltages on the leads and connections below it will fall. The voltages at the connections and leads at *D*, *E*, and *F* will fall by a greater percentage than those on the other side.

If the connection at *D* opens, all the leads and connections above it will rise to 20.0 V. The voltage at *D* (the trace, not the lead), and at the connections and leads of *F* and *E*, will fall by a greater percentage than those on the other side.

If the connection at *E* opens, a balanced bridge will have no change. In an unbalanced bridge if the normal voltage at *E* is higher than at *H*, the voltages on the *E* side will rise above normal. The voltages on the *H* side will fall below normal.

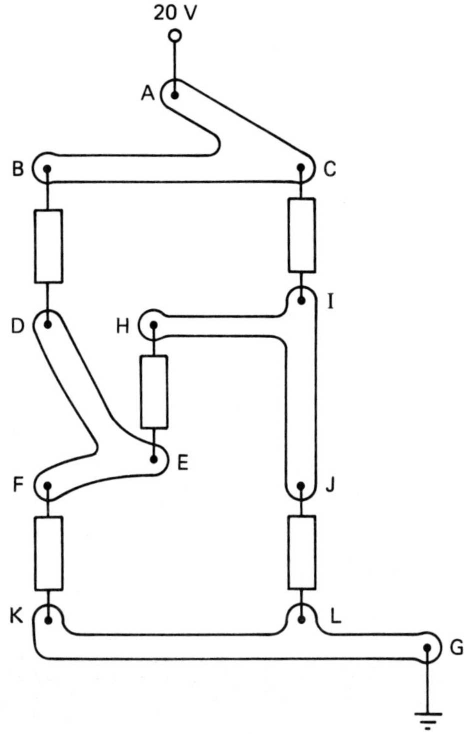

Figure 4-15 Note all the connections in this bridge circuit. There are several connections where a schematic may show only one. Many voltages can be changed when a single connection opens. See the text.

The *H*-side analysis is similar. The ground-side connections are similar to those of the simple series–parallel circuit in the preceding section. We leave it up to the reader to complete the analysis of this side in a problem at the end of the chapter.

SUMMARY OF SELECTED TIPS

Tip 4-3. If the resistance across an in-circuit part is higher than its rated out-of-circuit resistance, it is bad.

Tip 4-7. In a series-parallel circuit: The voltage drop across a bad series part will change by the greatest percentage. "Look for the voltage or voltage drop that has changed most drastically."

All the voltages above a partial short will tend to fall and all the voltages below it will tend to rise. All the voltages above a partial open will tend to rise and all the voltages below it will tend to fall.

All the voltages in a good branch will change by the same percentage. The voltages in the bad branch will not change by the same percentage.

If all the voltages in the branch parts change by the same percentage, the single nonbranch part is bad.

Tip 4-8. A short in a circuit block may tend to pull down the voltage above it and pull up the voltage below the block.

An open in a circuit block may tend to pull down the voltage above the block. (Coils, capacitors, and balanced bridges excepted.)

WHAT YOU LEARNED TO DO

In this chapter you learned:

1. That it is impossible to find out which part in a parallel circuit is bad with the voltmeter or ohmmeter alone.
2. How to cut open a parallel circuit to find out which part is bad with an ohmmeter or voltmeter. You also learned to do it with a series test lamp.
3. How to troubleshoot a simple series–parallel circuit with a combination of voltmeter, ohmmeter, and circuit-disconnecting procedures.
4. How to troubleshoot series–parallel circuits with more than three parts, including the pi circuit, by using only a voltmeter.
5. How to troubleshoot a bridge circuit, with a voltmeter, by noting that a defective side part shows the greatest percentage voltage change. You learned various ways of finding a defective bridge part in a balanced circuit.
6. How to work on circuit blocks by treating them as single parts.
7. The meaning of "loading down a circuit."
8. To find open printed-circuit connections with a voltmeter.
9. To "look for the voltage or voltage drop that has changed most drastically."

QUESTIONS

1. A parallel circuit with eight parts is connected as in Fig. 4-3. To check the circuit, the supply line is cut in half. An ohmmeter reading across the fourth part shows 0 Ω. In which half does the defect lie? How many parts are probably good and need not be checked? Where should you make the next cut?
2. Fig. 4-5, Section 4-2, shows a simple series–parallel circuit. Test point *A* has 0 V. What can you do next to find out whether P_1 is open, or whether a part below is shorted?
3. The top-side voltage in one branch of a series–parallel circuit (Fig. 4-8) has gone down from a normal 40 V to (20 V). The voltage below it has gone down from a normal 30 V to (15 V). The middle voltage on the other side has gone down from a normal 20 V to (10 V). Which part, if any, is bad and has its resistance increased or decreased?

4. The low-side voltage equals the high-side voltage of a branch part in the series–parallel circuit (Fig. 4-8). The high-side voltage has decreased, and the low-side voltage has increased. The voltages in the other branch are below normal. Which part is bad, and is it shorted or opened?
5. The voltage just below the series part in a series–parallel circuit (Fig. 4-8) has dropped from 6 V to 5.33 V, about an 11% drop. A side voltage has gone down from 4 V to 2.68 V, about a 33% drop. On the other side, the voltage has gone down from 2 V to 1.78 V, about an 11% drop. Which part is bad? Has it decreased or increased resistance?
6. Voltage *A* and *B* in a pi circuit (Fig. 4-9) are equal. Which part is likely to be bad, and what is wrong with it?
7. If both voltage points in a pi circuit go down by the same percentage, which part is bad? Has it increased or decreased resistance?
8. In a bridge circuit (Fig. 4-11), which side is bad—the side that has the greater percentage voltage change from normal, or the side with less? Which part is bad—the one with the greatest percentage change in its voltage drop, or the one with the least?
9. If a part within a circuit block opens, will the voltages above the block be pulled down, or up?
10. What will a shorted load do to the source voltage?
11. Does a "heavy" load have more resistance than a "light" load or less?
12. Cite what happens to the voltages in the bridge circuit in Fig. 4-15 of Section 4-7 as each of the connections *F*, *J*, *H*, and *G* open up?

PRACTICAL TROUBLESHOOTING EXPERIENCE

A. Here is some work that will help you develop some valuable skills. Obtain a cast-off circuit board and remove the parts by desoldering them. Use a solder sucker. Be as quick and neat as possible. You will have less chance of damaging the copper foils (traces). After the board is depopulated, clean it off with some special board-cleaning fluid. You can now solder in resistors to connect up the circuits we studied in this chapter.

1. Obtain a resistor and a small coil.
   a. Find a place on the board where you can insert and solder the parts in parallel. You may have to solder in a jumper wire.
   b. Measure the resistance across either part. Compare this resistance with that of the coil alone. Are they nearly the same?
2. Solder into the board a simple series–parallel circuit. The three resistors should each have at least 1 kΩ in value. No resistor should be more than three times greater or less than any other resistor. Make sure that the ground terminals of the pair are connected. Jump over foils and insert wire connectors, if necessary. Connect the circuit to a 5- to 15-V power source. You

might tack-solder them to a 9-V battery. You can also use jumper wires with alligator clips. All voltages are measured with respect to ground.

a. Measure the normal voltage at all connections with respect to ground and record them. (Recording voltages is good troubleshooting practice.) As a beginner, you might go over them twice just to make sure that you have the right voltages. A dirty test prod, part lead, or bad solder connection with excessive rosin can give you a bad reading.
b. Jump the single-series resistor with a resistance of equal or comparable value to simulate a reduced resistance value. You can tack-solder the shunt resistor or use jumper wires with alligator clips. Which voltages fell, and which did not?
c. Do the same as above with R_2.
d. Desolder and remove the bottom lead of the series resistor. This simulates an open resistor as well as having an actual open connection. Check the voltages at all the connections. Did the voltages rise or fall partially, or completely?

   Clean the pin hole and reinsert the lead, but do not resolder. Check the voltage on the lead against the voltage on the foil around it. Move the lead if necessary to get a voltage. What is the difference between these two voltages?
e. Reconnect and resolder the connection. Desolder and remove the high-side lead of one of the branch parts. Clean out the pin hole and reinsert the lead. Compare the voltages on the two sides. Move the lead if necessary to obtain a voltage that shows an open.
g. Reconnect and resolder the connection and desolder and remove a ground lead on one of the parts. What voltages do you get now?

B. Obtain several schematics that will show enough series and branching resistors and coils, and voltages, for carrying out the instructions below.

1. Find a resistor path, not necesarily in series, between two voltage points. Can you tell whether the resistors are dropping voltages?
2. Identify a series, resistor, and coil path between two voltage points. (A coil drops little or no voltage.)
3. Find branching paths consisting of resistors or coils. Are the resistors dropping voltages?
4. Can you find a resistor–coil path (series or branching) from a voltage point to ground?
5. Find a resistor with parts in series with it or branching off below it. What will happen to the voltage immediately below it if a part below it shorts? What if it opens?

5

In-Circuit Checks on Basic Electronic Parts

EXPERIENCE REQUIRED

You can take advantage of the tips in this chapter if you know how to:

1. Find opens, shorts, and changed-value resistances in series and parallel circuits.
2. Use the multimeter.
3. Identify resistors, capacitors, coils (inductors), transformers, diodes, and transistors.
4. Read resistance, capacitance, and inductance values.
5. Use the resistor color code.
6. Identify the schematic symbols for the basic electronic parts. You have heard of integrated-circuit (IC) chips.
7. Desolder and solder parts from printed circuit boards and solder-terminal lugs.

5-1 OUT-OF-CIRCUIT CHECKS OF BASIC ELECTRONIC PARTS

As far as possible, we try to check the parts in-circuit before removing them for out-of-circuit checks. Indiscriminate removal of parts takes up a lot of time, and the chances of damaging the circuits are a lot greater. There are times, however, when we do have to remove them. For example, we have to verify the in-circuit checks by re-checking a part out-of-circuit. Also, it is a good idea to check the new replacement parts out-of-circuit before you connect them into the circuit. There are also times when we have to pull out a suspected part as a last resort to test it. Anyway, if we learn how to check them out-of-circuit, we'll know what to look for when we check them in-circuit.

5-1-1 Parts Substitution

Substituting a good part for a bad one is usually the most reliable test you can make. Open parts can be checked by jumping good ones across them without removing them from the circuit. For shorted parts, however, you have to unhook one lead. (In Chapter 2 we learned that only one lead of a two-terminal part had to be disconnected to take it out of a circuit—as in Fig. 2-7.) If jumping a part brings the circuit back to life, that part is bad. Solder-in the substitute. You save a step by jumping in a good part without directly testing the bad part.

You do not always have to use the exact-value part for a substitution test. If the exact replacement is unavailable, use as close a value as you can find. After the test, you can solder in the exact replacement. Engineers use exact values to make sure that the circuit will operate under worst-case conditions. We usually work under conditions that permit the use of off-value parts for substitution tests.

Experience will tell you what to look for with an off-value replacement. If you lack experience, you can be fairly sure of one thing: If a circuit does not work at all, but works with the off-value replacement, the replaced part is almost certain to be bad. If a circuit that works badly works better, the replaced part is probably bad and is worth a trial replacement.

> TIP 5-1
>
> Substitution is the surest test for parts. You need not always use an exact-value part for test purposes as long as you use exact or recommended values for the final replacement.

5-1-2 Checking Resistors Out-of-Circuit

All parts have resistances (some have infinite resistance), but the resistor specializes in it. Its main purpose is to drop voltages or to hold them up. We can say that resistors are like the structural beams of a building. They hold things in place. They have other functions, too, but as troubleshooters we are mainly interested in the fact that they hold dc voltages in place.

Out-of-circuit resistors are usually checked with ohmmeters that have built-in batteries for pushing current through the part measured. This current is displayed as a resistance value. Resistors are also rated according to their power-handling capabilities. The higher-power resistors, above 2 W, are usually wirewound. The lower-power carbon or film resistors, from $\frac{1}{4}$ to 2 W, are color coded for resistance and tolerance.

5-1-3 Checking Capacitors Out-of-Circuit

Basically, a capacitor is a device that can store electrical charges or voltages. This ability is called *capacitance;* it is measured in microfarads, nanofarads, or picofarads. Capacitors are best checked by substitution. However, the multimeter is always on hand, and you can make fast preliminary checks before jumping in a substitute.

Although you cannot directly measure capacitance with the multimeter, you can use it in some cases for finding opens. This test makes use of the battery in the ohmmeter to charge the test capacitor. The charging current is displayed as resistance by the meter. At first the current rushes into the capacitor, so the meter shows a low resistance. As the capacitor fills up, the resistance increases. When the capacitor is fully charged and no current can flow, the resistance hits a maximum (overrange or infinity). If the capacitor is open, no charging takes place and you immediately have a maximum reading.

You will have to check to see whether this method will work with your particular digital meter. The test may work on some digital meters for medium- and large-value capacitors, about 0.15 μF and up. If you reverse the leads and take a second reading you may able to read as low as 0.1 μF. This is because the discharge from the first reading will be added to the charge of the second reading.

TIP 5-2

Certain digital ohmmeters can be used to check for open capacitors out-of-circuit. Check your meter on known good capacitors to see if this can be done. This check may work for values above 0.15 μF.

To perform the test, touch the probes to the capacitor leads. You will momentarily obtain a resistance reading; it will then increase toward maximum. For the lower capacitance values, this will happen quickly. For the higher capacitances this may take longer—in some cases, too long.

If you reverse the leads and take a second reading, you may read as low as 0.1 μF.

An open or low-value capacitor will give no indication at all. You will see the overrange indication on the digital display.

TIP 5-3

Medium- and large-value capacitors from 0.01 μF on up may be given the kick test with an analog ohmmeter. Touch the two capacitor leads with the ohmmeter set to the highest range. A 0.01-μF capacitor will give a slight kick of the needle. Larger values kick more. For low-value capacitors that are somewhat less than 0.01 μF, reverse the leads to get a bigger kick on a second try. An open or low-value capacitor will not kick.

Internal ohmmeter voltages are usually too low to check capacitors for partial shorts (leakages). However, if you do find a nonelectrolytic capacitor with less than infinite resistance reading, it is definitely bad.

Electrolytic capacitor leakage is very hard to check with an ohmmeter. The higher values take too long to charge up. A certain amount of leakage resistance is permissible. The readings will vary from ohmmeter to ohmmeter. A 5000-μF electrolytic may show maximum resistance on some digital ohmmeters. On others, a 5000 μF electrolytic might show 150,000 Ω leakage resistance, if you can wait long enough.

TIP 5-4

Ohmmeter tests for capacitor leakage resistance are not reliable, but if you get anything less than infinity or maximum reading for a nonelectrolytic capacitor, it is definitely bad.

For electrolytics, a certain amount of leakage is okay. Not all digital meters can be used to check electrolytic capacitors for leakage. Compare your readings with readings on good capacitors. Compare the rapidity of the change from low to high resistance.

Figure 5-1 shows a capacitor being tested for leakage with a voltmeter. One lead of the capacitor is still connected to a voltage point in the circuit. The voltmeter is connected to the other lead to check for leakage voltage. There should be no leakage with nonelectrolytic capacitors. A milliammeter could be used to check the leakage current, but a voltmeter is safer.

When testing electrolytic capacitors, do not let the charging current mislead you into thinking that you have excessive leakage. For a truer reading, charge the capacitor first.

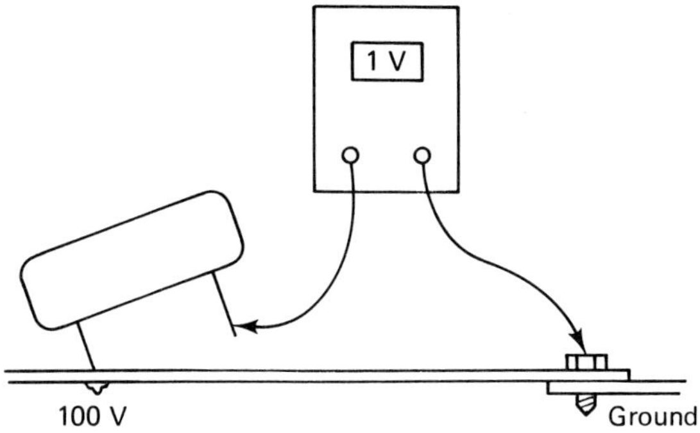

Figure 5-1 Checking a capacitor for leakage. This one is leaking 1 V dc. (Using a voltmeter is safer than using a milliameter.) For nonelectrolytic capacitors, any leakage at all is bad.

To be on the safe side, do not charge the capacitor by closing the open lead while the power is on. Close it while the power is off. Reopen it after the power is turned on. If you do not take this precaution, the sudden surge of the charging current may damage something.

This test will show a certain amount of leakage for good electrolytic capacitors. The permissible leakage depends on the meter resistance, the capacitance, the capacitor voltage rating, and the actual voltage on the capacitor. Compare your reading with a good capacitor connected to the same point.

In your spare time, check your meter with good electrolytic capacitors connected to various power-supply voltages. For example, a reading showing 30% of test voltage could indicate a bad electrolytic.

With nonelectronic voltmeters, the resistance of the meter varies with the range. The lower ranges have less resistance and will pull down the voltage. They also take less charging time. For example, 1% of the test voltage, on the same range for which you read the test voltage, may be okay.

TIP 5-5

A capacitor connected to a voltage point may be checked for leakage by disconnecting its low-side lead to check for voltage on it. There should be no leakage voltage for nonelectrolytics (see Fig. 5-1).

Electrolytics may show some permissible leakage. You may have to charge up the electrolytic first, before using the voltmeter. A reading showing 20% of the high-side test voltage may be okay on some digital meters. With nonelectronic voltmeters, about 1% is okay on the same range for which you read the test voltage.

The checks above are preliminary checks. If they cast doubt on a capacitor, connect a new one into the circuit; this is the final and best test.

If more accurate checks are needed, capacitor checkers of various types are available. They check for leakage as well as capacitance. Capacitance bridges are very accurate.

5-1-4 Checking Coils and Transformers Out-of-Circuit

Coils are easily checked out-of-circuit for opens with an ohmmeter. Usually, they either open up completely or they do not. There are no in-between values to worry about.

Smaller coils or larger heavy-wire coils are difficult to check for shorts with an ohmmeter. Their normal resistances are zero or near zero ohms. You may have to use an inductance bridge or a substitute coil.

> **Interesting Sidelight.** Let us look briefly at the characteristics of coils. They have the ability to store electrical energy. When you run a fluctuating current through a coil, it produces a fluctuating magnetic field around it. In effect, it is storing the current in the magnetic field. This ability to store energy is called inductance. Hence coils are called *inductors,* although sometimes they are called *chokes.* Inductance is measured in henries. In Chapter 6, we find out how chokes are used in power supplies.
>
> Coils also have the ability to pick up electrical energy from the fluctuating magnetic field. This ability is made use of in transformers. The basic transformer has two windings. One winding produces a fluctuating magnetic field. The other winding then picks up energy from this fluctuating field and outputs electrical current.

Figure 5-2(a) shows the basic transformer circuit. The transformer is made of two sets of coils or windings. One is called the primary and the other the secondary. The electromagnetic properties of the coils allows electrical energy to be passed from the *primary* to the *secondary* circuit. Only fluctuating electrical current can be transferred. A transformer will not work with direct current.

Figure 5-2(b) shows three transformer checks that can be made with an ohmmeter. Like coils, they can be checked for opens with an ohmmeter. Sometimes a short between the primary and secondary windings can be detected with an ohmmeter. Anything less than infinity indicates a short or partial short. The big transformers have iron cores that increase their power transfer capabilities. A short between a winding and the iron core of a transformer may also be detected with an ohmmeter. A short between the turns of the same winding cannot be detected with an ohmmeter. We will cover this when we take up power supplies.

Power transformers may not show a short until after they warm up. This problem is handled in Chapter 6. Most transformer checks are made in-circuit.

Figure 5-2(c) shows an autotransformer. The primary and secondary windings are tied together in series. This transformer is checked out like a coil. There should be continuity between all its terminals.

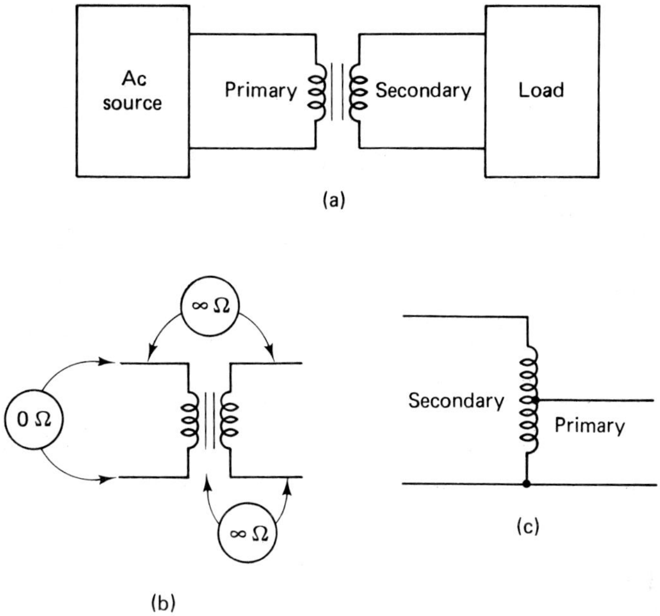

Figure 5-2 (a) Basic transformer circuit. (b) Checking a transformer out-of-circuit with an ohmmeter. (c) An autotransformer has dc continuity between the primary and secondary.

TIP 5-6

Coils and transformers may be checked for opens out-of-circuit with ohmmeters. Generally speaking, the good ones have low resistance values.

They should have infinite out-of-circuit resistance between the primary and secondary terminals. Transformers with iron cores should have infinite resistance between the windings and the core.

TIP 5-7

Coils as well as capacitors are usually checked by substitution. If parts are not readily available, special coil and capacitor checkers are used.

We have more on coils and transformers in the following chapters. They deal with alternating currents, power supplies, and signal circuits. Coils are also used with dc current for electromagnets, solenoids, and relays. If you find these electrical devices in an electronic circuit, treat them like coils when checking them out of the circuit.

5-1-5 Checking Diodes Out-of-Circuit

For testing purposes we should know that diodes and transistors are devices that pass dc current from one terminal to another in one direction only. Most diode types can be roughly checked with an ohmmeter. Certain types, such as high-voltage rectifiers and tunnel diodes, cannot be checked with an ohmmeter. We are concerned here only with diodes that can be checked with ohmmeters.

Ohmmeter checks for diodes and transistors give the best results when done out-of-circuit. A diode can be identified by the band that is printed around one end of it. This end is called the *cathode* (*K*; see Fig. 5-3). The unmarked end is called the *anode* (*A*). The diode symbol shown next to it in the illustration has an arrow pointing from the anode to the cathode. Practically speaking, conventional current can flow only in the direction of the arrow, from anode to cathode. It has a low resistance in this direction and a very high resistance in the other direction.

Figure 5-3 also shows the two steps required for checking a rectifier diode. Step 1: The meter shows a low-resistance reading. Step 2: The leads are reversed, and the meter shows a high-resistance reading.

If you do not get these high and low (hi–lo) readings, the diode is definitely bad. (Not all diodes can be checked with ohmmeters—high-voltage and tunnel diodes, for instance.)

If you do get these readings, the diode is probably good. To make sure, you can install a new diode, since you have already taken out the old one. You can also use an

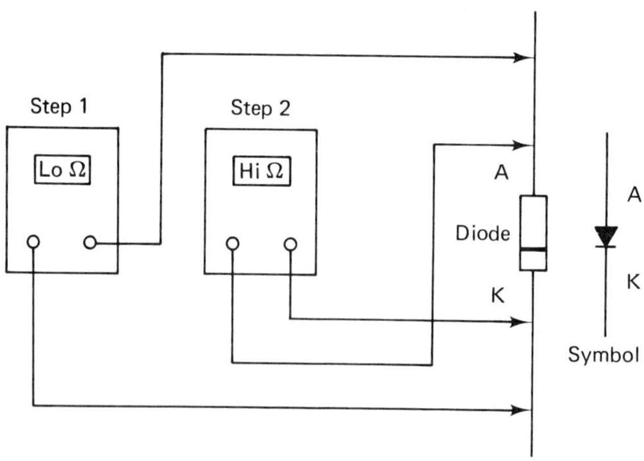

Figure 5-3 Two steps for checking a diode with an ohmmeter.

accurate diode checker. The ohmmeter test is mainly for finding bad diodes, not good ones. In troubleshooting, we focus on the bad parts.

> **TIP 5-8**
>
> To check a diode out-of-circuit with an ohmmeter, take two readings. Reverse the leads for the second reading. A good diode will give a hi–lo reading, in this case a near-infinite resistance for one reading and a low reading for the other. If you are using a deflection meter, use the ×100 or ×1000 ohm range. The other ranges may have damaging voltages. The ohmmeter is used mainly for finding bad diodes, not necessarily for good ones.

5-1-6 Checking Zener Diodes Out-of-Circuit

Figure 5-4 shows a zener diode, Z_1, being tested. Note its symbol. Zeners are special diodes that can be used for regulating voltages. We learn about them in Section 10-2-1. Unlike regular diodes, they can pass current when reverse biased, but it can only do so at a certain voltage. This voltage is called the zener voltage. Figure 5-4 shows a power supply delivering 12.5-V dc to a resistor that drops the voltage to 10-V dc. This 10 V is the zener voltage, V_Z; it will not change as long as there is a higher voltage above the dropping resistor, R_1. The power-supply output can vary greatly above 10 V, and the zener will hold the voltage to 10 V as long as the current does not exceed a specified limit. Zener diodes are available with various zener voltage ratings and maximum current or power ratings.

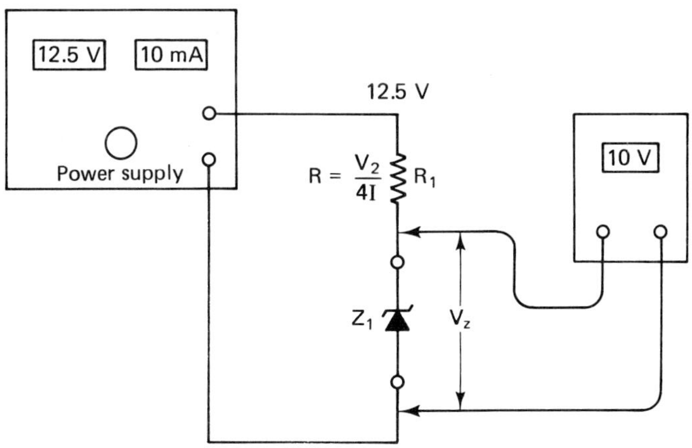

Figure 5-4 Checking a zener diode. The voltmeter voltage must not exceed the zener value at the rated current.

An ohmmeter can be used to see if a zener is bad but not to see if it is good. If it does not have hi–lo readings, it is definitely bad. If it does have hi–lo readings, it could be good, and you may decide to look at other parts in the circuit before making further checks on it.

You can make further tests on it with the setup in Fig. 5-4. Start from 0 V and see if you can crank up the voltage until the rated, 10-mA current is reached. If you then read the rated zener voltage across the diode, it is good. The rated current may be obtained by dividing the power rating by the zener voltage. The value of the dropping resistor, R, should equal about one-fourth the zener voltage V_Z divided by the current rating I, or $R = V_Z/4I$. The power rating should be at least one-fourth the power rating of the zener diode.

5-1-7 Checking Bipolar and FET Transistors Out-of-Circuit

Transistors are covered more thoroughly in Chapter 8, but it will not hurt to learn how to make a simple ohmmeter check at this time. The most common type of transistor is known as the bipolar transistor. It has three terminals called the base B, the collecter C, and the emitter E. Note the arrow at the emitter in Fig. 5-5(a); this is the customary way of identifying the emitter.

For illustrative purposes, Fig. 5-5(b) shows an equivalent transistor diagram with two diodes pointing toward the base. One is pointing from E to B, and the other is pointing from C to B. These can be checked as regular diodes. They should have hi-lo readings across their terminals. *Note:* Do not confuse these diode paths with transistor paths, C to E and B to E. You will learn more about these later.

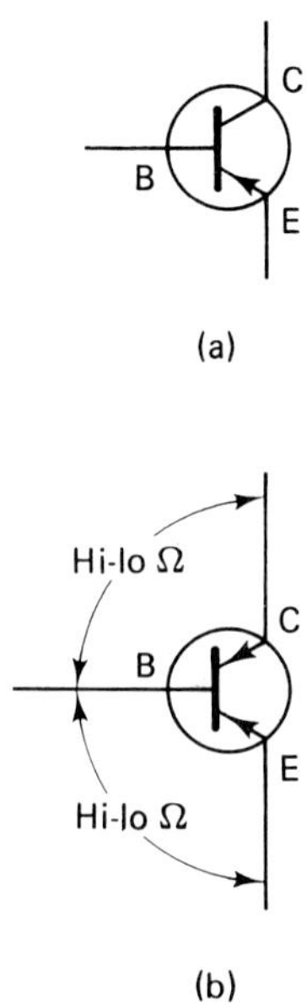

Figure 5-5 (a) A transistor has three terminals, B, C, and E. (b) The BC and BE terminals or junctions can be tested like diodes.

TIP 5-9

To check transistors, the *C*-to-*B* terminals and the *E*-to-*B* terminals can be treated as diodes. You should get hi–lo readings for each of these diodes.

TIP 5-10

To check a transistor out of circuit, you need only remove the base lead from the circuit. You can cut the lead or the PC foil; or you can remove the solder around the hole with a solder sucker and carefully measure the resistances while making sure that the lead does not contact the circuit.

There is another type of transistor that can also be roughly checked with an ohmmeter. It is called the junction-field-effect transistor (JFET). There are other types of FET transistors, but they cannot all be checked as fully as the JFET with an ohmmeter. This transistor usually has three terminals, the drain *D*, the source *S*, and the gate *G* [see Fig. 5-6(a)]. We discuss it more fully in Chapter 8.

TIP 5-11

A JFET transistor should have hi–lo readings between *G* and *D* and also between *G* and *S*. It should read lo–lo between *D* and *S*. (*D* = drain, *S* = source, and *G* = gate.)

Another transistor type that can be checked with the proper precautions is the metal-oxide-semiconductor field-effect transistor (MOSFET). There are two kinds of MOSFETs: the depletion type [Fig. 5-6(b)] and the enhancement type [Fig. 5-6(c)]. The MOSFET is susceptible to static charges, so care must be taken when handling it. It is stored with its leads shorted in conductive foam or metal foil. Ground your fingers before unpacking it. This will discharge any static charges you may have on your body. Also ground your soldering iron before unsoldering the MOSFET. Always turn the power off before you remove or insert a MOSFET.

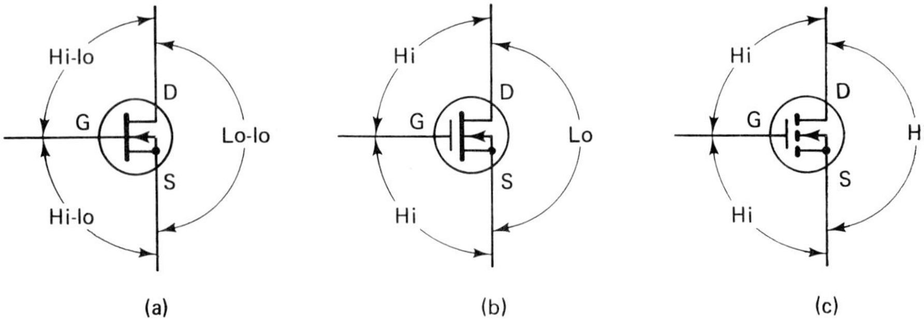

Figure 5-6 How to check various types of FET transistors with an ohmmeter. (a) JFET. (b) Depletion-type MOSFET. (c) Enhancement-type MOSFET. (Precautions in the text must be followed when testing MOSFETS.)

As a further precaution, it is a good idea to use an antistatic mat on you bench as well as a grounding strap on your wrist. They must both be grounded through leads that are in series with 1 MΩ resistors to avoid electrical shock. You should also have antistatic floor and seat coverings. Other coverings might generate static electricity when you rub them with your shoes, or clothing.

> **TIP 5-12**
>
> MOSFET transistors should have almost infinite hi–hi resistances between its *G* terminal and its *S* or *D* terminal. Ground your fingers and the soldering iron before handling a MOSFET. Make sure that the power is off when the device is being removed or inserted.
>
> To check MOSFETs, the *G* terminal must be shorted to the *S* terminals.
>
> Enhancement-type MOSFETs have almost infinite hi–hi resistances between *D* and *S*. The depletion-type MOSFET will have lo–lo resistances between *D* and *S*.

We have now gone over the quick out-of-circuit checks for the basic electronic devices. If for any reason, you need more accurate checks, use bridge instruments for the resistors, capacitors and coils, and gain and leakage checkers for the transistors and diodes.

5-1-8 Checking Silicon-Controlled Rectifiers Out-of-Circuit

Figure 5-7 shows a silicon-controlled rectifier (SCR) in a test setup. First of all, just what is an SCR? Briefly, it is a semiconductor switch. It is turned on by a positive

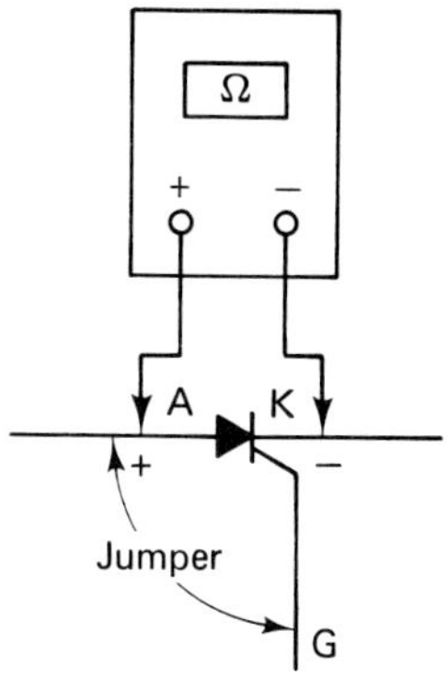

Figure 5-7 Checking an SCR diode. The diode path, *A–K*, should remain open until *A* and *G* are jumped. It should then close. After the jumper is removed, it should remain closed. It should open only after the meter is removed and remain open after the meter is reconnected.

voltage on its gate (*G*) terminal; the anode and cathode terminals must be forward biased. That is, the anode must be more positive than the cathode.

> *Remember:* A forward-biased anode-to-cathode path will not conduct unless it is triggered by the gate voltage. On the other hand, a loss of gate voltage cannot switch off the SCR. The SCR can be turned off only when the anode-to-cathode voltage is turned off.

The characteristics above allow us to use the setup in Fig. 5-7 to roughly test small SCRs. Use the $R \times 1$ range of the ohmmeter. First check for continuity from anode to cathode without the jumper that is shown in the illustration. The lead with the positive voltage is connected to the anode terminal. You should have an infinite resistance reading. Next trigger the gate terminal by shorting it to the positive voltage on the anode; you should now read about 15 to 50 Ω. Next remove the short, and the reading should not change, since the loss of gate voltage should not switch the SCR off.

For a better test, you must use a dc power supply with meters for reading the anode and cathode currents. You also need a voltage source and meters for reading the gate voltage and current. You can use a variable resistor to obtain the proper trigger voltage.

Summary for This Section. Again we say that for the final test, we usually substitute known good parts. Usually, if the circuit works after the substitution, we have saved a few steps. However, it pays to have specialized equipment in the shop for more positive readings when needed.

5-2 IN-CIRCUIT DC CHECKS ON BASIC ELECTRONIC PARTS

In Chapters 2 and 4 we learned to locate bad parts without regard to their specific characteristics. In actual practice, all parts have characteristics that affect dc voltages and resistances in certain ways. They send us distress signals that guide us toward them. The tips in this chapter take advantage of these signals or clues, but don't go overboard. You should not try to memorize all the tips. Eventually, as you gain more experience, you will develop your own bag of tricks.

The most important thing to learn from this chapter is the type of thinking that is required to isolate the defective part. Try to see how the fundamental rules (or mathematics, if you prefer) are applied in each case. The tips are listed primarily to get you started. Nevertheless, they will come in handy when you get stuck on a job.

5-2-1 Checking Resistors in DC Circuits

We have already learned how to check resistors in-circuit. The rules developed in Chapters 2 and 4 are for all kinds of parts. All parts have resistance, but resistors specialize in it.

Taking resistance measurements in-circuit is difficult because other parts in the circuit will affect the readings. Electrolytic capacitors connected to the supply line are especially troublesome. Capacitors are a little like batteries. They can be charged up with the voltage in your ohmmeter battery. The charging current goes through the meter and throws off the reading. High-value resistors in the meter and in the circuit slow down the charge time. You may have to wait forever before the reading reaches a steady level.

In some cases the capacitors will retain some residual voltage after the power is turned off. Reversing your ohmmeter leads will show that you are getting a false reading. In the case of analog meters, the indicator may "peg" below zero. Digital ohmmeters will "max out" in one direction and show lower-than-true resistance in the other.

> **Caution:** Ohmmeters can be damaged by charged electrolytic capacitors even after the power is turned off. Grounding the test point with a jumper wire through a 100-Ω resistor for a second or two to discharge the capacitor will protect your ohmmeter.

Get to know your ohmmeter, and see how it behaves in various circuits. With digital meters, note how fast the numbers change. They slow down toward the end. With analog meters see how the speed of the indicator slows down as it approaches the final resting point. If you know your meter, you will not have to wait forever to make a judgment.

You can also remove the electrolytic capacitor to remove its effect, but this can be messy. A better way, oddly enough, might be to short out the capacitor with a jumper. (You might want to discharge it through a 100-Ω resistor first to avoid shock and sparks.) In Fig. 5-8(a), the resistors you want to measure are R_1 and R_2. The capacitor is shorted out with a jumper connected to points B and C. If you measure each resistor separately, you will see that they are not individually shorted.

Actually, the jumper has placed the two resistors in parallel. Both ends of each resistor are tied together at A, and the other side is connected together by the jumper

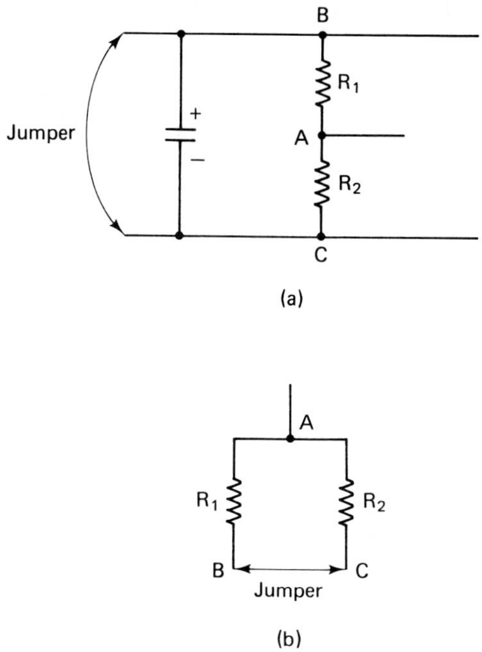

Figure 5-8 (a) Jumping the capacitor will keep it from interfering with the resistance measurements across R_1 and R_2. (b) The circuit redrawn. The resistors are in parallel.

through ground. Redrawing the circuit as in Fig. 5-8(b) makes it easier to see. The jumper has shorted out the capacitor and all other circuits connected to the supply line. Note the letters *A*, *B*, and *C*. Compare their locations in Fig. 5-8(a) with those in Fig. 5-8(b). We can now estimate the parallel resistance of the pair as we learned in Chapter 4 (see Tips 4-1 to 4-3).

TIP 5-13

If the power-supply electrolytic hinders the measurement of two resistors in a voltage path, short the supply line to ground (with the power turned off). The capacitors will be shorted out but not the individual resistors. They will be placed in parallel.

The following tips may be more helpful when dealing with the effects of electrolytics.

TIP 5-14

Removing the power-supply output fuse eliminates the effects of the power-supply capacitors on resistance readings. Don't forget, there may be other electrolytic capacitors elsewhere on the supply line.

TIP 5-15

Low-value resistances that can be read on the ×1 to ×100 range are relatively unaffected by capacitors with values less than 500 μF.

This is true because the lower ranges of your meter have less resistance, and the capacitor will charge up faster to give you a final value.

Interesting Sidelight. The time it takes to charge up a capacitor through a resistance is calculated in time constants TC. A 1-second TC equals 1 Ω times 1 F, or TC = $R \times C$. It takes about 6 TC to fully charge or discharge a capacitor for practical purposes.

5-2-2 Checking Resistors Connected to Capacitors

Figure 5-9(a) shows a resistor connected to a low-value capacitor. Low-value capacitors discharge almost immediately. A 1.0-μF capacitor takes only 60 milliseconds to discharge through a 10-kΩ resistor. That is not a long wait, so there is no problem here when using the ohmmeter.

There is one nice thing about this connection. The capacitor isolates the resistor from other circuits. It does not pass the ohmmeter voltage to the other circuits.

TIP 5-16

A low-value capacitor in series with a resistor will protect the ohmmeter reading from outside influences. You will be able to measure the true resistance value of the resistor.

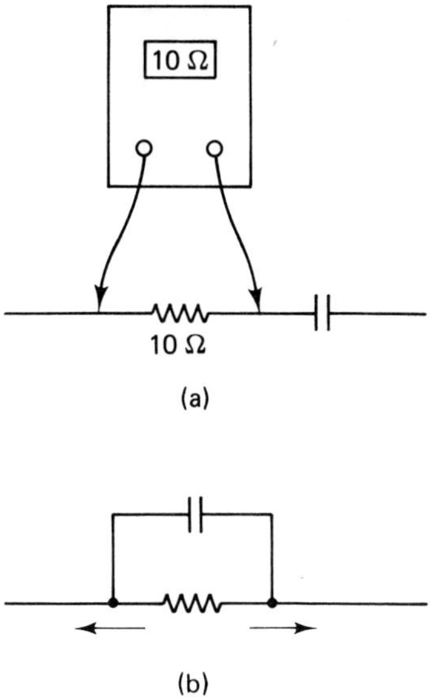

Figure 5-9 (a) A low-value capacitor permits a good resistance reading. (b) Only an open resistor can cause the voltages to pull apart.

Perhaps we should take note of something here. If the capacitor is good, the resistor should drop no dc voltage, but this is true of bad resistors, too. In other words, we have an exception to basic Tip 2-1, which says that a good dc voltage probably indicates a good circuit.

> TIP 5-17
>
> If all the dc voltages in a defective circuit test good, check the resistances or ac voltage drops across all the parts that are in series with capacitors.
>
> Also check those in series with varactor diodes (capacitor diodes) or other diodes having similar voltages on them (Section 5-2-3).

> TIP 5-18
>
> If the dc voltage drop across a resistor–capacitor shunt combination is above normal, the resistor has probably increased value.

Figure 5-9(b) shows a resistor shunted by a capacitor. The voltage drop across it is too high, indicating a possible open. Since a good capacitor is already open to dc, only the resistor can cause the abnormal voltage drop by opening up. The open will show up on an ohmmeter, too.

5-2-3 Checking Resistors Connected to Semiconductors

A resistor connected to a diode can often be checked in-circuit if the diode isolates the resistor from other circuits. Figure 5-10(a) and (b) show two circuits in which the resistor can be checked with an ohmmeter. In both cases the ohmmeter current can go through the resistor only. If you cannot readily determine the correct polarity, take two readings and use the higher reading.

Some digital ohmmeters will not be affected by semiconductors connected in either direction with the suspected resistor. Their internal voltage is not high enough to overcome a certain "barrier voltage" that semiconductors are known to have.

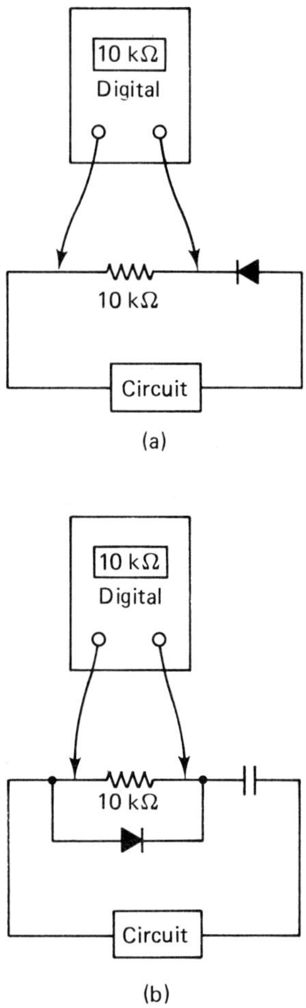

Figure 5-10 (a) The diode can isolate the resistance reading from the other parts around the circuit. (b) The capacitor isolates the ohmmeter reading. The diode may show a low reading. Take the higher of two readings.

Check your meter in many types of semiconductor circuits to see if it can show readings in either direction.

Some meters have special low-voltage settings for checking semiconductor circuits. In the low-voltage setting, the ohmmeter voltage source is not high enough to overcome the barrier voltages of the semiconductors.

TIP 5-19

Resistors can often be checked with ohmmeters even if there are transistors or diodes in the circuit. Just take a hi–lo resistance reading across the resistor. The higher of the two readings is the true reading. However, if other semiconductors are in the circuit, then both readings may be affected and you will not get a true reading.

With certain low-voltage ohmmeters, you will be able to read resistances on the low ranges under any of the circumstances above with just one reading.

5-2-4 Checking Capacitors in DC Circuits

We stated before that in troubleshooting, we proceed as if any bad part will produce a bad dc voltage. Here is another exception—an open capacitor. Capacitors react differently to dc and ac. When you check them in-circuit with dc voltmeters and ohmmeters, they appear to be open. On the other hand, when you check the same capacitors with an oscilloscope, you find that they are not open at all; they pass the ac signals. We will cover "scopes" and ac signal voltages in Chapter 7. Coils also have split personalities; we'll find out more about them later in this chapter and in Chapter 7.

TIP 5-20

If all the dc voltages in a circuit are okay but the circuit is not working, suspect an open capacitor or shorted coil. See Tips 5-17 and 5-25.

Since a capacitor is completely open to dc, a resistor in series with it should not drop any dc voltage. If a capacitor shunted to (in parallel with) a resistor shows a short, it is very likely to be shorted since resistors seldom short.

TIP 5-21

When checking for an open capacitor, jump it with a good one of comparable value. If normal circuit operation is restored, the capacitor is open.

TIP 5-22

A resistor in series with a capacitor should drop no dc voltage. If it does, the capacitor is shorted.

TIP 5-23

If a capacitor shunted across a resistor shows a short, the capacitor is more likely to be shorted than the resistor. (But don't forget the effect of the other circuits.)

5-2-5 Checking Coils in DC Circuits

We previously noted that shorted coils and good coils can show no difference in resistance and dc voltage drop. As far as dc is concerned, a coil is nothing more than a length wire. You must be aware, though, that when it comes to ac or signal voltages, there is a big difference. We cover ac circuits in Chapter 7.

Coils usually have much lower resistances than the other parts connected to them. This means that the other parts can be ignored when taking coil resistances. Suppose that we have a coil shunted by a 10-Ω resistor. How much effect will the resistor have on the resistance measured across the coil? Calculations will show that a 0-Ω coil will measure 0 Ω. A 1.0-Ω coil will measure 0.99 Ω. A 2.0-Ω coil will measure 1.67 Ω. Even in the latter case, an open coil will easily be detected. We cited an extreme case. Coils are usually connected to much higher resistances.

TIP 5-24

Coils are easily checked in-circuit with ohmmeters. Coil resistances are so low that their readings are hardly affected by other parts in the circuit. Their resistances, in-circuit or out, usually run from about zero to a few hundred ohms. In a few cases they'll run into the 1000-Ω range.

TIP 5-25

Coils should drop zero or very low dc voltages. If you do not get a low voltage across a coil, suspect an open. If you do not have normal circuit operation and the dc voltages are normal, suspect a shorted coil or open capacitor (see Tips 5-17 and 5-20).

TIP 5-26

Coils usually open completely when they open. They are not likely to increase in resistance.

TIP 5-27

Small coils with thin wires are more likely to open than to short. They often open at the terminals and are repairable.

Speaker coils may open at the connections on the cone. They can sometimes be resoldered.

TIP 5-28

Grounded coils should have very little or no dc voltage on them. They may have ac voltages on them.

TIP 5-29

Some small single-layer coils tend to lose a little of their inductance as the windings spread apart with age. They can be "retuned" by squeezing them back to their original axial length.

TIP 5-30

When checking for an open coil that drops very low signal and dc voltage, you can jump it with a length of ordinary wire. If operation is restored, the coil is open. See the ohmmeter and voltmeter checks, Tips 5-24 and 5-25.

The tip above does not work for transformers and "tuned circuits." You'll learn more about these later.

5-2-6 DC Transformer Checks

As mentioned before, transformers are basically ac devices. Their main function is to transfer ac voltages from one circuit into another. You will learn more about this in the next two chapters. Troubleshooters are also interested in their dc (resistance) characteristics.

Shorted turns in a coil usually cannot be detected with an ohmmeter, but shorts between primary and secondary may. The voltmeter may also be used. There should be no dc transfer from one winding to another.

TIP 5-31

Open transformer windings can be checked in-circuit with ohmmeters, because their normal resistances are very low compared to the other parts in the circuit.

In Chapter 3 we learned to troubleshoot a power input circuit. We started off by measuring the resistance across the ac plug prongs. If you have a transformer in the input circuit, you would be measuring its primary resistance in addition to the fuse resistance.

Following is a list of the primary resistances of power-supply transformers of various output powers. The power rating of a transformer is about the same as the power rating of the equipment itself. You can find this rating on the back cover of the equipment. These ratings should give you an idea of what kind of resistances to expect. We're referring to 120-V ac equipment.

POWER RANGE	PRIMARY RESISTANCE RANGE
1500 W, high power	0.5 Ω
190 to 500 W, high power	1 Ω
100 W, medium power	10 Ω
10 W, low power	200 Ω

More Useful Information. Here is some more information that may be helpful. The secondaries of power transformers for low-voltage (semiconductor) circuits have lower secondary resistances than their primaries. The secondaries of high-voltage transformers for video or TV display tubes (CRT) should have several hundred ohms, up to 1 kΩ for older color TV sets.

Secondaries of audio and CRT vertical output transformers have low ohms, from a fraction to 10 Ω. The primaries are 200 or more ohms.

5-2-7 Checking Diodes In-Circuit with an Ohmmeter

Bad diodes may be found with an ohmmeter, even though they may be connected to other parts. When checking diodes, the ohmmeter must show a high reading and a low reading. When the reading is high, it will be a reduced by the circuit resistances. However, it is still greater than the low reading, which is also reduced to some extent by the circuit. This check is almost always satisfactory. It may be followed up by other checks when necessary. If you get a lo–lo reading, check the circuit for low-resistance parts that are in parallel with the diode. You can use analog or digital meters for these checks.

TIP 5-32

Diodes may often be checked in-circuit with an ohmmeter. The high reading will be affected, but usually not enough to prevent a satisfactory indication.

5-2-8 Checking Diodes with a Voltmeter; Diode Switches

When checking diode circuits, remember that conventional current flows through the diode only in the direction of the arrow and only from a higher positive to a lower positive (or more negative) voltage. To check a diode with a voltmeter, find out which terminal has the higher positive voltage on it. If the anode has the higher voltage, as in Fig. 5-11(a), the diode is said to be *forward biased*, and it acts like a short. It will conduct current and drop very little voltage. The silicon diode mentioned in the tip below is the most common type. The germanium type is used less often.

TIP 5-33

A forward-biased silicon diode acts like a short; the voltage drop is low, about 0.7 V (0.3 V for germaniums). If it drops much higher than this, say 1.5 V, it is bad. If it drops much lower than this, say 0.2 V when forward biased, it is bad.

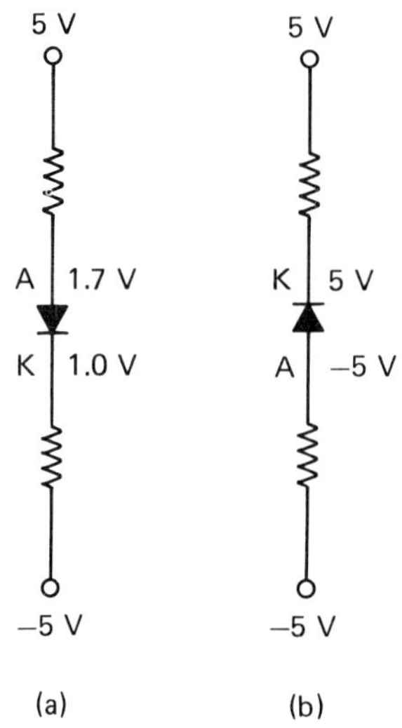

Figure 5-11 (a) Forward bias voltages for current flow. The anode has the higher positive voltage. (b) Reverse bias voltages for cutoff. The cathode has the higher positive voltage.

There is one other voltage check. When the cathode has the higher positive voltage, as in Fig. 5-11(b), it is said to be *reversed biased;* it acts like an open. It will not conduct, and it will drop the maximum voltage allowed by the circuit.

> **TIP 5-34**
>
> A reversed-biased diode acts like an open; it should have the maximum voltage that the circuit can allow on its terminals. Otherwise, it is bad.

When checking diode circuits, you may have difficulty determining whether a diode has the right voltages on it with respect to ground. It may help if you just remember what it takes to make it conduct. The diode conducts only under two circuit conditions: when the cathode is less positive, or when it is more negative than the anode. Pehaps you should remember that more positive means less negative and more negative means less positive. A study of the following examples may help.

When a diode is forward biased, we say that it is "switched on," and when it is reverse biased we say that it is "switched off." Figures 5-12(a) to (e) show examples of diodes that are switched on or off. Figures 5-12(f) to (i) show good and bad diodes. You can tell that a silicon diode is bad if it does not drop a maximum voltage when it is reversed biased, and if it drops much more or much less than 0.7 V when forward biased. Figure 5-12(i) is tricky. The 0.7-V drop is not the result of the forward-bias drop; it is the result of the low-resistance short in the defective diode.

5-2-9 Checking Zener Diodes In-Circuit

Figure 5-13 shows a typical zener diode regulator circuit. The zener is holding the lower load circuit voltage to 10 V dc. Check this voltage first. If it is too high, the zener is bad. If it is too low, check the supply voltage. If the supply voltage is okay, either the dropping resistor has an increased resistance, or the zener or load circuit is partially shorted to ground.

Turn off the power and check the dropping resistor with an ohmmeter; if the resistance is higher than the marked resistance, it is open. If it is less, a number of things could be wrong, but the easiest check is to take hi–lo readings on the diode. If you have lo–lo readings, the diode could be shorted. (A hi–hi reading would not give you the lower-than-normal output voltage you now have.)

Next, if it does not look too difficult, look for a drastic short in the load circuit. If you cannot find a drastic short, or if it is going to take too long to find one, you can start disconnecting things.

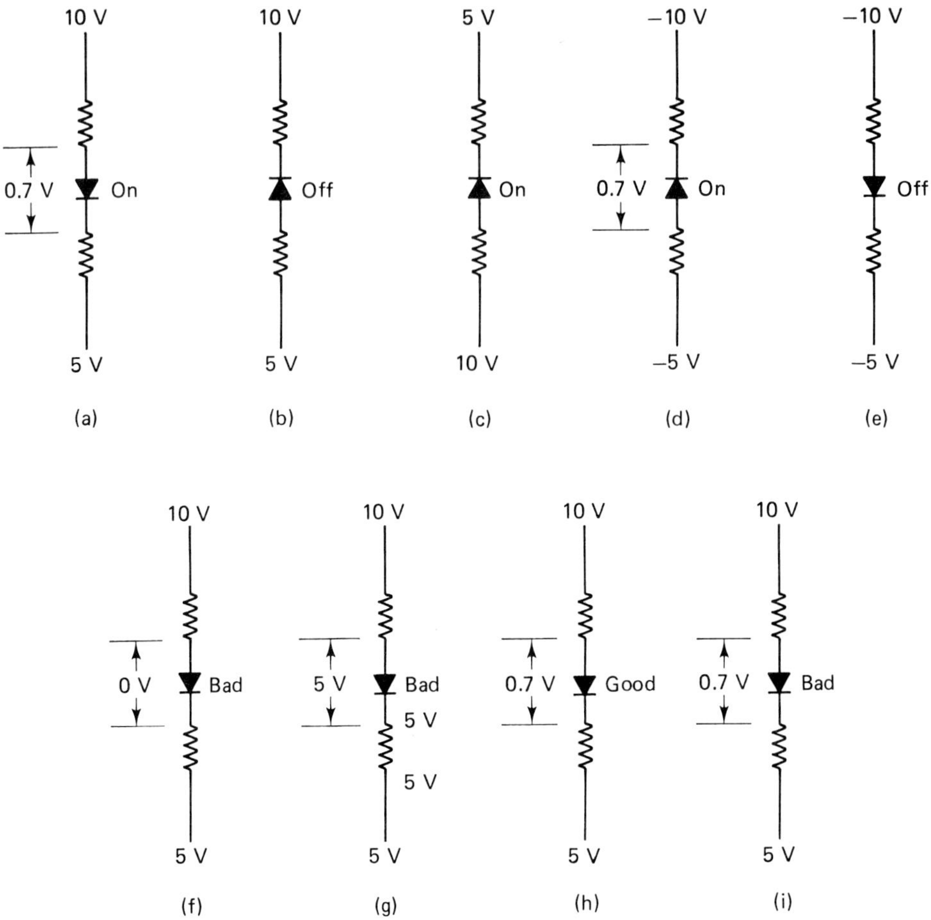

Figure 5-12 See (a) to (e) for voltages that switch the diode on and off. See (f) to (i) for voltages that show good and bad diodes.

Turn off the power and disconnect the zener. Turn on the power. If the voltage rises above 10 V, the zener is partially shorted. If the voltage does not rise above 10 V, the load circuit is partially shorted.

5-2-10 Checking Transistors In-Circuit (Ohmmeter Only)

A preliminary ohmmeter check may be performed on transistors in-circuit as you did with out-of-circuit transistors in Section 5-1-5. This check works for transistors that are completely shorted. The effects of other parts in the circuit have to be considered as you did with diodes. Also, this test may not work for partially defective transistors that cause weak or unstable operation. In Chapter 8 we show how to make more positive checks with a voltmeter.

Substitution is the best check for transistors. Out-of-circuit checkers are available that give good results, and they have uses; but if you are going to unsolder a

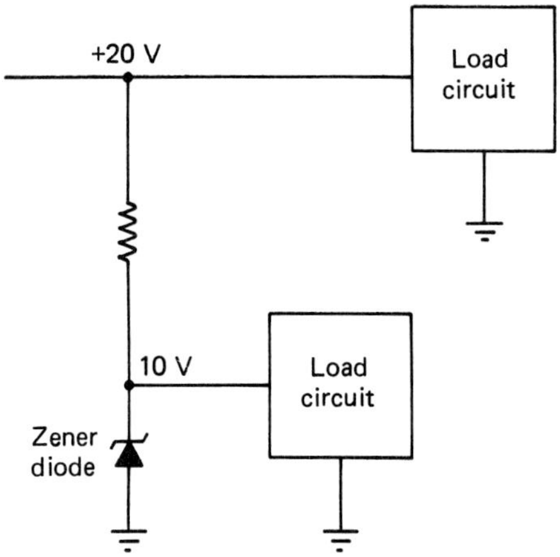

Figure 5-13 Checking a zener diode in-circuit. If you have the zener voltage, 10 V in this case, on the cathode, the diode is probably okay.

doubtful transistor and then solder it back into the circuit, the chances of damaging it becomes greater. You may have more peace of mind if you just go ahead and put in a new one.

TIP 5-35

Check in-circuit transistors as you would out-of-circuit transistors. If you obtain hi–lo readings, the transistor is probably good. Check the rest of the circuit to make sure. If you do not have hi–lo readings, it could be bad, especially if there are no coils and diodes in the circuit. See Chapter 8 for further tests.

TIP 5-36

Preliminary checks on FETs of various types can be done in-circuit for hi–lo resistance readings. Your high readings will be decreased by other parts in the circuits. Your low readings should be low.

5-3 MORE ON SWITCHES, CONNECTIONS, AND CONDUCTORS

A firm understanding of switches, connections, and conductors is absolutely essential for good troubleshooting. VCRs and other types of electronically controlled mechanical equipment contain many types of switches for both sensing and controlling movement. Electronic switches have replaced mechanical switches in many applications, but their basic principles remain the same.

For every electronic part there are at least two connections. For every connection there is at least one conductor. As for conductors, with the advent of integrated-circuit chips, there are a whole lot more of them than there are of parts.

Modern troubleshooting is largely a matter of checking switches, connections, and conductors. You still have to use the fundamental rules to check them.

> **TIP 5-37**
>
> A high-side switch should pull its low-side voltage up when it is on, and drop it when it is off [see Fig. 5-14(a)].

> **TIP 5-38**
>
> A low-side switch should pull its high-side voltage down when it is on and push it up when it is off [see Fig. 5-14(b)].

> **TIP 5-39**
>
> Connections should have no voltage drop. If the high-side voltage is higher than the low-side voltage, the connection is defective.

> **TIP 5-40**
>
> Connections and conductors are prone to short with other connections and conductors. Look for excessive solder and frayed wires. Check for shorts between conductors with an ohmmeter.

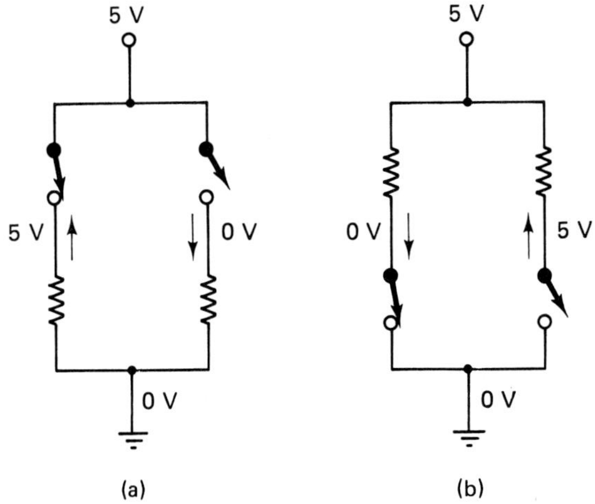

Figure 5-14 Note the ways in which to obtain high and low voltages with switches. In particular, note how a high voltage is obtained in (b).

TIP 5-41

There should be no significant voltage drop between any two points on a conductor. If there is, there is an open somewhere between the two points. This is very important in computer work.

TIP 5-42

Conductors are not always easy to follow. Very often, they can only be followed by checking the continuity from one point on the conductor to another, or from one connection to another.

TIP 5-43

Hidden parts may be checked by reading the exposed end of conductors and leads of parts to which they are connected.

5-4 TIPS BASED ON THE NATURE AND FREQUENCY OF PARTS FAILURES

Knowing how parts fail and how often they fail can provide valuable clues. In general, parts fail when placed under excessive voltage, heat, or mechanical stress. Many parts fail soon after they leave the factory. They tend to fail in batches.

> TIP 5-44
>
> If you find that one particular factory run has the same part failing repeatedly, do not waste too much time on voltage checks. Just note the symptoms carefully and replace the part. Bad parts like these are reported occasionally in factory service literature.

5-4-1 Transistor Failures

A fuller treatment of transistor circuits begins in Chapter 8. At this point we are interested in the transistor failure rate—it is very high. In fact, you can often find the cause of circuit troubles just by checking transistors with an ohmmeter or transistor checker.

Power output transistors are exposed to the maximum supply voltage. They are damaged by heat. They require good ventilation and are mounted on metal heat sinks for good heat dissipation. For this reason they are often mounted outside, on the back of the cabinet. They are connected directly to output devices that are exposed to the outside world. Anything can happen out there. They are subject to damage from the inside, too. Defective internal transistors and parts can place damaging voltages on their terminals.

Low-power transistors go bad for various reasons; prolonged exposure to heat, transient voltages, and cost-effective selection are the major reasons. Since they lie in the center of most voltage paths, they are subject to damage by shorted parts from all sides. Later you will learn that they are also damaged by open parts. High-power transistors are also subject to these stresses.

> TIP 5-45
>
> Transistors have a high failure rate. Give them high priority when checking out a circuit.

TIP 5-46

Transistors are damaged by heat. They require ventilation. Apply heat-conductive silicone grease when you mount them against a heat sink. Transistor equipment should not be operated for long periods in a hot environment.

TIP 5-47

Transistors often go out when defective parts place incorrect operating voltages on their leads. Check for other defective parts when you find a bad transistor.

TIP 5-48

Power transitors usually fail in catastrophic ways. They blow fuses, trip circuit breakers, and damage power rectifiers (see Chapters 6, 10, and 11).

5-4-2 Capacitor Failures

Capacitors will fail when exposed to heat and excessive voltages. They tend to develop high-resistance shorts that change the operating voltages in sensitive circuits. Low-voltage capacitors fail most often, probably because of less stringent quality control. Capacitors also open up quite frequently. In fact, they are more likely to open than to short. They are made of aluminum foil, which cannot easily be connected to copper leads. Repeated exposure to voltage, temperature, humidity, and mechanical changes will loosen up and corrode the connections. Electrolytic capacitors are also made of aluminum and tend to open for the same reasons. In addition, they dry up and lose capacitance with age. Capacitors also short or develop partial shorts, or as they say, they get ''leaky.''

TIP 5-49

Capacitors have a high failure rate. Small low-voltage electrolytics fail most often. Other types of low-voltage capacitors fail very often, too.

TIP 5-50

Electrolytic capacitors tend to open more often than they short. This means that jumping all the electrolytics when something goes wrong is not a bad idea. Use one or two large-value, high-voltage capacitors for checking all the capacitors. The ratings should be higher than that of the capacitors tested. Turn off the power when connecting the jumper capacitors.

TIP 5-51

Electrolytic capacitors may fail if a part failure places incorrect polarity or ac voltages on them. Electrolytics have positive and negative terminals. They must have positive voltages on the positive terminals with respect to the negative terminals.

TIP 5-52

While defective nonelectrolytic capacitors will often cause other parts to fail, they are not usually damaged by failures of other parts in the circuit as long as the polarity is okay.

Because of their high failure rates, a good troubleshooting procedure might be to check the connections, transistors, electrolytics, and low-voltage capacitors before looking for other defective parts. At any rate, they should be given high priority in any troubleshooting procedure.

Whereas nonelectrolytic capacitors seldom fail because of the stress placed on them by other part failures, this is not so with electrolytics. Electrolytics are polarized; they have positive and negative terminals. The positive terminal must always have a positive voltage on it with respect to the negative terminal; otherwise, it will be damaged.

> TIP 5-53
>
> Open capacitors can sometimes be found by wiggling them and making their internal leads establish or disestablish good internal contact.

Some technicians wiggle parts indiscriminately, especially when things are not going right. This is a nasty habit that can cause more damage than it does good. Wiggling is an art.

5-4-3 Resistor Failures

Resistors fail for a number of reasons. Sometimes, low-value, low-power resistors are damaged by excessive voltage and current over long periods of time. At other times, the internal resistive material does not make good electrical connection with the metal leads. Many times, long exposure to heat and high voltages can gradually increase their resistances.

Sometimes, resistors abnormally increase value as they warm up and return to normal after they cool off. These resistors can be found by carefully warming them up with a soldering iron tip held near them and observing the symptoms. After the circuit has warmed up, defective resistors can also be found by cooling them off with freeze spray and noting the symptoms.

Carbon resistors very seldom short out. Two-watt carbon composition resistors have been known to short; they can be carbonized by high voltages. High-resistance ceramic types can develop arc paths in high-voltage circuits. This can reduce their resistance.

The old carbon-composition types are the worst. Whether they increase value or decrease value depends on the nature of the stress they undergo. If they go through the normal cycle of warm-up, cool-off, and voltage stress, they will increase value over time. If they undergo drastic overvoltage and overcurrent, they tend to decrease value. However, they may also show discoloration and burn marks. The burning can also break and open up a resistor.

Generally speaking, if a defective resistor shows no signs of burning, it is more likely to be open than shorted. Hairline cracks and corroded internal contacts can open them up. Expansion and contraction over a period of use can eventually open up the contact between the resistive material and the copper lead.

Wirewound resistors normally run hot—hot enough to sear your fingers. The heat tends to deteriorate their connections and mounting material. Constant expansion and contraction eventually cracks their ceramic insulation and breaks the resistance wire. The wire itself is subject to stress, corrosion, and crystallization that cause them to open up.

TIP 5-54

Quarter- and half-watt resistors that do not show discoloration from overheating usually have increased resistance.

TIP 5-55

Discolored carbon resistors of any size occasionally short out or have decreased resistance.

TIP 5-56

A low-power, low-value resistor may burn out completely and open up when a diode, transistor, or capacitor in its path shorts. It may be acting as a fuse, so do not replace it with a higher-power resistor.

TIP 5-57

A resistor can burn out if a parallel part, usually a coil, opens and increases the voltage drop across the resistor [see Fig. 5-15(a)].

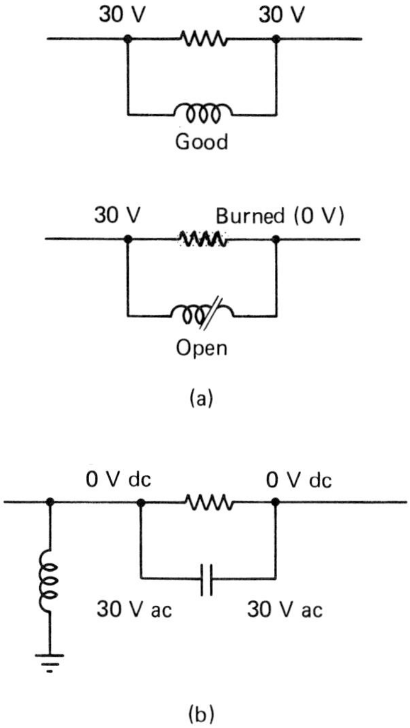

Figure 5-15 (a) An open coil increases the voltage across the resistor to 30 V, which is excessive. (b) An open capacitor will increase the ac voltage across the resistor from 0 V to 30 V and burn out the resistor.

TIP 5-58

A resistor can burn out, even if there is 0 V dc across it, when an electrolytic capacitor bypassing it opens. The resistor may not be able to handle the resultant high ac voltage that was normally bypassed [see Fig. 5-15(b)].

TIP 5-59

Resistors are often okay when the circuit is first turned on, but as they warm up, they may gradually increase in value. These resistors can be located quickly by heating them and then cooling them with freeze spray.

TIP 5-60

High-value carbon resistors that drop large voltages tend to increase in value with age.

TIP 5-61

Wirewound resistors do not change value, nor do they short internally. They only open completely.

TIP 5-62

Wirewound resistors normally run hot. A cold resistor is definitely open if current is supposed to be flowing through it.

5-4-4 Coil and Transformer Failures

Coils do not fail as often as other parts. Their copper windings are durable and make good solder connections with the leads or lugs that connect to the outside. Occasionally, they do open. Their weakest point is at the connections where the coil wire is soldered to the terminal. The heat during the soldering tends to weaken the copper wire. This is especially true of thin-wire coils. Traces of enamel insulation may cause a poor internal solder joint. Age and continual expansion and contraction can break the thin wire that is connected to an immovable terminal. In transformers the side with the higher voltage has the thinner wire. Coils with thick wires are subject to cold-solder joints that open up.

Power chokes are well constructed and carefully insulated, because a short to the grounded iron core can be costly and dangerous. They seldom short to ground, but occasionally they do.

TIP 5-63

If the markings on an open wirewound resistor cannot be read and no schematic is available, cut the resistor in half and measure the good half. The substitute should have twice the resistance of the good half. The power rating can be figured out by dividing the square of the resistor voltage drop by its resistance and adding 50% for safety ($P = V^2/R$)

TIP 5-64

Carbon resistors seldom short or decrease value. When they fail they seldom cause other parts to go bad. Once in awhile a grounded hold-down resistor may open up and cause a transistor to fail (see Chapter 8).

High-voltage transformers often fail. In consumer equipment, it would be too costly to make them 100% damage-proof. As it is, they are very expensive. Power transformers fail occasionally. In older equipment the big ones run warm and are subject to heat damage. They have thick wires that are difficult to solder.

TIP 5-65

Coils fail less often than do capacitors and resistors.

TIP 5-66

Small, low-voltage coils seldom short out; they open up. They tend to open at the terminals and are often repairable.

TIP 5-67

Coils with medium-sized to thick wires rarely fail, but they tend to "uncoil" with age and lose their tuning. Their turns can be pressed together to retune them (see Chapter 7 for tuned coils).

TIP 5-68

High-voltage transformers fail often. They can damage the high-voltage output transistors (see Chapter 10).

TIP 5-69

Power transformers fail occasionally. Their thick ground leads may be poorly soldered to the metal chassis. Their thin internal wires may break at the terminal lugs.

5-5 STUDYING A SCHEMATIC

Figure 5-16 shows a small section of a four-page schematic for an RCA CTC-123 color TV set. We are looking at the U_{701} luminance/chroma integrated-circuit (IC) chip. We are not ready for ICs yet, but we do know a few things about the parts around the chip. Let's study the schematic to see what can happen when these parts go bad. An experienced troubleshooter would check the signals on the chip and know which pins to check. We are not ready to do this yet.

> **Caution:** Do not work on the actual circuit. We are only studying the schematic. A beginner might do more harm than good. R_{722} shown on the right and above the chip is shaded. This means that you must use a factory replacement if it goes bad. It is the part of a circuit that prevents the formation of harmful x-rays. It is connected to a dangerous high-voltage circuit.

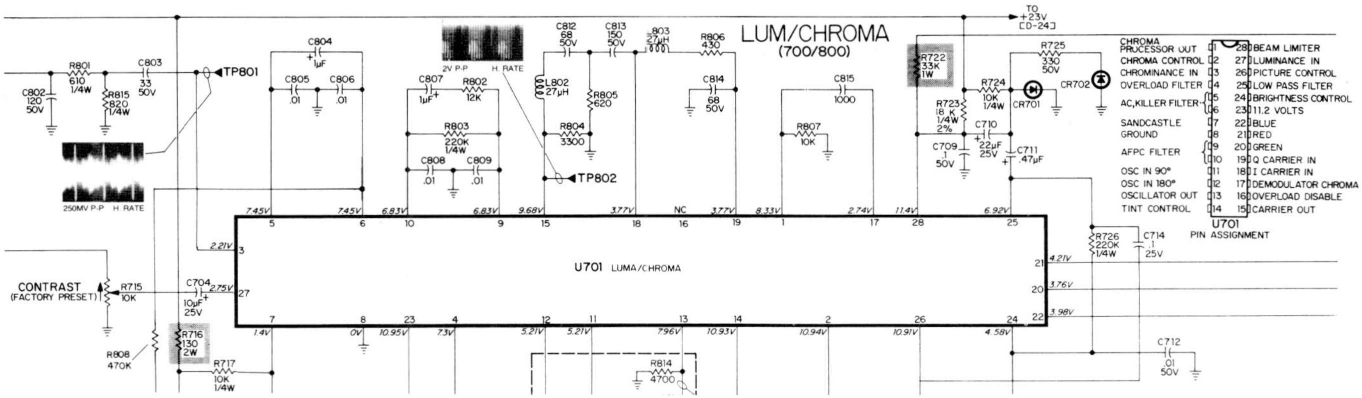

Figure 5-16 RCA color luminance circuit. With the voltages given in the schematic, you can check the parts around the schematic using the tips in this chapter. (Reprinted courtesy of Thomson Consumer Electronics)

We will consider only dc voltage and resistance changes. It should be noted, however, that bad parts also affect signal voltages, which in turn may cause the internal IC circuits to change the dc voltages. We know nothing about these circuits, so not all of our dc checks are conclusive. Also recall that a 0-V or 0-Ω measurement may not be exact but something close to it. Furthermore, a part that is thought to be open or shorted may really have an excessively high or low resistance.

With these considerations in mind, let's look at the parts along the top edge of U_{701}, starting at pin 5. (The pins are not drawn in order.) If the voltage on pin 5 is too low, a shorted C_{805} could be pulling it down. A similar thing could be said about pin 6 and C_{806}.

A shorted C_{804} may not affect the voltages on pins 5 and 6 because they are normally equal. This is only a guess because we have no idea of what is going on inside the chip. So far, we have not had any definite checks. If nothing else can be found, C_{804} is suspect. Generally speaking, low-voltage electrolytics have a high failure rate.

Moving over to the pins 10 and 9, we find something more definite. If C_{807} is good, R_{802} should have 12 KΩ across it. R_{803} should not have more than 220 KΩ across it. If pins 9 and 10 have unequal voltages, R_{802} would drop 0 V if C_{807} is good.

If pin 15 voltage is too low, C_{812} could be shorting out. A defective L_{802} may have no affect on this dc voltage. L_{802} itself should not drop any voltage and have 0 Ω. If pin 15 voltage is too high, R_{804} could have increased resistance. R_{805} should have *exactly* 0 V above it unless either C_{812} or C_{813} are shorted.

Moving over to pins 18 and 19, L_{803} should measure 0 Ω. If pins 18 and 19 have unequal voltages, an open L_{803} would drop some voltage. If pin 19 voltage is lower than pin 18 voltage, C_{814} could be shorted.

Looking at the next two pins, if pin 1 voltage is too high, R_{807} could be open. If pin 17 voltage is too low, C_{815} could be shorted. If the voltage on pin 28 is bad, R_{722} or the circuit connected to it could be bad. Let's look at the parts that are closer to pin 28. If its voltage is too low, C_{709}, C_{710}, or C_{711} could be shorted; or R_{723}, which is connected to the 23-V supply line, could be open.

An open R_{724} will lower the voltage below it. This voltage is held or "clamped" to 0.7 V by diode CR_{701}. If this voltage is too high, diode CR_{701} is open.

R_{725} should drop no voltage unless CR_{702} is completely shorted, which is unlikely. In other words, a bad R_{725} would be hard to find with voltage checks alone. Since it is low valued compared to the other resistances, you can check it with an ohmmeter. It should measure 330 Ω. Make sure that your ohmmeter battery does not affect the reading.

At pin 25, a shorted C_{711} should raise the voltage on it. R_{726} has too high a resistance to affect this voltage significantly (ignoring the internal IC actions for the moment). The same thing can be said about C_{714}. If C_{714} itself shorts, it might raise the voltage on pin 25 because it is connected to a higher-voltage point. Again, we cannot be sure of this.

This completes our short study of dc voltages on a schematic. You should do the same on your own.

WHAT YOU LEARNED TO DO

In this chapter you learned to check specific parts in- and out-of-circuit. You learned how to:

1. Combine the basic rules with the specific characteristics of resistors, capacitors, coils, and transformers to isolate these parts in-circuit.
2. Check diodes in-circuit.
3. Check connections, conductors, and switches in-circuit.
4. You learned which parts fail most often and what conditions make them fail.

QUESTIONS

1. Why is it better to check parts in-circuit rather than out?
2. Why is it possible to check a part out-of-circuit when one of its leads is still connected to the circuit?
3. Suppose you find a series part that has the maximum supply voltage across it. Is it open or shorted? What is the best way to confirm your suspicion?
4. What resistance value should a nonelectrolytic capacitor have when checked with an ohmmeter out-of-circuit? How would you evaluate a resistance check on an electrolytic capacitor?
5. Do coils have high or low resistances compared to other circuit parts? How does this fact help the troubleshooter?
6. You find that a diode has the same resistance reading measured both ways out-of-circuit. Is it good?
7. The anode terminal of a diode is connected in series with a grounded resistor. The cathode has 2.0 V on it. How many volts should the resistor have on it?
8. A 4-kΩ resistor is in series with a capacitor. You measure 8-kΩ across the resistor. What other check can you make to verify that it is bad?
9. If the circuit above has a voltage drop across the resistor, which part is bad?
10. A circuit having a 10-kΩ resistor connected across a capacitor has a greater-than-normal voltage drop across it. Which part is bad?
11. If the circuit above has a below-normal voltage drop, which part is bad?
12. Two equal-value resistors are connected in series between supply and ground with a diode in between the resistors. The diode arrow is pointing toward ground. The supply voltage is 30 V. What voltages would you expect at each test point?
13. If the arrow of the diode in Question 12 is pointed toward the supply, what voltages would you expect?

14. If the arrow is pointed toward ground and the supply voltage is −30 V, what voltages would you expect?
15. If the arrow is pointed toward supply and the supply voltage is −30 V, what voltages would you expect?
16. You find a small grounded coil with 14.3 V dc on it. Is it good, shorted, or open?
17. The secondary circuit of a transformer has dc voltages in it. The only dc source is on the primary side. What could be wrong?
18. If you measure 1.0 Ω across the ac plug prongs of a a low-power test instrument, what, if anything, is wrong with it?
19. Name, in order of failure rate, the parts that are most likely to fail.

PRACTICAL TROUBLESHOOTING EXPERIENCE

Find a well-populated scrap PC board with all the basic parts represented. You will also need a power supply that can output 9 V dc. Preferably, it should have a meter to display the current. A 9-V battery will do. Make sure that the battery voltage does not drop lower than 8.5 V when loaded. Cut out some branch supply lines until you get close to 9 V.

1. With the power off, check a few resistors.
   a. Did you have any problems with readings that do not settle down? If so, tack-solder short lengths of jumper wires across the big electrolytics.
   b. How many of the resistance readings equaled the coded values of the resistor across which the measurements took place? (Check your meter against known-good out-of-circuit resistors.) If you are using a low-voltage ohmmeter, you'll probably get a higher percentage of readings that equal the coded value of the resistors.
   c. If a resistor shows a value equal to its coded resistance, desolder one end of it to see if it is truly good. In actual practice you will assume that it is good and come back to it later if necessary.
2. How many resistances had more than their coded values? If you found one, is it bad? You may have to check all the resistors. More likely than not, all the resistors are good. Unsolder the ones that seem bad and recheck them out-of-circuit.
3. Can you find a resistor in series with a capacitor? If you can, measure the resistance across it. What value should you measure if it is good?
4. Find a resistor in parallel with a capacitor and measure the resistance across it. Open up one end of the capacitor. Measure across the resistor again. Did its value change?
5. All in all, from the experience above, would you say that measuring resistances is a good way to find a bad part?

6. Find a series resistor and capacitor combination and connect it to a 9-V dc power source through a milliameter. The resistor should be at least 1 kΩ. Do not let the current exceed your maximum power-supply rating. If you have a variable supply, bring up the voltage slowly and watch the ammeter meter. Do not allow the current to exceed the maximum.

   Measure the voltage drop across the resistor. How much voltage do you get? What part are you checking, the resistor or the capacitor, and is it good or bad?
7. With an ohmmeter, check all the coils that you can find on the board. What was the highest in-circuit resistance measured? Measure this coil out-of-circuit. Was there a substantial difference between the in-circuit and out-of-circuit measurements? Was it necessary to remove the coil to verify the in-circuit check? Would you say that in-circuit resistance tests of coils are fairly reliable? What can be wrong with resistance checks? How long would it take to test all the coils if you check only whether they are open or not?
8. If there are any small transformers on the board, check the resistances between all terminals. If it has two separate windings, the resistance between the primary and the secondary could be high. Remove the transformer and remeasure the terminals. Do you get the same readings as before? Was it necessary to remove the transformer to verify the in-circuit check?
9. Look for diodes and check them in-circuit and out. Compare the results. Did you find an in-circuit measurement that looked good but turned out to be bad when measured out-of-circuit? Did you find one that looked bad but turned out to be good?
10. Check the transistors. Check all terminals for hi–lo readings. You should get at least two hi–lo readings with one of the terminals involved in both readings.
11. Are there any FETs on the board? How can you tell which are FETs? (At this point the only way you can tell is by the resistance checks and the schematic or transistor manual.) Later, voltage checks along with the resistance checks will let you know whether or not you have an FET.

6

Basic Power Supplies

EXPERIENCED REQUIRED

Before you can learn the procedures for troubleshooting a basic power supply, you should know how to:

1. Check the basic electronic parts out-of-circuit.
2. Check all the basic parts except transistors in-circuit.
3. Connect batteries in series so that their voltages add up.
4. Use ac adapters for battery-operated equipment.
5. Check branch circuits.
6. Check connections, including ground connections.
7. Take the proper safety precautions (see Chapter 3).

For the experienced reader, some oscilloscope readings will supplement the dc checks. However, these are not absolutely necessary for working on basic power supplies. For beginners it is important that you get hands-on experience early. You will learn just enough about ac to get you started. In the next chapter we will provide more detail on ac circuits.

6-1 THE POWER SUPPLY

A high percentage of troubles that require bench work are associated with the power supply. Compared to other electronic circuits, basic power supplies are relatively simple, and beginners can do a lot of work on them.

Working on a power supply requires some knowledge of ac. Direct current (dc) is smooth, steady, and flows in one direction. Alternating current (ac) fluctuates and alternately reverses direction many times in 1 second. The rapidly fluctuating voltage that comes out of electrical outlets is okay for light bulbs, motors, and long-distance power transmission, but it is not good for directly powering electronic circuits. Electronic circuits generally require smooth-flowing dc power sources.

Our first concern in troubleshooting something new is to find out what it does. The power supply does two things. It converts ac to dc, and it supplies the required amounts of voltage and current. If we know what it does when it's good, we'll know it's bad when it doesn't.

6-2 BLOCK DIAGRAM OF A POWER SUPPLY

Our next step is to find out what each section of a power supply does. The one thing that all electronic parts, sections, and systems do is to modify voltages. Figure 6-1 shows a block diagram of a power supply. Each block represents a section having a particular function—a job. Each section has an input and an output. One of the most important things in troubleshooting is to know the input and output voltages of the various sections within a system.

6-2-1 The Input Section

The input section connects the ac power lines to the power supply and changes the voltage to the level required by the system. In this example the voltage is decreased from 120 V ac at the input to 20 V ac at the output. It also contains the power on-and-off switch and a protective fuse.

6-2-2 The Rectifier Section

The rectifier section in Fig. 6-1 converts the ac voltage that fluctuates in two directions to a type of dc voltage that pulses in one direction. The rectifier is a one-way valve through which the fluctuating current is pumped into the filter section. You will not see much pulsing at the rectifier output, because it is filtered out by the following section.

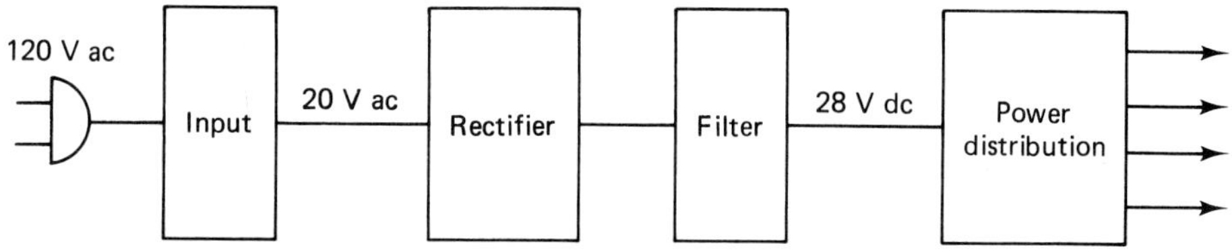

Figure 6-1 Block diagram showing the main sections of a power supply.

6-2-3 The Filter Section

The action to this point can be compared to an air compressor that has a one-way valve that allows air to pulse into a tank and keeps it from backing out. The tank smooths out the pulses. That is exactly what the filter section in Fig. 6-1 does with electrical pulses. It smooths them out.

There is one other thing you should know about the filter section. It can output a dc voltage that is about 1.4 times higher than the ac voltage at the rectifier input. In our power supply the 20 V ac voltage ends up as 28 V dc. We'll discover why in Section 6-3-3 when we look at what the parts in the filter section are doing.

6-2-4 The Power-Distribution System

The power-distribution system shown in Fig. 6-1 does just what its name says it does. It distributes suitably filtered dc voltages of different levels to various circuits. Some circuits require more completely filtered dc voltages than others. The power-distribution system may provide additional filtering.

6-3 WHAT EACH PART IN A BASIC POWER SUPPLY DOES

Thus far we divided the basic power supply into several sections and saw what each did to the input voltage. Now we have to look at the parts in each block and see what type of voltages they input and output or maintain.

6-3-1 The Transformer Input-Circuit Parts

Our basic power-supply input circuit, shown in Fig. 6-2, has a step-down transformer, *T*, which reduces the voltage from 120 V ac to 20 V ac. The transformer primary

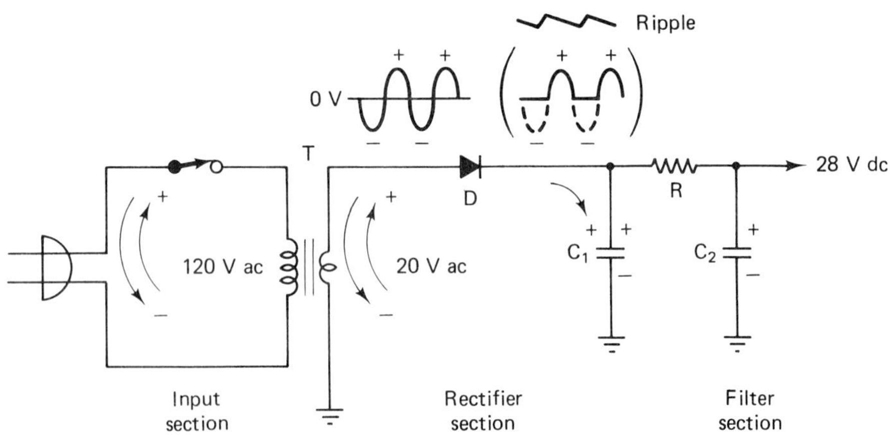

Figure 6-2 Schematic of a basic power supply showing the input and output voltages at key points. The arrows show changes in the direction of current. The waveforms represent what can be seen with an oscilloscope.

circuit is similar to the power input circuit in Chapter 3. The transformer windings "step down" the 120 V ac at the input to 20 V ac at the output. This voltage, which is flowing alternately in both directions, is pumped into the rectifier.

6-3-2 The Rectifier

Recall that diodes pass voltages or currents in one direction only. In Fig. 6-2 we see a diode, *D*, passing current into a capacitor and resistor. Ahead of the diode we see a wave that represents the alternating voltage being impressed on the diode. This wave is called a sine wave. The tops of this wave represent the peak positive voltages. The bottoms represent peak negative voltages. The centerline represents points where the voltage passes through 0 V. Note the arrows at the transformer secondary. They represent one full cycle of alternations.

> *Note:* The two *alternations* make up one *cycle* of the wave. The number of cycles per second is called the *frequency* of the wave. It is measured in *hertz*. In this country the standard frequency for ac power-outlet voltages is 60 Hz.

The alternating voltage tends to push the current through the diode. The diode, however, can pass current in one direction only. As a result, we have a series of pulses all going in the same direction. We show that only the downgoing arrow can pass through the rectifier.

Also see the half-wave pulses depicted at the output of the rectifier. The dashed halves show the switched-off alternations of the wave. The solid halves show what can pass through the rectifier if it were not filtered at this point. In reality, capacitor C_1 in the next section partially filters them out, so that we have the ripple shown above the half-wave waveform.

6-3-3 The Input Filter Capacitor

We mentioned that the filter section acts like a compressor tank that outputs smooth air pressure. Another comparison can be made. Capacitors are somewhat like automobile batteries; they store voltages. Auto batteries are charged through rectifiers, and so are these capacitors. The fact that a capacitor cannot hold its charge very long is no problem. The rectifier pulses keep recharging the capacitor 60 times a second, which is fast enough to keep it from losing its charge.

Now we can see why one-way pulses are needed to fill up the capacitor "tanks." If both pulses are allowed to pass through the rectifier, one pulse will charge the capacitors and the second pulse will discharge them. The net result would be zero volts.

Our rectifier keeps pumping these one-way pulses into the capacitor fast enough to keep it from discharging. It can only discharge through the resistor, and this takes time. Within a fraction of a second, the capacitor "tank" is filled to the top. It reaches the peak voltage the rectifier is capable of delivering—28 V. The capacitor will maintain this peak charge as long as the power is on.

So now we can see why 20 V ac produces 28 V dc. When we say 20 V ac, we are actually talking about the power-delivering ability of a fluctuating voltage that has a peak of 28 V and goes through zero volts at times. The mathematical name for this capability is called the *root-mean-square (rms)* voltage. It is also called the *effective voltage,* because a 20-V rms voltage delivers as much power as a 20-V dc voltage. It has the same effect, even though its peak voltage is nearly 28 V.

Note: The equation for converting a peak voltage to rms is

$$V_{rms} = 0.707 \times V_{peak} \text{ and } V_{peak} = 1.414\ V_{rms}$$

In our example: $V_{rms} = 0.707 \times 28\ V_{peak} = 19.8\ V_{rms}$

Electrolytic capacitors are used for power-supply filters because they can have very large values, around 50 μF to several thousand mirofarads. For many applications, the input filter capacitor alone cannot do a complete filtering job. Another filtering section is needed.

6-3-4 The Filter Resistor and Output Capacitor

The filter resistor, R_1 in Fig. 6-2, may drop very little dc voltage, but the output filter capacitor causes it to drop almost all the pulse voltage. The output filter capacitor has the capability of shorting out the pulses but not the dc (see Chapter 7). The final result is almost pure dc at the output.

Coils called chokes were used in the older supplies for additional filtering. They have the ability to pass dc and block pulsing currents (more on this in chapter 7). The filter section has been largely replaced (and improved) by electronic circuits. We will see these circuits in Chapter 11.

6-3-5 The Power Distribution System

In Chapter 1 we stressed that the power supply was one of the gateways into the system. A look at the power distribution system will show you why. It provides various voltage levels for nearly all the circuits in the system. The block diagram of the distribution system shown in Fig. 6-3, indicates that we can apply all the basic rules we learned in preceding chapters—in particular, review Tip 3-8 on series–parallel circuits. The power supply is considered to be "above" the point being measured and everything else is "below" or toward ground. The voltage points that are nearer to the power supply have the higher voltages, and those nearer to ground have the lower voltages.

R_4 in the illustration serves to hold down and stabilize the 20-V supply line. It is called a *bleeder* resistor. In higher supply-voltage systems, it "bleeds" off the dangerous voltages that are stored in the filter capacitors after the equipment is turned off.

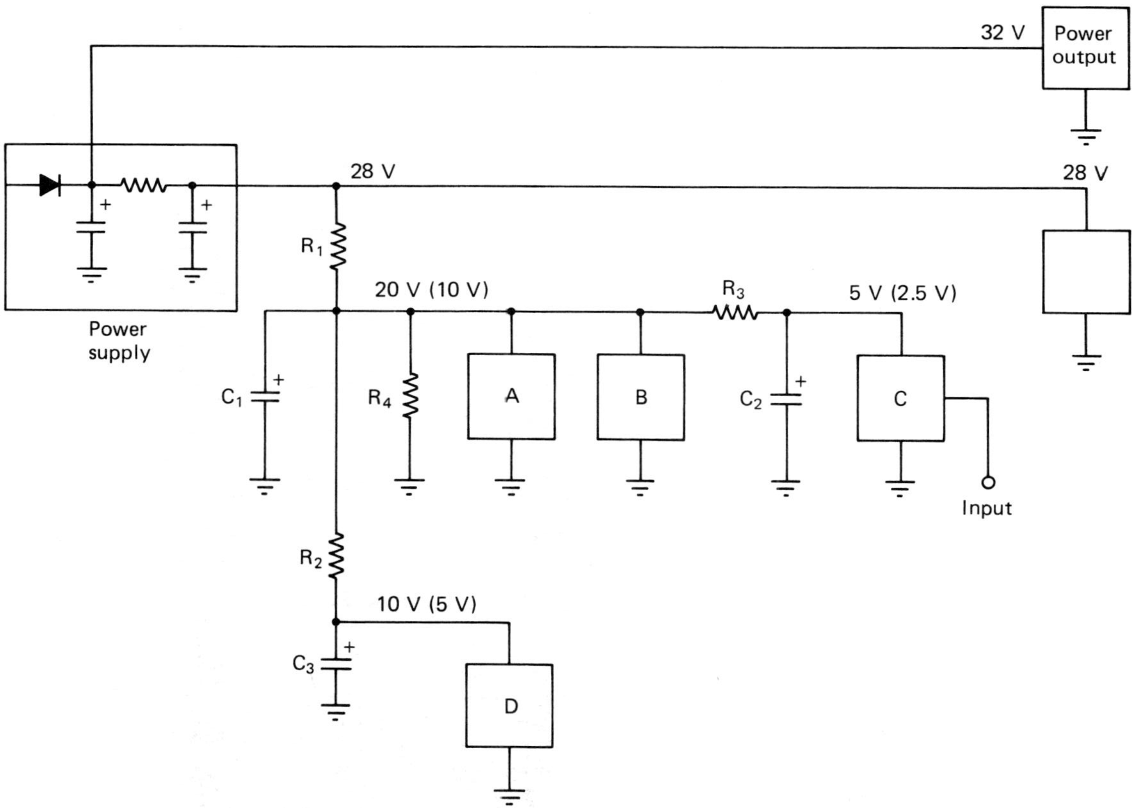

Figure 6-3 Power-supply distribution system. The resistors drop the voltages as required by the various load circuits. The capacitors filter out any signals that may be generated by these circuits.

Electrolytic capacitors are stationed throughout the distribution system at strategic locations. Their purpose is twofold: First, they provide extra filtering for the more sensitive circuits; and second, they filter out interference signals that may get into the supply line from various sources. In this respect they are called *bypass capacitors,* since they bypass unwanted signals to ground. They are also called *decoupling* capacitors because they prevent circuits from coupling (passing) unwanted signals to each other through the supply lines.

The system output circuits require the least filtering. Sometimes they are fed right off the power-supply rectifier, as shown in Fig. 6-3. Note the 32-V line. The only filtering it gets is from the power-supply input filter capacitor. On the other hand, the signal input circuits require the most filtering. Note the input circuit, block *C*, in Fig. 6-3. It has a low voltage, which is typical. Its dropping resistor also serves as a filter resistor. The rule of thumb for most applications is to cut down the ripple to within 2% of the dc supply line voltage. Of course, for high-quality equipment, more filtering is needed.

One problem with these supply-line electrolytics is that they are often hidden in the most remote and inaccessible locations in the chassis and in the schematic. Chassis designers and draftsmen seem to be playing hide and seek with us.

A Very Important Case History. The power supply is a gateway into the system. Remember Fig. 1-1? The power-distribution system has pathways that lead you to the defective sections. You will work on these pathways in the following example. You will be repeating this example many times in troubleshooting. Study it carefully. Refer to Fig. 6-3. The chassis we are looking at has a weak and defective output. Generally speaking, one of the first things we do is to check the supply voltages to the various circuits.

Our checks show that the main supply-line voltage is okay at 28 V, but the 20-V supply point has dropped to (10 V), as shown in parentheses. This means that either resistor R_1 has partially opened or there is a short below it (Rules 2-2 and 2-3). None of the blocks has a complete short. R_1 is a wirewound resistor; therefore, it is not likely to open partially. Hence we conclude that there is a short below. One of the four blocks or capacitors below is partially shorted.

Next we see that the 10- and 5-V supply points have gone down proportionately with the 20 V point. They have all fallen by 50%. This leads us to the following line of thinking.

Capacitor C_3 is definitely not shorted—probably not even partially. If it were, we would see a greater drop than the proportional 50% drop from 10 V to (5 V). R_2 is not partially open. If it were, the voltage above it would have gone up instead of down to 10 V (Rule 2-2). It is also a wirewound type that is not likely to open partially. If it opens, you'll have 0 V below it.

We have a similar situation with respect to C_2 and block *C*. They are probably not partially shorted. R_3 is a carbon resistor which could open partially, but this would raise the voltage above it (Rule 2-2) instead of lowering it.

Therefore, we should be looking into blocks *A*, *B*, C_1, and R_4 to find out why the voltage above them has fallen. Capacitor C_1 could be partially shorted (Rule 2-3), or blocks *A* and *B* could have a shorted part in them. R_4 is less likely to short than C_1

or the circuit blocks (see Section 4-1 for more on circuit blocks). If the blocks do not have too many circuits, your best bet would be to check the circuits within them for a drastically shorted part. If you can understand this example, you will make a good troubleshooter.

Our knowledge of the power-distribution system shows that several different circuits with different functions could be on the same line. If that line goes bad, all these circuits would also go bad. This gives us the following important tip.

> **TIP 6-1**
>
> If the output symptoms show that several functions are defective, look for a defective supply line that feeds the circuits responsible for these functions.

In summary, a good knowledge of the four power-supply sections will let us know where to look for the trouble. For instance, in audio equipment, if you hear excessive hum, the power-supply output could be pulsating: Check the filter. In video equipment, if you have a display but no sound, check the power-supply distribution line to the sound section. These faults and others are treated in more detail in the following sections.

The power supply is not that difficult to work on. Just concentrate on the voltages at each point. The voltages are usually on the schematic. You should be able to identify the power supply and its parts and to know where and what to check when voltages go bad. We cover this next in more detail.

6-4 NORMAL VOLTAGES AND CIRCUIT RESISTANCES

A good reason why you should know typical voltages and resistances is that the schematics are not always correct. Notice how schematics are constantly being revised? This is not always because some circuit changes have been made; sometimes a draftsperson or someone else has made an error.

6-4-1 Normal Voltages

Figure 6-4 shows a schematic of a very basic power supply. It consists of an input circuit, a transformer, a rectifier, a filter, and an output fuse. Note the input and output voltage relations between the sections.

1. For safety the 120-V ac input circuit voltage is not measured. Unplug the line cord and use an ohmmeter to check this section. This circuit is similar to the lamp circuit in Chapter 3. The troubleshooting methods are the same.

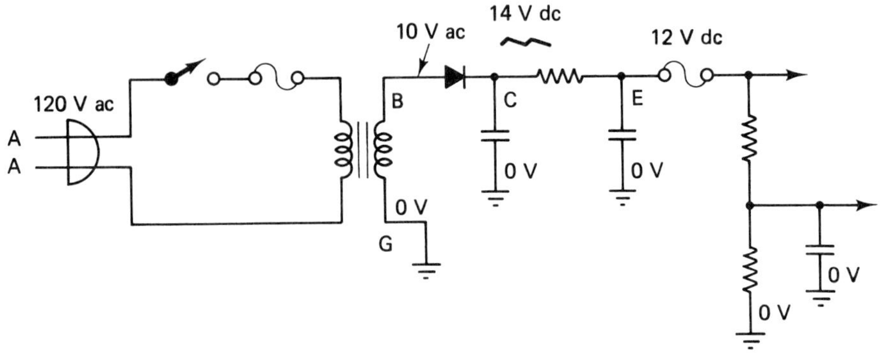

Figure 6-4 Power supply showing typical input and output voltage relationships. The text also shows the normal resistances in this circuit.

2. The transformer output voltage may or may not be shown on the schematic. It should be an ac voltage of about 70% of either the rectifier or filter dc output voltage.
3. The dc output voltage at *E* is usually given in the schematic. It should be about 1.4 times greater than the ac voltage at point *B* when using the multimeter.
4. You should not use an oscilloscope to check these voltages unless you are familiar with the oscilloscope safety precautions (see Chapter 7). Sometimes the schematics show the "scope" waveforms and their peak-to-peak ripple voltages. These ripple voltages will read considerably less on your multimeter ac function. At the input filter the oscilloscope voltage can be quite large, about 5 to 10% of the dc voltage is typical at this point. At the output filter, it is about 1% or less for most equipment.

If you do not have an oscilloscope or the type of meter that reads peak-to-peak voltages, that's okay; there are other, faster ways of dealing with ripple (see Section 6-6). Our primary aim is to get rid of the trouble.

5. There should be zero voltage at every ground lead in the power supply. Do not overlook checking for this.

6-4-2 Normal Resistances

We start with the input resistance of the circuit as shown in Fig. 6-4. The resistance across the ac plug prongs A–A′ takes in the transformer primary resistance and the switch and fuse—all at once. This overall resistance should be very low since the resistances of these parts are low.

The resistances in the primary circuit give you better information than do the secondary circuit resistances. The normal secondary circuit resistances will vary from meter to meter. We can still give you some helpful information. Let's look at the transformer secondary resistance *B–G* in Fig. 6-4.

TIP 6-2

The input resistance across the ac plug prongs is very low—from only a fraction of an ohm for equipment using high-power transformers, and several hundred ohms for low-power transformers.

TIP 6-3

The secondary resistance of a power supply transformer is lower than the primary resistance for most solid-state equipment. For low-power, high-voltage equipment, it could be higher.

Resistance *C–G* has a high reading and a low reading. Switch the leads after the first reading. Wait until the capacitors charge or discharge. You should have the transformer resistance plus the diode forward resistance. Get to know your meter. Try it on good diodes in various circuits.

To measure the filter resistor you may have to use a low-voltage ohmmeter or take two readings, one with the leads reversed. Otherwise, the rectifier will affect your reading. Reverse the leads for the second reading. Also wait for the capacitor charging and discharging to reach a steady state. If the rectifier is good, the higher of the two readings should give you the resistor value.

The resistors in the distribution system should be measured with a low-voltage ohmmeter. Of course, if you have very bad shorts or opens, any ohmmeter will show it.

6-5 ABNORMAL VOLTAGE AND RESISTANCE READINGS

Now that we have a fairly good idea of what types of voltages to look for at various points in the power supply, we can start looking for bad voltages and their causes. First we will see what shorts can do to voltages.

6-5-1 Catastrophic Effect of Shorts on Voltages

If a short occurs in a power supply, you may not be able to read the voltages. As soon as you turn on the system, the fuse or something will blow and leave you no voltages to read. You have to use another method. Either take resistance readings or use the series-lamp method described in the next section.

In addition to the basics given in Chapter 2, here are a few other things you should know about voltages in a shorted power supply:

TIP 6-4

A partially shorted power-supply transformer, besides running hot, will give you low output voltage. It may also buzz and vibrate before it blows the fuse.

TIP 6-5

A shorted power-supply rectifier will put an ac voltage on the electrolytic filter capacitor and damage it. The capacitor may even sizzle and explode. The transformer will overheat as a result.

The replacement rectifier rating must be at least 50% more than twice the input ac peak voltage.

Interesting Sidelight. The peak voltage across a power-supply rectifier equals the capacitor voltage plus peak input voltage. On the reverse cycle, the top of the ac source has a negative voltage and its bottom has a positive voltage. This voltage is added in series with the capacitor, which has a negative polarity at its bottom and a positive on top. Think of the reverse-voltage source and the capacitor as two batteries connected in series. For the reason given above, power-supply diodes are given *peak inverse-voltage* ratings.

6-5-2 Effect of Opens on Voltage Readings

The input circuit is best checked for opens with an ohmmeter, but sometimes the trouble shows up only when voltage is applied. You can unplug the circuit and clip the voltmeter probes to a test point and plug the circuit back in. Repeat this for each check. This may be slow, but it is safe.

The secondary circuits can be checked safely without unplugging it for each measurement. The transformer isolates the secondary from the dangerous ac line. To give you maximum options, we include voltages for all sections. Refer to Fig. 6-4 for the following voltages. All dc voltages are with respect to ground.

1. Troubleshooting the ac input circuit is basically the same as for the lamp circuit in Chapter 2. If you have to measure voltages, unplug the circuit and clip your voltmeter leads across the part before plugging it back in for the check. Any part in the input circuit that opens up will develop 120 V ac across it. If measured with respect to ground (A'), you will have 120 V ac on one side of the open and 0 V ac on the other.
2. An open at any point in the primary circuit will cause a loss of voltage at the secondary output, point B.
3. An open anywhere in a basic transformer will cause a loss of voltage at the secondary output, point B.
4. An open rectifier will cause a loss of dc voltage at its output and leave the ac voltage at its input unaffected.
5. An open input filter capacitor C_1 will cause a substantial loss of dc voltage and an unstable or erratic performance at the output of the equipment. A very large ripple voltage may be read at point C with a meter or oscilloscope. This could be normal; the input capacitor is not always expected to filter out the ripple completely. It usually gets a lot of help from the rest of the filter circuit parts.

TIP 6-6

An open input filter capacitor will cause a loss of power-supply output voltage, large ripple, and serious deterioration of circuit operation. Check it with a substitute capacitor shunted across it. If normal operation is restored, the capacitor should be replaced.

6. An open filter resistor will cause a loss of voltage at point E and all points after it. The voltages ahead of it will be left intact or slightly higher.
7. An open output filter capacitor will place ripple voltages on all points after point E. Not only will the power-supply ripple get into the various circuits, the circuits themselves will feed ac voltages into the distribution system. The result will be erratic and often unpredictable. There will be no substantial loss of supply voltage. The main function of the output capacitor is to filter out the ripple. The symptoms produced by open output filter capacitors will be shown in Section 6-6-3.

TIP 6-7

An open power-supply output filter will cause ripples and other unwanted signals that can be seen on the supply lines with a scope. The result will be erratic and often unpredictable operation of various circuits. The dc voltage may not go down substantially. The capacitor can be checked by shunting it with a capacitor of equal or higher value to see if normal operation is restored.

8. An open distribution resistor will cause the voltages above it to rise and the voltages below it to fall. The voltages below will drop to zero.

TIP 6-8

If an open power-supply ground connection has more than zero volts on it, all the voltages above it, toward the supply, may rise. Operation may be erratic.

TIP 6-9

An open supply-distribution line capacitor may cause erratic operation, strange symptoms, and pulses on the line. A popular rule among troubleshooters is: "When you're stuck, jump the electrolytics" (turn off the power when you shunt an electrolytic capacitor).

6-5-3 Effect of Shorts on Resistance Readings

In general, a shorted part will result in reduced resistance across its terminals. In some cases the trouble may not show up until voltage is applied. However, resistance checks are easy and safe. The following readings are taken with the parts in circuit. The numbers in Fig. 6-5 correspond to the item numbers.

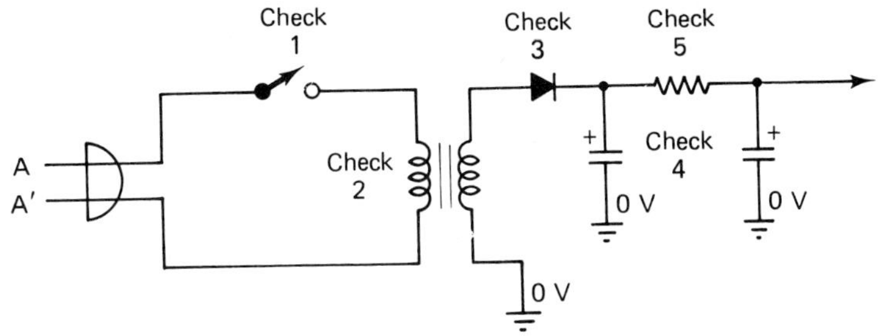

Figure 6-5 Checking resistances in a basic power supply. The numbers indicates checkpoints in the text.

1. A switch shorted to ground is a dangerous matter. It is not likely to occur. If it does, something may burn or explode and you will be able to spot it visually. If there is no visual indication of a short to ground, an ohmmeter check may reveal it, provided that the transformer primary resistance is not too low. Otherwise, you may have to disconnect a lead between the transformer and switch to isolate them from each other. A short between the two switch terminals will measure near-zero ohms in the off position.
2. A shorted transformer input winding may not show a conclusive ohmmeter reading unless the schematic gives you the resistance value. The normal resistance may be too low to make a guess. This is especially true of high-power equipment. Low-power transformers for small instruments may give you more positive indications for shorts. They are normally about 100 Ω or more, so if you see something like 2 or 20 Ω you may be looking at a short.
3. A shorted rectifier will give you near-zero ohms both ways.
4. A shorted filter capacitor may be detected with an ohmmeter if it shows very low resistance—less than the combined resistances of all ground-return resistor paths in the system. Of course, the rectifier (or fuse, when there is one in the primary circuit) would be blown if the filter is shorted. Remove the rectifier before checking the capacitor.
5. A filter resistor is not likely to short; but if it does, it will measure less than its nominal resistance. This resistor usually has a low value.

6-5-4 Effect of Opens on Resistance Readings

In general, an open part will give you a higher-than-normal resistance reading across its terminals. In-circuit resistances for various parts are given below. Refer to Fig. 6-5 again; disregard the numbers.

1. Any open part, connection, or wire in the input circuit will give an infinity reading at the ac plug, *A–A′*. We learned this in Chapter 2.
3. An open switch will have no continuity when turned to the on position.

3. An open rectifier will not give you a hi–lo reading. A good one will.
4. You cannot obtain useful readings on filter capacitors. Other parts in the circuit will affect your readings. As stated in Chapter 5, the best way to check for an open capacitor is to jump it with a good one.
5. An open filter resistor will give you a higher resistance value than its nominal value. This is because it has a much lower resistance than the other parts in the circuits.
6. There should be zero resistance between any ground lead and ground. Do not overlook this resistance check.

6-6 PRELIMINARY POWER-SUPPLY CHECKS

Now that we have seen what all the parts can do when they are good or bad, we are ready to troubleshoot them. Before checking the parts themselves, you should do the easy and more obvious things first.

6-6-1 First Steps

The first steps are the same as for the lamp and power input circuits in Chapter 2. Determine whether there really is something wrong with the equipment. Sometimes the user does not know how to turn it on. Try the switch and other user controls. Note whether the equipment is plugged in. You will be surprised how often this happens. Check the outlet with a lamp or any other appliance at hand. Wiggle the plug. Observe the safety rules. Don't touch metal. Squeeze the cord toward the plug to see if you can make the inner broken wire ends touch and get some kind of reaction from the system. Do the same with the other end of the cord.

After seeing that something is really wrong with the equipment, look for dripping tar from a shorted transformer or ooze from an electrolytic. Sniff for burned shellac from the transformer and other odors from burned parts. Check the fuse if it is easily accessible. During this preliminary stage, you may discover clues that indicate which part or section is bad.

One of your major objectives in this preliminary stage is to determine whether the supply is open or shorted. Shorted power supplies may require special procedures because you cannot apply power to them.

6-6-2 Symptoms Resulting from Shorts

By symptoms we mean the faults that can be detected without test instruments: the things that we can see, hear, smell, and feel. We cannot take voltage readings when there is a short, so we rely heavily on the symptoms. A power-supply short usually burns something. The following tips may prove helpful.

TIP 6-10

A shorted ac plug or line cord will blow the building fuse. Look for possible burn marks near the ac outlet receptacle or ac plug.

Shorts in unfused equipment may blow the building fuse or trip the building circuit breaker. Shorts in low-power equipment (around 5 W) may not blow any fuses or circuit breakers, but the power transformer will heat up.

A fuse is not very likely to short. If it is bypassed or replaced with one of higher rating, the results can be catastrophic when the equipment shorts. The power supply may blow and damage the rectifier, sizzle the electrolytics, or even burn the transformer. The building fuse may blow.

Note: Using Ohm's law ($R = V/I$), we have the following table, which will give you a feeling for shorts that can blow 120-V ac line fuses.

RESISTANCE OF SHORT	RATING OF BLOWN FUSE
8 Ω	15 A
6 Ω	20 A
4 Ω	30 A
3 Ω	40 A
2.5 Ω	50 A

As a reference point, a short of one turn in the input winding of a low-power transformer can overheat it, but its resistance is not likely to be much less than 200 Ω.

TIP 6-11

If a switch shorts internally or externally across its terminals, you will not be able to turn the equipment off.

When low-voltage circuits fail, since they usually feed signals into the higher-voltage circuits, the entire system may seem inoperative. Later, you will learn how to use substitute output signals in place of the dead low-voltage circuit outputs.

TIP 6-12

If any hot-side terminal, conductor, or part in the primary circuit shorts to the ac ground-side terminal or conductor, the input fuse will blow.

A shorted rectifier, filter capacitor, or transformer winding may cause the transformer to buzz and blow the input fuse.

A shorted transformer may also have a smell of burned shellac and drip melted tar.

TIP 6-13

A short in the power supply will leave behind a good power-supply output fuse and a blown input fuse. This may indicate a good power supply and a short in the output circuit of the equipment. Although the fuse is rated to protect the rectifier, the rectifier may be damaged also.

TIP 6-14

A shorted low-voltage supply line may not blow the fuse but may cause loss of operation in the low-voltage circuits only. Look for smoke and sparks. A burning resistor may be detected by turning off the lights.

Turn off the power and sniff around. Burned carbon resistors have a recognizable odor.

6-6-3 Symptoms Resulting from Opens

The most obvious and general symptom for an open in the power supply is a completely dead system with parts that are cold to touch. The indicator lights may not light. Unplug the equipment before touching the parts.

The following tips may give only generalized symptoms for the system outputs. More specific descriptions and troubleshooting procedures associated with these symptoms are given in later chapters and sections. Some but by no means all of these chapters and sections are referred to in the tips.

The symptoms described apply to nearly all types of equipment and circuits because nearly all of them have power supplies.

TIP 6-15

A shorted capacitor in the medium-voltage circuits may not blow the fuse but may cause a series resistor to smoke or discolor.

Smoke and sparks can be very deceptive. A sudden spark of short duration will reflect light off a shiny part and mislead you as to where it actually is. A wisp of smoke will curl around some corners, catch an air current, and suddenly appear in a location nowhere near its origin.

If only the circuits having medium supply voltage are inoperative, look for an open at the medium-voltage takeoff point in the power-supply distribution system.

TIP 6-16

Any open in the input circuit will cause a complete loss of operation. The power-indicator lamp will not light, and the circuit parts will be colder than normal. Unplug the equipment before checking the temperature with your fingers. Touch lightly at first so as not to burn your fingers; there could be a short instead of an open.

Open connections in the input circuit are likely to cause intermittent operation (see Chapter 16).

6-7 ACCESSIBILITY AND FAILURE RATES

Some parts in the chassis are easier to get at than others. If it is going to save time, do the easy things first. The power supply, in particular, has some check points that are much easier to get at than others. The plug, cord, and sometimes the fuse can be accessed without removing the chassis. Parts that are externally mounted, such as the power and regulator transistors, are usually accessible. Indicator lights may be accessible. Sometimes they are not so easily accessible, but you may still have to account for the time spent replacing the lights.

TIP 6-17

An open in the secondary-circuit dc voltage path will cause complete loss of operation, and the parts may be colder than normal, although the transformer may hum very slightly. Unplug the equipment before touching the parts.

An open rectifier will cause complete loss of system operation, and the parts will be colder than normal.

An open filter resistor will cause loss of operation and the parts below it will be cold. The resistor itself will be cold.

If a circuit takes its voltage from the input side of the filter resistor, look for voltages and symptoms that show it is working.

An open dropping resistor in the distribution lines will be cold. The symptoms may show that the circuits below it are not working and that the other circuits are working.

An open bleeder resistor will raise the voltages in the lines above it. This may cause arcing in circuits that develop high voltages (see Section 11-6). Look for symptoms associated with circuits that can be controlled by changing voltage levels. Some circuits are specifically designed to be controlled by changing voltage levels (see Sections 11-2, 12-4, and 12-5).

TIP 6-18

An open input filter-capacitor will cause weak and unstable operation and other symptoms associated with weak voltages and with ripple voltages in the supply lines. These ripple voltages are often called hum voltages (see Section 11-4 and Chapter 13).

An open output capacitor will introduce even more ripple or hum voltages into the circuits but will not reduce the voltage appreciably. There may be no reduction of output power, but there will be an increase of instability and other symptoms associated with ripple or hum.

The interference signals from various circuits may get into the supply lines and cause all sorts of strange troubles (see Sections 11-4 and 12-1).

An open capacitor in the distribution lines will cause similar unstable and erratic operation.

Some parts fail more often than others. Many times it is better to check these parts first (see Section 5-4). The fuse and rectifier often go out, and you should check these as soon as you can get to them.

Wirewound resistors and fuses open up when they fail; they seldom short. Do not waste too much time trying to figure out whether a shorted power resistor is pulling the voltage the wrong way. Try some other things first.

6-8 A TROUBLESHOOTING PROCEDURE FOR THE POWER SUPPLY

One of the first problems a beginner encounters when working on the power supply is to find it. You have to find out where it is in the chassis and also in the schematic. Most problems elsewhere in the system involve finding the supply-voltage source, so the tips given here, though very simple, are worth going over.

> **TIP 6-19**
>
> To find the power supply in the schematic, look for the ac plug or a supply line. Either of these will lead you to the power supply. The power switch, fuse, transformer, power rectifiers, and filter section are easy to spot. They are often grouped together in one corner of the schematic.

> **TIP 6-20**
>
> To find the power supply in the chassis, look for the big electrolytics, the transformer, and the red wires running from the transformer to the rectifiers. These wires are usually a twisted mess, so you will have to use a pair of long-nose pliers to pry them apart gently and carefully tug at a lead. You will catch sight of it responding to the tug at some distance away. If necessary, repeat the tugging over several lengths of wire until you find the other end. You can also use an ohmmeter to trace out the connecting wires and printed-circuit conductors.
>
> The power transformer normally runs warm and even hot in some cases, so it may be located where the heat will not affect the other parts. It is also very heavy, so it may be placed to obtain a good balance.

TIP 6-21

If you cannot find the output terminal of the power supply, look for various electrolytics and power resistors at random. If you cannot find voltages on any of them, the power supply is probably defective.

The procedures given below need not be followed exactly. There is no definite sequence for troubleshooting a power supply, for several reasons. First, the parts are so few that it may not really matter what you do first. Second, the accessibility of the parts differs in each case, so it might be best to check the easy things first. Third, there are two procedures for troubleshooting power supplies: one for opens and one for shorts.

6-8-1 Troubleshooting for Complete Opens

The following checks will help you find an open in the power-supply circuits (refer to Fig. 6-6).

Check 1. Check the symptoms: If symptoms indicate an open (Section 6-6-3), proceed to the next step. If not, see the procedure for shorts.

Check 2. Check the input circuit as you did for the lamp circuit in Chapter 3.

Check 3. Check the power-supply output fuse since it is easily located and likely to be blown. It will also tell you whether the fault is in the power supply or the load circuits. If it is blown, there is a short in the load circuits fed by the power supply. If it is good, there is an open in the power supply.

If there is no output fuse, check for voltage at the output load. Then check for voltage in the power supply. If you have voltage at the rectifier but not at the output load, there is an open between the two points. Possibly there is a fusible resistor between the two points.

An open fuse will have a voltage on its input side but none at the output side. The same is true of a burned-out distribution resistor. If you do find either of them open, the trouble is not in the power supply but is probably in the load circuits.

If you do not have voltage on the power-supply side, proceed to Check 4 to check the rectifier.

Check 4. Check for an open rectifier, since it is the next-most likely part to go out. Check for ac voltage at the rectifier input.

If you have normal ac voltage at the rectifier input and no dc voltage at the rectifier output, the transformer is good and the rectifier is bad. If you do not have ac voltage at the rectifier input, check the transformer secondary circuit (see Check 5). If it is more convenient, you can check the rectifier before you check the fuse.

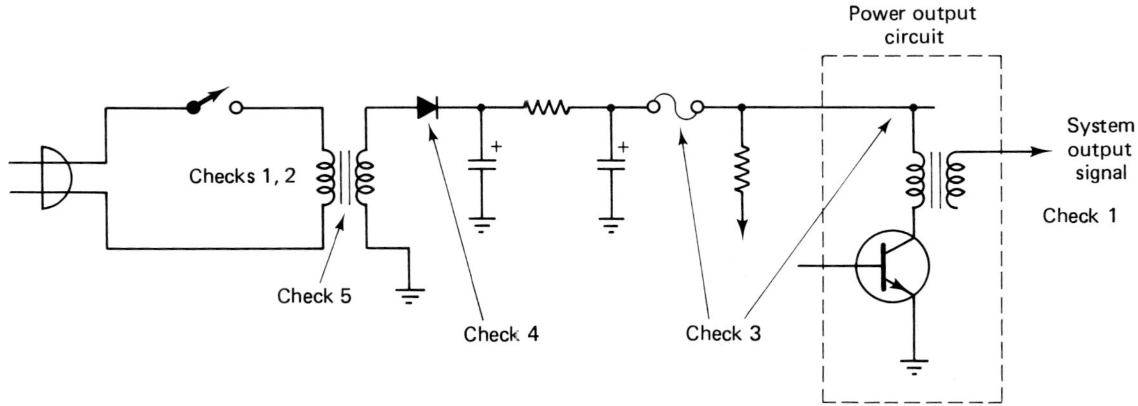

Figure 6-6 Key checks for locating opens as shown in the text.

Check 5. Check for an open in the transformer circuit: Check for ac voltage at the transformer secondary connections. No ac here indicates an open transformer if the input circuit and the secondary ground connection are good.

Check the transformer ground connection. If you have voltage on the transformer ground lead, the connection is bad. Be very careful, or be aware that operation might be restored while you are making this test. If you are lucky enough to notice that operation was restored just when you touched the ground lead, you will know for sure that the connection is bad and should be resoldered. Otherwise, you may not know what restored proper operation.

The transformer primary circuit should be checked for opens with an ohmmeter. It is similar to the lamp circuit we worked on in Chapter 3.

6-8-2 Troubleshooting Dead Shorts with an Ohmmeter

A dead short in the power supply can have catastrophic results. You cannot just replace the fuse and turn the system back on; the fuse will blow again. If there is no fuse, something else will blow. You will have to use the ohmmeter; the voltmeter is useless in this situation.

Generally speaking, the ohmmeter has its limitations. First, as we learned before, the filter capacitors cause false readings. Second, shorts may show up only when the parts are warm. Third, manufacturers' schematics usually do not show in-circuit resistances. Nevertheless, the ohmmeter can detect dead shorts (near zero ohms) safely and quickly. Refer to Fig. 6-7 for the following ohmmeter checks. They need not be done in the exact order given. Do what is easiest.

Check 1. If the output fuse is blown, the output circuit is shorted. If there is no fuse, check the rectifier. If the rectifier is blown, check the output circuit for a short (see Chapter 10).

Check 2. A rectifier short is almost always a near-dead short, which is easily checked with an ohmmeter. If the rectifier is defective, something after it may have shorted and burned it out; do Check 3. If it is good, something before it is shorted; do Check 6.

Check 3. If the rectifier is shorted, remove it and check the filter capacitors. A shorted filter capacitor will damage the rectifier, causing it to short out. Also recall that a shorted rectifier may damage the filter capacitor.

To check the filter capacitors, take the resistance readings to ground on both sides of the filter resistor. The side with the lower resistance has the short. We are checking for near-dead shorts. If the input side is shorted, the input capacitor is shorted. If the output side is shorted, either the output capacitor or the load circuits are shorted.

Check 4. To determine whether the output capacitor or the load circuits are shorted, check to see if you have a power-supply output fuse. If you have a good fuse, the filter capacitor is probably shorted. A short in the load circuits would blow the fuse; this would show up as an open when you first check out the equipment.

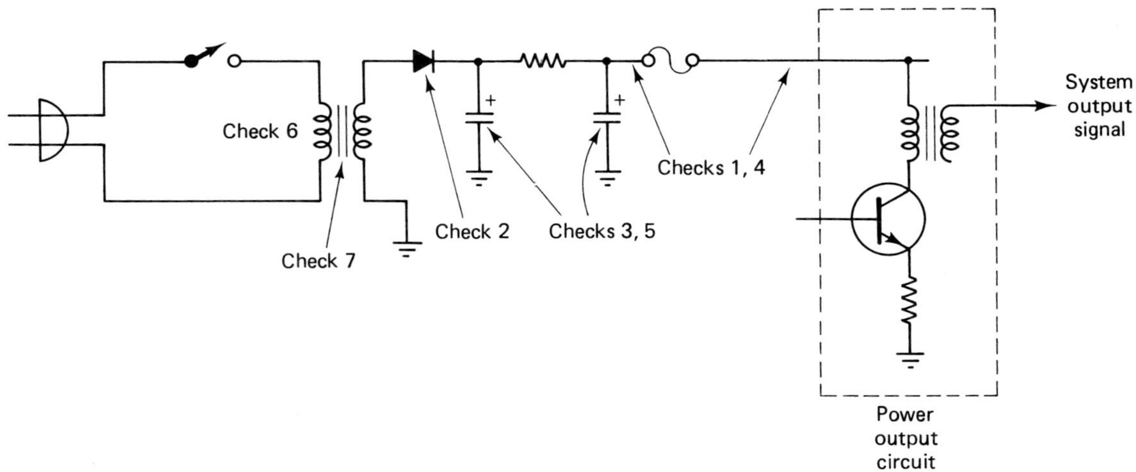

Figure 6-7 Key checks for finding dead shorts with an ohmmeter as shown in the text.

If you have no fuse, disconnect the lead to the power-output circuit. Now you can check either side for a short to ground. (The other leads from the power supply are not likely to show a dead short). This step will also make it easier to check out the power supply, because the external load circuits will be isolated, so they cannot affect your readings. Leave the output connection disconnected for the following checks.

Check 5. Repeat Check 3. Now that the output load circuits have been cut out, a low resistance across the capacitors shows that they are shorted. Give the capacitors time to charge up.

Check 6. To check the input circuit, see Chapter 3. The only parts that are likely to short in the input circuit without showing burn marks are the switch and the transformer primary. Any hot side-to-ground side short will show burn marks. A shorted switch will not be able to turn the system off.

Check 7. A shorted transformer primary will give you a lower-than-normal reading at the plug prongs. This resistance may be from ten to several hundred ohms for medium- to low-power equipment. A dead short will give you a fairly positive indication. High-power transformers may measure only a fraction of an ohm. If you feel you are getting a lower-than-normal reading, just note it and go on to the next step in search of a more positive indication.

The secondaries of step-down transformers are difficult to check with the ohmmeter unless you are given exact resistances. The resistances of step-up secondaries should be greater than the primary resistances. See Section 5-2-6 for more on transformer resistances. See the next section for a better way to check transformers.

6-8-3 Checking for Power-Supply Shorts with a Series Lamp

In Chapter 3 we showed you how to connect a lamp in series with a parallel circuit in order to test it for shorts. We also used the series lamp for checking the power-supply transformer. We can apply the same method to the power supply itself. This is a time-honored method inherited from appliance repairmen. It is used in power circuits because it is nearly impossible to take voltages when the power supply is shorted.

The power rating of the test lamp should be about the same as the equipment being tested. You can clip it across the input fuse holder as shown in Fig. 6-8, or you can rig up a series test lamp that can be connected directly to the ac plug. Jump all the fuses (except the one to which the test lamp is connected). With the lamp connected, turn the equipment on and check for the following conditions and make the follow-up checks.

1. If the lamp does not glow at all, you have an open. With the lamp still connected, make voltage checks to see where you are losing the voltage. You should get normal or higher-than-normal voltage at every point above (before) the open, and zero voltage below (after) the open.

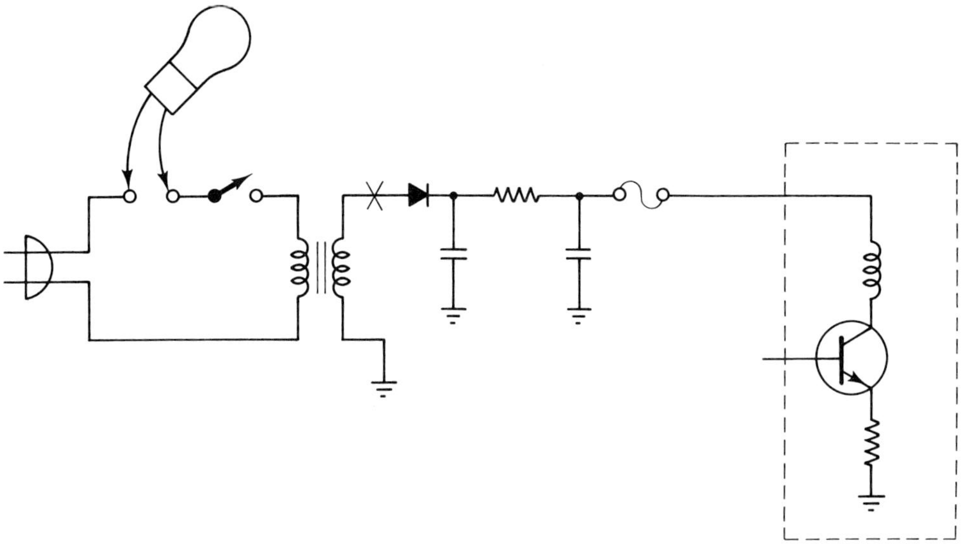

Figure 6-8 A series test lamp can be used for finding shorts in the power-supply and power-output circuits. A short in these circuits will cause the lamp to glow brightly. Removal of the shorted part or circuit will dim the lamp.

2. If the lamp glows bright, you have a dead short in the power-supply or power-output circuit. Follow the checks in Section 6-8-4.
3. If the lamp glows dim, the circuit may be okay, you may have a partial short, or you may have a complete short in a section other than the power-supply or power-output section. Follow the checks in Section 6-8-5.
4. If the lamp glows to medium brightness, follow either Section 6-8-4 or Section 6-8-5, or both.
5. In high-power equipment, the lamp may glow brightly even if the equipment is not shorted. In this case you can check the brightness by shorting some point in the power supply to ground and comparing the two brightness levels. (Observe safety precautions). If there is no change in brightness, you have a dead short. Otherwise, follow either Section 6-8-4 or Section 6-8-5, or both.

6-8-4 When the Lamp Glows Bright

If the lamp glows brightly (indicating a short), take the following steps. Refer to Fig. 6-8 again.

Check 1. Disconnect the power and cut the power-supply lead to the power-output section. If the lamp does not light when the power is turned back on, the power-output section is shorted. If the lamp still glows bright, the power supply is definitely shorted.

If the power supply is shorted, leave the output leads disconnected and do the next checks.

Explanation for Check 1. We are looking for a "dead" short. That is, we are looking for anything that would blow a fuse or possibly damage a rectifier or power transformer. The two circuits that are likely to cause this type of catastrophic short are the power supply and the power-output section.

If any other circuit shorts, the power-distribution resistor that delivers voltage to that circuit will either open or cut down the current to the lamp. In either case, the lamp will not glow to full brightness. (Ohm's law says so, too: $I = V/R$.)

Check 2. Turn off the power and check the rectifier. If it is good, the short lies before it (see Check 5). If it is defective, the short lies after it (see Check 3).

Explanation for Check 2. If the lamp glows bright there is a short before or after the rectifier. If the short occurs after the rectifier, it will be damaged. If the short occurs before the rectifier, it will be bypassed and not damaged.

Check 3. Replace the rectifier, turn on the power, and check the lamp. If the lamp glows bright, one of the capacitors is shorted. Do Check 4 to find out which.

Explanation for Check 3. This check cannot be done with a bad rectifier. An open rectifier will not pass voltage to the electrolytic filter capacitor, and a shorted rectifier will put ac voltage on it and possibly damage it. Electrolytics require dc voltages of the proper polarity.

Check 4. If you want to know which of the two capacitors is shorted, measure their voltages. If there is little difference between the two voltages, the input capacitor is shorted. If there is a large difference, the output capacitor is shorted.

Check 5. If the lamp glows bright after Checks 1 to 4, the transformer or the input circuit is shorted. Leave the secondary disconnected and let the transformer warm up. If it gets warm or hot, it is shorted. A good transformer should not have normal warmth with the secondary disconnected and the lamp cutting down the voltage.

If the transformer tests good, the input circuit is shorted. It is best checked by visual inspection and ohmmeter checks. If the input fuse had blown, the short must be after the fuse. (It is jumped for these checks). The most likely cause is damaged wire insulation or a frayed wire at a connection. If the fuse is not blown, the most likely causes are a shorted line cord or ac plug.

6-8-5 When the Lamp Glows Dim

If the series test lamp glows dim, the circuits could be operating normally; or there could be a partial short, particularly in a low-voltage circuit; or the trouble may occur only when full voltage is applied. Anyway, the lamp is now dim and not dropping the full voltage. This leaves us some voltages in the circuits with which to work.

TIP 6-22

Shorted power-input and power-output transformers may be tested by running them at reduced voltages with their secondaries disconnected. If they heat up excessively or develop hot spots that can be felt by moving your fingers over the surfaces, they are shorted. **Caution:** Turn the power off and touch lightly; shorted transformers can be very hot, both electrically and in terms of heat.

Check 1. With the series test lamp dropping the voltage, measure voltages on the power-supply filter resistor (see Fig. 6-8). If the output voltage has fallen by a greater percentage than the input voltage, there is a short below. The wire-wound power resistor is not likely to have increased resistance.

Check 2. Measure the dc voltages on the power-distribution resistor that outputs the second highest voltage after the filter resistor (see Fig. 6-3). If the output voltage of R_1 has decreased by a greater percentage than the input, look for a short below.

Check 3. Test for ac voltage at the filter resistor. If you have ac, the rectifier diode is shorted and the electrolytic filter capacitors are drawing just enough current to make the lamp light dimly. The filter capacitors should be checked with a good rectifier in place.

Check 4. Measure the transformer primary and secondary ac voltages. If the lamp brings down the secondary voltage by a greater percentage than the primary, there is a short somewhere, possibly in the transformer.

Check 5. To check the transformer further, turn off the power and disconnect one of the secondary leads, then recheck the voltage. If the percentage decrease of the secondary voltage is now equal to that of the primary (or even less), the transformer is good and there is a short below. If the secondary voltage is still too low, the transformer is defective.

If the transformer seems to be hotter than normal, disconnect its secondary (either lead is okay). If it still gets hot, it is partially shorted. If it runs cooler or the lamp goes out completely, the transformer may be okay. As a final check, repeat this test without a series lamp. A short could develop at the higher voltage.

Check 6. If any part gets hotter than normal, look for a short in its path.

6-8-6 Shorts in Multiple Secondaries

In Fig. 6-9 the power transformer has many secondary leads. Shorts and opens can be determined, as before, by the brightness of the lamp. If the lamp glows bright, the shorted branch can be found by disconnecting the leads one at a time. If the lamp ceases to glow bright when a branch is cut out, that branch is shorted.

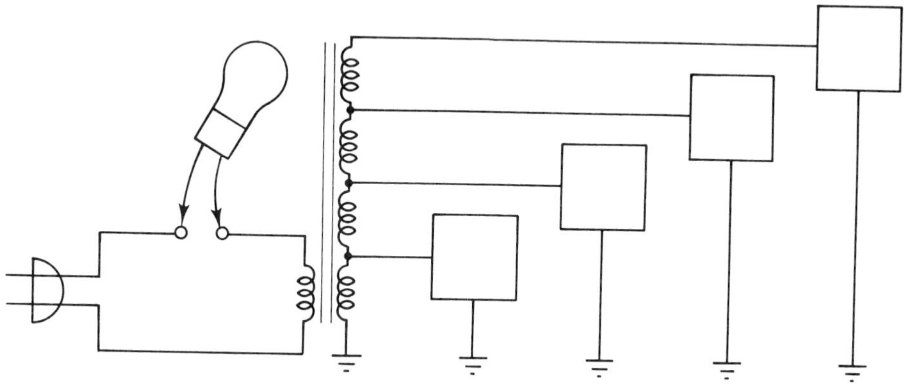

Figure 6-9 A short in any one of the blocks fed by the transformer secondary will cause the lamp to glow brightly. Removal of the shorted block will dim the lamp.

When the lamp glows dim, it is difficult to tell whether the system is good or bad. All the secondary lead voltages will be pulled down by nearly the same percentage. However, there will be voltages left in the circuits to work with. We are concerned about a short that blew the fuse when full power was applied. Therefore, look for shorts in the power-supply and power-output circuit as you did with the single secondary transformer. If these seem okay, check the power-distribution resistor that feeds the highest supply-voltage line. The next priority is to check the next-highest voltage line.

If these voltage points are too difficult to get at, you can cut out the secondary leads one at a time. When a bad branch is cut out, the brightness level may change considerably and the remaining circuits may be more operative.

> TIP 6-23
>
> To find a short in a system fed by a multiple-secondary transformer, one of the things you can do is to disconnect the secondary leads one at a time until more normal voltages are obtained at the various outputs. The short is in the circuit that held down all the voltages.

6-9 OTHER RECTIFIER CIRCUITS

We have covered the basic power supply that used a half-wave rectifier. There are other types of rectifier circuits that we should learn how to check. Let us examine some of them.

6-9-1 The Full-Wave Rectifier

You might come across power supplies with two rectifiers which have their cathodes tied together and their anodes connected to the opposite ends of the transformer secondary, as in Fig. 6-10. The transformer secondary has a center tap that is connected to ground. This circuit is called a full-wave rectifier with center-tapped secondary. You should be able to identify it and know the normal voltage relationships in it.

Both ends of the secondary should have the same ac voltage with respect to ground. The output dc voltage of the rectifiers should be about 1.4 times their ac input voltage measured to ground.

If you use an oscilloscope, the normal ripple voltage as shown in Fig. 6-10 will vary at a rate of 120 Hz rather than 60 Hz, with the half-wave power supply as shown in Fig. 6-2. This information will help you interpret power-supply symptoms in certain equipment. (The use of the scope is taken up in Chapter 7.)

Without going into the details of circuit operation, we can give you some voltages and faults to look for.

1. If one of the secondary windings or leads opens, you will lose the ac voltage on that side. The good side will deliver a 60-Hz ripple voltage at the rectifier output. The voltage will be somewhat lower because the filter capacitor is for the normal 120-Hz dual-rectifier output. For the same reason, you will also have excessive ripple.
2. If one side of the secondary shorts, the fuse may blow. If the short is partial, you may have lower ac voltage on that side. The transformer may also overheat. The dc voltage may be lower.
3. If one rectifier opens, the effect will be the same as when one side of the transformer secondary opens. You will still have ac voltage at input of the good diode. You will have a lower dc output voltage and more ripple.
4. If one rectifier shorts, the other will be damaged. The filter capacitor also may go, especially if one rectifier shorts and the other opens. In most cases the fuse

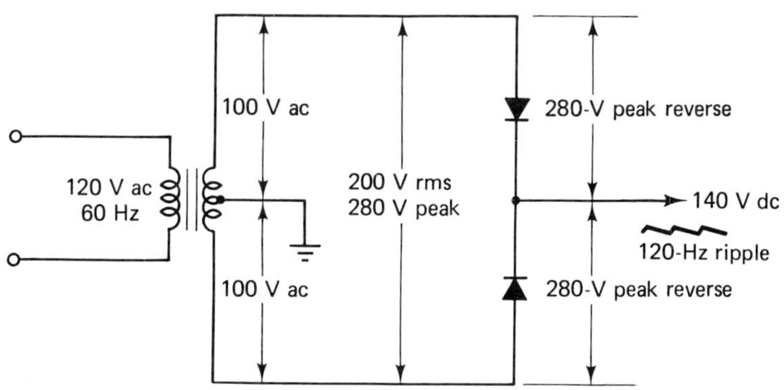

Figure 6-10 Schematic showing the voltage relations found in a full-wave power supply with a center-tapped transformer.

will blow and save the other rectifier and the capacitor. Some technicians will take no chances and replace both rectifiers automatically. This is a good idea.

5. When replacing rectifiers in this circuit, make sure that their voltage ratings are about 50% greater than *twice* the dc output voltage. (This is called the *peak inverse voltage rating.*)

Interesting Sidelight. For the half-wave rectifier, we saw that on the reverse cycle, the filter-capacitor voltage and the ac input voltages were in series. This doubled the inverse voltage on the rectifier. The full-wave rectifier in Fig. 6-10 would have this doubled inverse voltage on each rectifier even without the capacitor.

Let's look into this. The voltage from either secondary output is 140 V peak. The full secondary voltage is 280 V peak, as shown in Fig. 6-10. The voltage to ground on each rectifier input is 140 V peak. But the voltage *across* a rectifier when the other is conducting (shorting) is 280 V peak, the full secondary voltage. The rectifier must be able to withstand this voltage.

6. If the rectifier circuit is defective, look for a short in the filter section or load circuits.
7. If the rectifier circuit is good, look for trouble in the transformer and input circuits.

6-9-2 The Bridge Rectifier

The bridge rectifier in Fig. 6-11(a) has replaced the full-wave rectifier to some extent. It, in turn, is largely being replaced by the oldest of them all, the single-diode, half-wave rectifier. Modern regulator circuits and very-large-value capacitors allow the use of the single-diode circuit. We look at regulated supplies in Chapter 10.

The arrows in the illustration should give you an idea of how each alternation is rectified. The solid arrows show the current path during the positive alternation, and the dashed arrow shows the path during the negative alternation. They both end up going in the same direction at the output. Note also the solid diode symbols and the white diode symbols; one handles the current coming out of the top of the secondary, and the other handles the current coming out of the bottom.

Figure 6-11(b) shows the same circuit redrawn so that it is easier to see how each alternation is fed to the load and returned to the transformer. Note the correspondence of points *a*, *b*, *c*, and *d* in both figures. When the black diodes are "on," the white diodes are "off." When the white diodes are on, they "switch" the transformer leads, and the black diodes are off.

Let's see what we can do with a voltmeter (see Fig. 6-11 again). The secondary leads should have equal ac voltages to ground, but they will be half-wave voltages and cannot be read directly with your multimeter. You can check them with a scope.

TIP 6-24

Diodes in a bridge rectifier are easily checked in-circuit with an ohmmeter. The two diodes with their cathodes tied together should have the same readings, and the two with their anodes tied together should have the same readings.

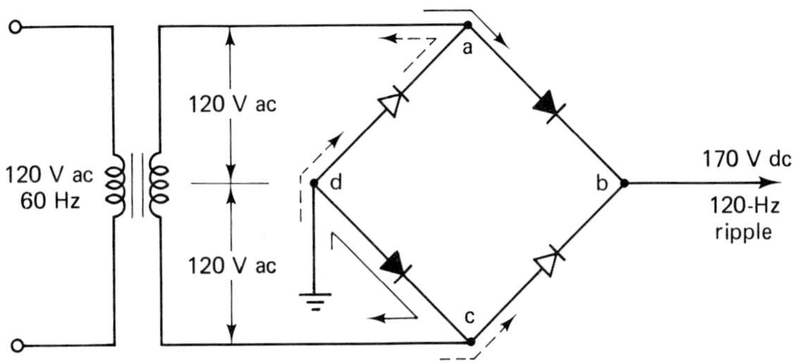

(a)

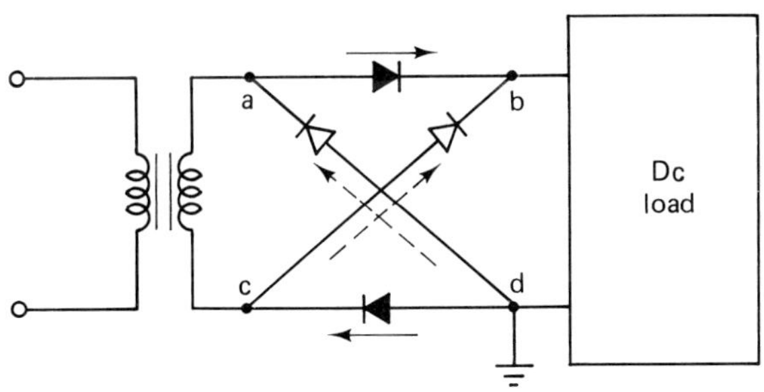

(b)

Figure 6-11 Full-wave bridge rectifier. (a) The solid and dashed arrows show the alternating directions the current takes toward the same output. (b) The same circuit redrawn to show current going in and out of load.

Notes

1. The ac voltage to bridge rectifier must be read across the two ac input terminals of the bridge. It cannot be measured from ground. It can be measured across the secondary of the transformer if there is one.
2. The ground connection should have zero volts ac or dc. The rectifier output voltage should equal about 1.4 times the ac voltage between the two secondary leads. If you have a scope, the output ripple frequency should be 120 Hz. This is a full-wave rectifier.
3. If any rectifier shorts, the fuse will blow. If any rectifier opens, the output dc voltage will fall a little. The ripple voltage will increase, and the operation of the system may be erratic or unstable. Hum may be heard in audio equipment and so-called hum bars seen in video equipment. You will see 60-Hz ripple on the supply line with an oscilloscope.
4. If the rectifier circuit is defective, look for a short in the circuits after it. If the rectifier circuit is good, look for a defect in the transformer or the input circuit.

6-10 TRANSFORMERLESS POWER SUPPLIES—SAFETY

Before working on transformerless power supplies, you should be familiar with the safety precautions in Chapter 3. We review one precaution here and let you know why it is so important in this situation.

6-10-1 Safety Precautions

Some lightweight equipment have no transformers. Their chasses are connected directly to the 120-V ac lines. This is a potentially dangerous situation. Someone can receive a shock by touching the chassis if it is connected directly to the hot side of the 120-V ac outlet. Take the safety precautions given in Chapter 3. The best safety measure is to use an isolation transformer.

Most transformerless equipment come with polarized plugs. A polarized plug has one large prong, so that it can fit into an outlet in only one way, as shown back in Fig. 2-2(b). This ensures that the chassis, which is connected to the large slot, is connected to the larger ground slot of the outlet. Someone can clip the ground prong to make it fit into a nonpolarized outlet. Use an isolation transformer.

Another safety feature may be found in some transformerless equipment. Instead of connecting ground to the chassis, it is connected to a common lug or thick wire to which all the circuits are grounded. Such a grounding arrangement is called a *floating ground*. When checking voltages, connect the voltmeter ground lead to the circuit ground lugs, not to the chassis. The chassis is connected to ground through a capacitor. For safety, use an isolation transformer.

Use an isolation transformer when you have to use a scope or signal generator. On this type of equipment, any test instrument that is powered by 120 V ac requires the use of an isolation transformer not only for safety, but also to prevent hum voltages that will interfere with the testing.

6-10-2 Voltage Doublers

Voltage doublers were once very popular in inexpensive, lightweight, transformerless TV sets. They had an output of about 275 to 285 V dc. They are half-wave rectifiers (see Fig. 6-12). It would be instructive to see how this half-wave voltage doubler works. It is easiest to look at the negative swing of the line voltage first. When this happens, D_1 charges C_1 in the direction of the dashed arrow. Note the polarity of the charge on C_1. When the line voltage swings positive, the charge on C_1 is added to it. Notice that the charged-up capacitor C_1 is in series with the ac plug. This doubles the voltage. It is a pulsing voltage and must be filtered by C_2. C_2 charges to the peak of the ac line voltage plus the C_1 voltage.

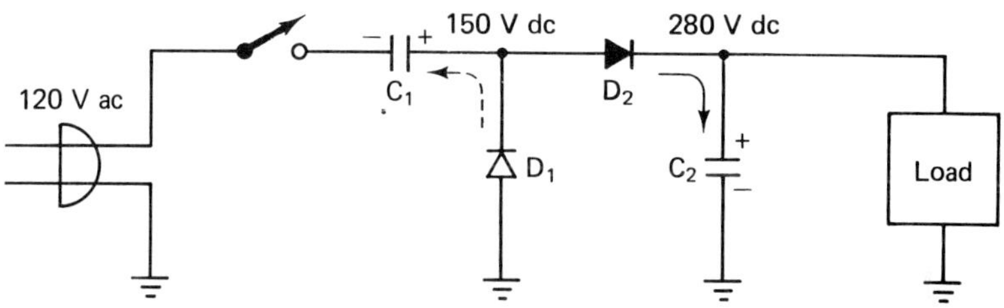

Figure 6-12 Voltage doubler. D_1 charges on one alternation, and the stored voltage in C_1 is added to the next incoming alternation to double the voltage.

When everything is working okay, you should have about 120 V ac at the C_1 input. C_1 outputs about 150 V dc when measured with a dc voltmeter. The junction of the two diodes, D_1 and D_2, is a good place to read this voltage; it is usually easy to spot. The output of D_2 should measure about 280 V dc. It may have a 60-Hz ripple voltage as high as 5.0 V peak to peak (p-p) on the scope. At the output capacitor C_2, it should have a ripple of about 1.0 to 2.0 V p-p.

Note: The doubler circuit is usually checked with an ohmmeter. Each diode should have hi–lo measurements. If the doubler rectifier circuit is defective, look for a shorted filter capacitor or load circuit. If it is good, check for an open input capacitor C_1. Also remember that a shorted rectifier can damage the electrolytic capacitors.

6-11 WHAT CIRCUIT BREAKERS CAN TELL YOU

Fuses are inconvenient because they must be replaced after a temporary condition blows them. Circuit breakers, on the other hand, can be reset merely by pushing a small button. Figure 6-13 shows a circuit breaker and its schematic symbols.

Everything we have learned about fuses in troubleshooting can also be applied to circuit breakers. For instance, if the circuit breaker trips but the power-supply output fuse does not blow, the fault is in the power suppply.

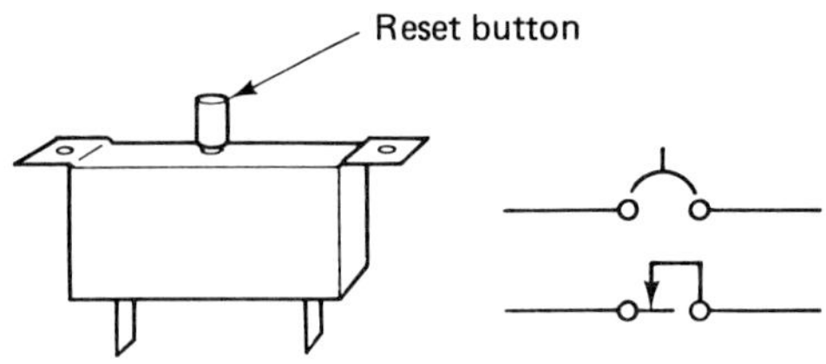

Figure 6-13 Circuit breaker and its symbols.

In addition to what we have learned about fuses, the location of the trouble can also be determined by the length of time it takes before the circuit breaker trips after the equipment is turned on. This is because they are tripped when the temperature in the breaker reaches a certain level. A shorted power supply or a shorted power-output circuit that is directly connected to it will trip the breaker immediately. Listen for transformer buzz or growl when this happens. This is a sure indication of a power-supply or power-output circuit short.

Other circuits will take some time before they trip the breaker—a matter of several seconds. These are the lower-voltage circuits that are fed through dropping resistors. It takes more time for the current to heat the breaker to the tripping point.

For longer, unpredictable periods of normal operation before cutoff, it is best to replace the circuit breaker without further ado. In many cases, the circuit breaker itself is at fault. They trip too early under high, but still acceptable line voltage conditions.

TIP 6-25

The length of time it takes for a circuit breaker to cut out after turn-on gives a clue as to where the trouble lies. Immediate cutoff indicates a short in the power-supply or power-output circuit. A few seconds delay indicates a low-power circuit.

Very long periods of normal operation before cutoff could mean that the circuit breaker itself is defective. Replace it. Chances are very good that this will solve the problem. If the problem persists, you have at least eliminated one possible cause.

6-12 CHECKING OUT THE POWER SUPPLY AFTER PARTS REPLACEMENT

After replacing the defective parts, you should check out the power supply before turning on the power. If the output circuit of the equipment under repair is not the cause of the power-supply failure, you can go ahead with the test in this section. (The power-output stage is taken up in Chapter 11.)

6-12-1 The Series Lamp

After replacing the defective part, you can use a series test lamp as in Section 6-8-4 and Fig. 6-8 to check out the power supply. This is a safe method for checking a good power supply. The lamp protects the circuit for two reasons. First, like any resistor, it will drop more voltage as the current increases due to a short or a sudden surge of current at turn-on. Second, its resistance increases as it gets hotter. After connecting the lamp, check the voltages at various supply points. Check to see if any power resistor is dropping excessive voltage (see Check 2, Section 6-8-5). Leave it operating for 30 minutes and see if the lamp increases brightness. Also recheck the voltages to see if there have been any changes. If there are no brightness and voltage changes that indicate a short, operate the equipment at full voltage and check the voltages as before. Turn off the equipment if you see indications of a short.

You might want to check further. Double the lamp wattage and check the voltages on the distribution resistors as before. Be on the lookout for any creeping voltage changes as the equipment warms up. If the creeping does not stop, something could be shorting out.

6-12-2 Using a Variable AC Transformer for Test Purposes

A variable ac transformer can also be used to check the power supply. It is connected to the the 120-V ac line and its output can be varied from 0 to 140 V ac. There is a knack to using it, however. You must monitor the dc power-supply voltage as you slowly bring up your voltage from zero. If at any point, as you steadily turn the knob, the voltage fails to come up with it, you may be developing a short. You might back down and try to bring it up again to make sure. The skill part is in not allowing the short to get out of hand. Do not let the voltage sink of itself. Bring it down quickly with a spin of the knob!

Connecting an ac ammeter in series with the variable transformer is an excellent idea. The current should not exceed about 70% of the input fuse rating. The current should go up at about the same rate as the voltage or the turn of the knob. If at any point the current begins to rise faster than the voltage or the turn of the knob, you may have a short.

Some manufacturers' procedures for testing power supplies call for the use of a variable transformer. This is another advantage over the lamp method. The beauty of the series lamp, however, is that it cuts down the voltage automatically when something shorts out.

6-13 MODERN POWER SUPPLIES

The simple power supply we checked out in this chapter is basically the same as the modern supplies in use today. These supplies still have the input, rectifier, filter, and distribution systems.

The rectifier section remains basically the same. The filter section has changed greatly. Electronic regulator circuits have greatly optimized the filtering and stability of the power supply. The filter choke has largely been replaced, but filter capacitors are still used.

In many cases the regulator circuits are encased in integrated-circuit (IC) chips. They are easy to find in the schematics and circuits. Checking these chips should not be difficult. Basically, if it has a dc input and no dc output, it is bad. However, it also has provisions for regulating the output. We take this up in Chapter 10.

If the regulator circuit is made up of separate (discrete) transistors, resistors, and diodes, you can treat the entire circuit as a block. Look for its input and output terminals. The input terminal is at the rectifier output or the input filter capacitor. There may be an output filter capacitor at its output. If you have trouble finding the regulator circuit output, check for dc voltages at random on various electrolytics and power resistors. If you cannot find dc voltages at any point, the regulator circuit could be bad. To find the regulator, look for the ac plug and trace the input circuit from there. It is fed by the rectifier.

WHAT YOU LEARNED TO DO

In this chapter you learned to troubleshoot a basic power supply. You learned the relations between the various voltages in the supply. You learned how to:

1. Check the voltages and resistances to find out which part in this simple power supply is bad.
2. Recognize the symptoms for shorted and open power supplies.
3. Establish a procedure for troubleshooting the basic power supply and doing the easiest things first.
4. Identify various types of commonly used rectifier circuits.
5. Check out the equipment after a power-supply repair.
6. Identify the full-wave and bridge rectifiers as well as the voltage doubler.

QUESTIONS

1. What are the first things to check if the equipment is completely inoperative?
2. What are the symptoms for a short in the power supply?
3. What are the symptoms for an open in the power supply?
4. If the power-supply output fuse is blown, what section is most likely to be shorted?
5. Draw a block diagram of a power supply.
6. If you are looking for the power-supply section in a schematic or the equipment, what parts would you look for?

7. What symptoms and defective voltages would you find when an input filter capacitor opens?
8. What symptoms and defective voltages would you expect if the output filter capacitor opens?
9. What are the symptoms for a short in the power-supply transformer?
10. What are the symptoms for a shorted filter capacitor?
11. If one of the filter capacitors in the basic power supply shorts out, how can you tell which one is shorted?
12. Name two types of full-wave rectifier circuits.
13. When using the series test lamp to find a short, how do you know that a short has been removed?
14. How can you tell where the short is by checking the circuit breaker?

PRACTICAL TROUBLESHOOTING EXPERIENCE

You will need a multimeter, a series test lamp, and optionally, an oscilloscope and variable ac power source (variac) for this job. Find a piece of cast-off electronic equipment that has one of the power supplies we studied in this chapter and follow the instructions below.

1. Remove the chassis and look for the power-supply parts.
2. Note what type of rectifier and filter circuit is used.
3. Take the proper safety precautions and check the equipment out with a series test lamp. Determine whether the equipment is shorted, open, or possibly good (see Sections 6-8-3, 6-8-4, and 6-8-5).
4. If the lamp glows bright, carry out the ohmmeter checks in Section 6-8-2 and the series test lamp checks in Section 6-8-4. If you have a full-wave or voltage doubler rectifier, check out the diodes individually with an ohmmeter. A bad rectifier circuit can be caused by a short after it. A good rectifier circuit points to a short before it.
5. If the lamp does not glow at all, you have an open in the supply or in a circuit operating directly off it. Carry out the checks for opens in Section 6-8-1. Also see instruction 4 above for full-wave and voltage doubler rectifiers, and regulator circuits.
6. If the lamp lights dimly, remove the lamp and take the voltage and resistance readings shown in Section 6-4. For full-wave and doubler rectifiers, refer to Sections 6-9 and 6-10. If you have a regulator, check its input and output waveforms.
7. *Optional:* Take the scope readings across the transformer output, the rectifier output, and the filter or regulator output.

7

Checking Signal Circuits

EXPERIENCE REQUIRED

All circuits in this chapter can be checked with simple dc voltage and resistance measurements and substitutions. The oscilloscope (scope), however, can really speed up your work. There is enough information in this chapter to help you get started.

Use of the scope requires a knowlege of the basic ac measurements: frequency, period, amplitude, reactance, and impedance. These measurements are explained briefly in this chapter.

Both dc and signal checks are given for each circuit when applicable. When a dc or signal check cannot be made, this is clearly stated.

7-1 SIGNAL CIRCUITS

The purpose of electronics is to process signals. In Chapter 8 you will see how transistors perform this job. They operate on dc voltages. Hence you will find both dc voltages and ac signal voltages in the circuits around them. Let's find out what this means to the troubleshooter. How can we take advantage of this?

We know that capacitor and coil circuits behave differently to ac and dc. Capacitors block dc and tend to pass ac signals. Coils do the opposite. They pass dc and tend to block ac signals. What happens when both ac and dc are on the same lines? Mathematically, you add the dc voltage to the ac voltage. For troubleshooters this is no big problem. They simply set their meter or scope to read the dc voltage on a line, and then set their meter or scope to read ac on the same line. In fact, the scope can read both ac and dc on the same line simultaneously.

The scope is the best instrument for reading ac signal voltages. The multimeter can be used, but it has limitations. For one thing, the frequency range of the meter is limited. Also, the scope can read very small signals that cannot be read with an ordinary meter. The scope can also reveal distorted signals that cannot be directly revealed by a meter. Furthermore, some meters cannot read ac voltages on lines having dc voltages.

Check your meter manual for the voltage and frequency ranges of your meter and whether it can read ac voltages on dc lines. The following tip deals with this last limitation.

TIP 7-1

Your meter can be used for checking ac voltages on dc lines if it reads 0 V ac when connected to a dc battery in either direction, or if it has a special input jack (called the "output" jack) for reading ac signals.

If your meter cannot read signal voltages on dc lines, take your reading through a series capacitor. Use a value that equals the highest-value coupling capacitor in the circuit. (For *coupling,* see the following section.)

There are also severe limitations with regard to reading pulse-circuit voltages with a multimeter. We will find out more about this when we take up pulse signals in Section 7-10.

7-2 SIMPLE OSCILLOSCOPE MEASUREMENTS

Let us go over some of the more basic uses of the oscilloscope. Figure 7-1 shows how it is used. Its input lead consists of a shielded wire connected to a probe. The outer

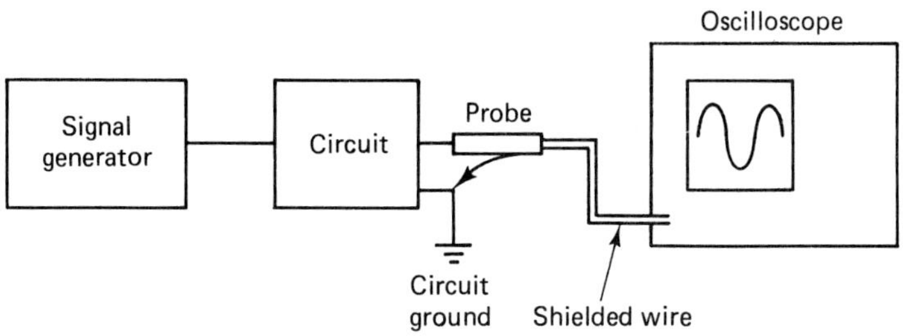

Figure 7-1 Set-up for checking a circuit with a signal generator and oscilloscope.

shield connects the circuit ground to the oscilloscope ground. The inner wire carries the signal to the scope.

A *signal generator* is also shown injecting a signal into the test circuit. If the circuit is good, the scope will pick up the signal at various points in it. Note the ac signal that is picked up and displayed on the screen of the scope.

7-2-1 Cycles and Amplitude

Figure 7-2 shows the graph of a sine wave. This graph can be seen on a scope. Many important equations are connected with it. We limit ourselves to those of most interest to a beginning scope user. The graph shows three *cycles* of an ac wave. The top peaks of the wave (touching line *A*) represent the peak positive voltage of the wave. The bottom peaks of the wave represent the peak negative voltage. The horizontal line that divides the wave equally between peaks *A* and *B* is the zero-volt line. The points where the wave crosses this line represent zero volts. The height of the wave from the zero-volt line is called its *amplitude*. The height from bottom to top (*B* to *A*) is called the *peak-to-peak* (p-p) voltage.

Converting from Peak-to-Peak to RMS. Signal voltages are usually given as peak-to-peak voltages in the schematics. Ac power-supply voltages, on the other hand, may be given as rms voltages. A brief explanation of rms voltage is given in Section 6-3-3. To compare these voltages, divide the peak-to-peak voltages by 2.818, or multiply the rms voltages by 2.818. (Section 6-3-3 deals with peak voltage, which is half the peak-to-peak voltage, or 1.414 times the rms voltage.)

7-2-2 Frequency and Period

Time is represented along the zero-volt line from left to right in Fig. 7-2. Every point to the right of a given point on the wave represents a later point in time.

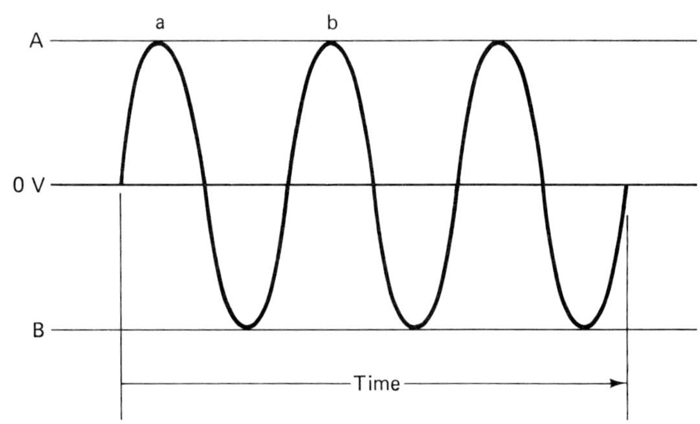

Figure 7-2 A graph of a sine wave shows how the voltage changes with time.

If it takes 1 second to complete the three waves, we say that it has a *frequency* of 3 Hz. If the same length represents one millionth of a second, we would have 3 million cycles per second, or 3 MHz.

This example shows that to measure frequency, you must adjust the scope so that the horizontal length of one cycle of the wave represents a certain amount of time. To do this, you must be able to convert from frequency to *period.* A period is the time it takes to complete one cycle. This period, *T,* is the reciprocal of the frequency.

$$T = \frac{1}{f}$$

Conversely, the frequency is the reciprocal of the time period for one cycle of *T* [press the reciprocal (1/X) button on your calculator]:

$$f = \frac{1}{T}$$

Very often the frequency is given in hertz, kilohertz, or megahertz in the schematics, and these have to be converted to periods when working with a scope. When dealing with periods on the scope, you will use the "time-division" (time/div) or "seconds-per-centimeter" (sec/cm) controls.

7-3 OSCILLOSCOPE ADJUSTMENTS

Caution: Before using the oscillosope, we should remind you that the scope cabinet and the ground lead of the probe are connected to ground. They are connected to the third prong of the ac plug. Do not clip the ground lead of your probe to a "hot" chassis. Use an isolation transformer. Otherwise, the scope input will burn and there is a danger of shock. Section 3-1 deals with safety precautions.

7-3-1 To Obtain a Display

You may have a blank screen when you first turn on a scope to look for a signal. If so, select the *auto mode* position on the *trigger mode* switch. If this does not bring in the signal, turn up the *intensity* or brightness control and ground the scope probe. This will cut down the signal and obtain a straight horizontal line. The line, however, may be lost beyond the top or bottom of the screen. Adjust the *vertical position* control to bring the waveform back down or up into the screen.

If you do not want to ground the probe, it may be quicker to turn the *input switch* to the ac position and turn the voltage control (volt/division control) to maximum so as to obtain as small a waveform as possible (which is often reduced to a straight line). Then, if necessary, adjust the vertical-position control to bring the waveform up or down and into view.

The *intensity* control should be readjusted to give you just enough brightness for eye comfort. Too much brightness will shorten the lifespan of the display tube. The *focus* and *astigmatism* controls should be adjusted to give you as fine a line as possible on all parts of a waveform.

7-3-2 Measuring Signal Amplitude and DC Level

The oscilloscope screen has horizontal and vertical lines called divisions [see Fig. 7-3(a)]. The horizontal lines are for measuring voltages. You can adjust the scope so that each division represents a certain amount of voltage. For instance, it can be set to 1 V per division, so that the 6 divisions shown in Fig. 7-3(a) would represent a 6-V p-p wave. If the scope were set to 1 mV per division, the same display would represent 6 mV p-p.

We must now find the dc level of this wave. This is the voltage with respect to ground of the zero line of the waveform. Is it 0 V? If so, the top peak would be +3 V and the bottom peak would be -3 V. Suppose that the line represents 10 V with respect to ground? Then the top peak would be 13 V and the bottom peak would be 7 V.

You can adjust the scope for measuring the dc level of the wave as follows. With no signal on the probe (you might have to ground the probe to kill the signal), set the

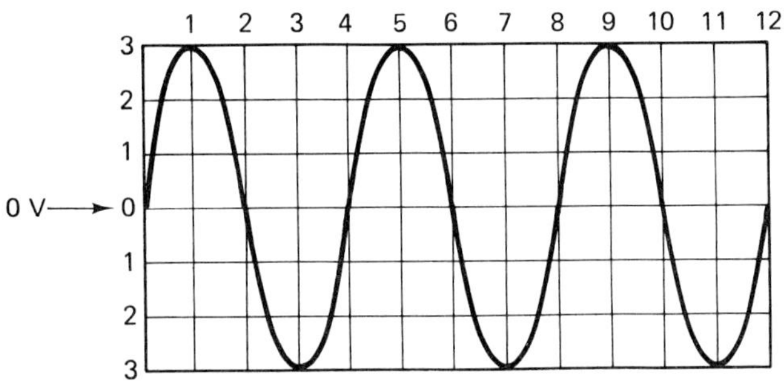

(a)

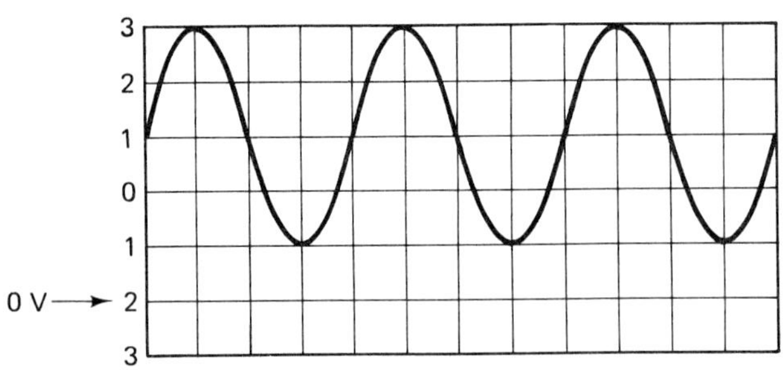

(b)

Figure 7-3 (a) Each height division on the oscilloscope screen represents a certain amount of voltage. Each division on the 0-V line represents a certain fraction of a second. (b) The 0-V line has been moved down 2 divisions. The dc level of the wave is 3 divisions above zero.

mode control to the "auto" position. This will give you a straight line across the screen, indicating 0 V *ac*. To obtain the *dc* level, set the *input switch* on the scope to the DC position. Without a signal input, the line across the screen now indicates 0 V dc and ac. The vertical position control allows you to move this line up and down as needed.

In Fig. 7-3(a) the 0-V dc level is positioned across the middle of the screen. Use the *vertical-position control* to do this. In Fig. 7-3(b) the 0-V dc level is moved down to accommodate a signal that has a dc level that is 3 divisions above 0 V. For example: If the the *volts/division control* is set to 10 mV per division, you are reading a wave that is swinging between 10 and 50 mV, a 40-mV p-p voltage swing. The dc level of this wave is 3 mV. We can say that the 40-mV p-p wave is "riding" on a 3-mV dc voltage. If we read this voltage with a dc voltmeter, we would measure 3 mV dc.

The voltage/divison control adjusts the height of the display to accommodate the input signal. The smallest range on your scope may be the 1-mV V/div position. When set to this range, each height division on the scope represents 1 mV.

The ability of a scope to read dc voltages can be a terrific advantage. A pure dc voltage will give you a straight line. Its height above the 0-V line indicates its voltage.

If you are interested only in the peak-to-peak voltage and not the dc level, switch to the ac position of the input selector.

> *Note:* When you change the height of the display, you are not changing the signal at the input. Beginners are often confused about this. You are merely changing the height of the display so that it can be measured. To reduce the height of the waveform so that you can see its top and bottom, switch to a higher voltage range.

7-3-3 Measuring Frequency and Period

You can adjust the scope so that each horizontal division represents a certain amount of time. Time is usually measured in milliseconds (ms) or microseconds (μs). Look at Fig. 7-3(a) again. One cycle of the wave covers four horizontal divisions. Let's say that the scope is adjusted so that one horizontal division represents 1 ms. This means that it takes 4 ms to complete one cycle. That is, the signal has a period of 4 ms. This represents a frequency of 250 Hz ($f = 1/T$).

The Time/Division Control. The time-period measurement adjustment is done with the seconds-per-division (time/division or time/centimeter) control. In the example above, the time/division control is set to the 1-ms position.

7-3-4 The Trigger Control

The purpose of the *trigger control* is to stop the wave so that you can measure it. You turn it one way or the other to stop the wave. It allows the input signal to cause the display waveform to start at the left of the screen. This ensures that the display waveform is always in time with the signal.

The waveform can be triggered from an outside source as well as from the signal itself. There is an extra input *external trigger* for this purpose.

There are other controls and features. You can read the scope manual to learn how to use them. A good way to learn how to use the scope is to connect it to a multifunction signal generator with a voltage indicator on it. Vary the output voltage, frequency, and waveform of the signal generator, and fiddle around with the scope controls.

Dual-Trace Oscilloscopes. Dual-trace oscilloscopes make it possible to see two waveforms on a scope. You can read both the input and output signals of a circuit at the same time. There are two probes, one for each *channel,* and there are two sets of controls. There is one trigger control for both channels. The signal on either channel can be used to trigger both waveforms. There is a switch for this purpose. There is also one external trigger input.

There are two ways in which the scope is able to show the two waveforms. One is the *alternate mode,* which rapidly switches between displaying one channel, then another. The other is the *chop mode,* which chops each waveform into hardly visible bits and alternately displays them to make two waveforms. The chop mode is used when the frequency is too slow and the alternately displayed waves can be seen blinking on and off.

7-4 CAPACITIVE AND INDUCTIVE REACTANCE

Strange things happen when ac voltages are impressed on capacitors and coils. Capacitors and coils can store electrical charges under compression and release them in ways that complement and oppose each other. The ability of capacitors to store and release charges is called *capacitance.* The math symbol for capacitance is C. The ability of a coil or inductor to store and release charges is called *inductance.* The math symbol for inductance is L.

7-4-1 Capacitive Reactance

For the troubleshooter it suffices in most cases to assume that capacitors block dc voltages and pass ac voltages and currents. Capacitors seem to pass ac currents because they charge and discharge with every alternation. Thus the charge and discharge current flows back and forth through the circuit as if it were passing through the capacitor.

A larger capacitor allows more charging current to flow. A smaller capacitor charges and discharges less, so it allows less current to pass "through." Thus we say that a smaller capacitor has a higher *reactance* than a larger capacitor.

A higher-frequency current reaches a peak and switches direction before it can fill up the capacitor and stop flowing. As a consequence, more current flows in the circuit with each alternation. Hence there is less reactance at the higher frequency. A lower-frequency current will fill up the capacitor and stop flowing before it reaches a peak and changes direction. Hence not all of it gets "through." Therefore, there is more reactance at the lower frequency. Reactance is measured in ohms. The equation for capacitive reactance, X_C, is

$$X_C = \frac{1}{2\pi fC}$$

The π (Greek lowercase letter pi) in the equation was put in by mathematicians to make the equation conform with the cyclic or circular nature of alternating current and waves in general.

In one respect, reactance is similar to resistance. Ohm's law works for reactance. That is, $I = V/X_C$. In fact, reactance is measured in ohms.

7-4-2 Capacitor Voltage and Current

When voltage is first impressed on a capacitor there is maximum charge current and minimum compression or voltage. As the capacitor charges up, the current approaches minimum, and the pressure or voltage builds up to a maximum.

Thus when alternating current flows in and out of the capacitor under electrical (electrostatic) compression and decompression, the voltage and current do not reach their peaks at the same time. When the current first rushes into the capacitor, there is no opposition, consequently, there is no pressure or voltage. The current leads the voltage. We say that the current and the voltage are *out of phase*. (The current and voltage across a resistor are *in phase*.)

7-4-3 Inductive Reactance

As far as dc measurements go, a coil is just like another piece of wire. Coiled or uncoiled, it drops the same amount of dc voltage. In the space around the coil, however, the dc current develops a magnetic field. (This happens with any length of wire, but not as much.) As long as you have a steady dc voltage and current, this magnetic field has no effect on the current in the circuit.

However, when the current increases, the expanding field generates a counter voltage that opposes the increase. When the current decreases, the collapsing field generates a counter voltage that opposes the decrease. Thus a coil develops a constant opposition to ac.

This opposition is called *inductive reactance,* X_L. It is measured in ohms. The amount of this reactance depends on the size of the coil, *L,* and the frequency.

$$X_L = 2\pi fL$$

Ohm's law and voltage division works with inductive reactance, too: $I_L = V_L/X_L$.

7-4-4 Inductor Voltage and Current

The voltage leads the current through an inductive current. When the voltage is turned on, an expanding magnetic field opposes and delays the current and maximum pressure or voltage is built up. It is only when the the field reaches a steady state that the delayed current begins to pass through. When the voltage goes down, the field collapses and aids the current and pushes it through. In other words, when the voltage is high, the current is low; when the voltage is low, the current is high—they are out of phase.

7-5 VOLTAGE DROPS IN SERIES *RC* AND *RL* CIRCUITS—IMPEDANCE

We encountered series *RL* (resistor–coil) or *RC* (resistor–capacitor) circuits in Chapter 5 when we learned to make dc checks on them. Now we can make other checks. Because capacitors and inductors can store charges, their voltage drops in series with resistances do not add up to equal the source voltage.

The total opposition that an *RC* or *RL* series combination offers to the current is called *impedance, Z*. It is measured in ohms. It is less than the arithmetic sum of X and L and greater than either of them. For instance, if $X = 4\ \Omega$ and $R = 3\ \Omega$, then $Z = 5\ \Omega$, not $7\ \Omega$. The actual equation is

$$Z^2 = X^2 + R^2$$

Fortunately for us, they drop dc voltages as we would expect them to (see Chapter 5). Hence they can usually be checked out with a dc voltmeter or ohmmeter, particularly when something goes wrong.

7-6 *LC* CIRCUITS—RESONANCE

Resonance is a very important phenomenon in electronics. It allows transmitters to generate signals and operate at their own assigned frequencies. It allows receivers to select stations. It helps to cut down interference signals. There are also undesirable oscillations that create interference. These are very important in troubleshooting.

7-6-1 The Series *LC* Circuit

The ac voltage drop across a series *LC* circuit can have nearly 0 V, even though the ac voltage across each part is, say, 10 V. This is because their voltages do not peak at the same time. Recall that the voltage leads the current across a coil and lags it across a capacitor. In fact, they cancel each other. This is why mathematicians give capacitive reactance a negative sign. That is, $X_L + (-X_C) = 0\ \Omega$. The frequency when X_L and X_C are equal and cancel each other is called the *resonant* frequency. We say that a series *LC* circuit has minimum impedance at resonance. We say "impedance" instead of "reactance" because the coil has some resistance in it.

7-6-2 The Parallel *LC* Circuit

In a parallel *LC* circuit, just the opposite happens. The circuit has maximum impedance at the resonant frequency. Simply put: Above the resonant frequency, the capacitor ceases to block the current; below the resonant frequency the coil ceases to block the current. Hence the impedance is highest at the resonant frequency. Actually, it is not this simple.

7-6-3 Oscillations

There is something else taking place in the parallel *LC* circuit at resonance. To understand the following, you must know that reactance, unlike resistance, has an elastic quality. It bounces current back to the source. Resistance absorbs energy and generates heats. Pure reactance bounces energy back to the source and does not generate heat.

At resonance, the capacitor under elastic (electrostatic) compression discharges current into the coil, and the coil reactance electromagnetically bounces it right back into the capacitor and recharges it—and the action keeps repeating itself. Incidentally, when the capacitor voltage is high, the coil voltage is low and ready to accept the capacitor discharge. When the coil voltage is high, the capacitor voltage is low and ready to accept the coil output. This bouncing back and forth can occur only at one frequency, the resonant frequency.

This repetitive action is called *oscillation*. The parallel *LC* circuit is often called a *tank circuit*. To lend credence to the explanation above, a resonant *LC* circuit will continue oscillating after the initial source voltage is removed.

> *Note:* The equation for resonance is derived from the fact that the inductive and capacitive reactances are equal when it occurs. That is, $2\pi fL = 1/(2\pi fC)$. This converts algebraically to $f_r = 1/(2\pi\sqrt{CL})$, where f_r equals the resonant frequency.

7-7 CHECKING SIGNALS

Having learned a few things about oscilloscopes and the nature of ac circuits, we can now begin to check signal circuits more speedily than with dc checks alone.

7-7-1 Signals in *RC* Circuits

Figure 7-4(a) shows a capacitor in series with a grounded resistor. The capacitor drops the signal voltage from 10 mV p-p to 5 mV p-p. This is a very small signal. There is also 5 V dc above the capacitor, which is blocked from the resistor.

If the capacitor shorts as in Fig. 7-4(b), the dc voltage on the resistor will rise and you can check it with a dc voltmeter. A scope will also pick up the short because the dc level will rise. In this case the scope must be switched to the dc position.

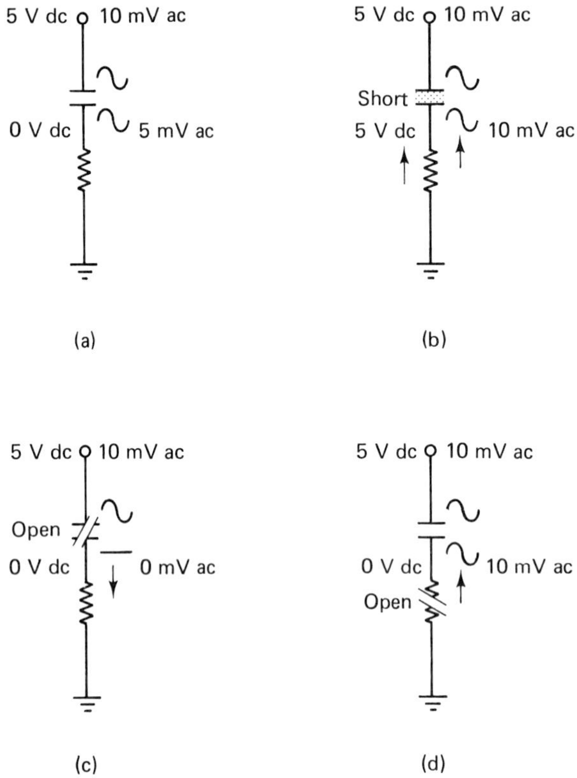

Figure 7-4 (a) Normal ac and dc voltages in an *RC* circuit. (b) A shorted capacitor pulls up both dc and ac voltages. (c) An open capacitor changes only the ac voltage. (d) An open resistor pushes up the ac voltage only.

If the capacitor opens, you would have no signal on the resistor [see Fig. 7-4(c)]. In this case an analog voltmeter will not be much help, because you would see 0 V ac on the meter even if the actual voltage is good. A normal 5 mV would read 0 V on an analog meter. You might be able to pick up the signal voltage with a digital meter, but it must be capable of at least indicating the existence of a signal as low as 1.7 mV ac, which is the rms value for 5.0 mV p-p. Be sure that you are not picking up noise or interference. This can be a problem.

You can do better with a scope. If the capacitor and resistor are good, you see a 5.0-mV p-p signal riding on 0 V dc, as was seen in Fig. 7-4(a). Even if you pick up noise and the waveform looks "grassy," at least you can measure the signal voltage. And perhaps the "grass" (noise interference) has something to do with the trouble you are looking for.

In Fig. 7-4(c), the capacitor opens. The dc voltage above it does not change, because the capacitor is already open to dc. Hence a dc voltmeter will not detect the open capacitor. On the other hand, the ac voltage will fall, and a scope will detect it. A subsititute capacitor can also be used here.

If the resistor opens, as in Fig. 7-4(d), the signal voltage will rise but the dc voltage will not. You will still have 0 V dc on it. The voltmeter may not detect it. You can check this with an ohmmeter if you think something is wrong, but the scope is faster.

Figure 7-5(a) shows another *RC* circuit. This time the resistor feeds a grounded capacitor. Normally, the resistor does not drop any dc voltage. It has 10 V dc above it and 10 V dc below it, because the capacitor blocks dc and builds up maximum pressure above it. On the other hand, it does not block the ac voltage completely, and the ac pressure above it drops from 1 V p-p to 0.5 V p-p.

In Fig. 7-5(b) the resistor shorts, but the dc voltage below it cannot be pulled up because it is already up. The ac signal voltage, however, is pulled up through the capacitor. The scope and ohmmeter will pick this up, but not the dc voltmeter.

In Fig. 7-5(c), if the capacitor opens, only the signal output will rise. This is because the capacitor is already open to dc. Only a scope or a substitute capacitor will pick this up.

Can a scope find defects in parallel *RC* circuits that cannot be found with a dc voltmeter? Figure 7-5(d) shows a capacitor and resistor in parallel. Both the dc and signal voltages are higher at the source side. If the resistor opens or shorts, completely or partially, both the dc and signal low-side voltages will change. If the capacitor shorts, both dc and ac voltages will rise. In all the cases above, both the dc voltmeter and the scope will detect a fault.

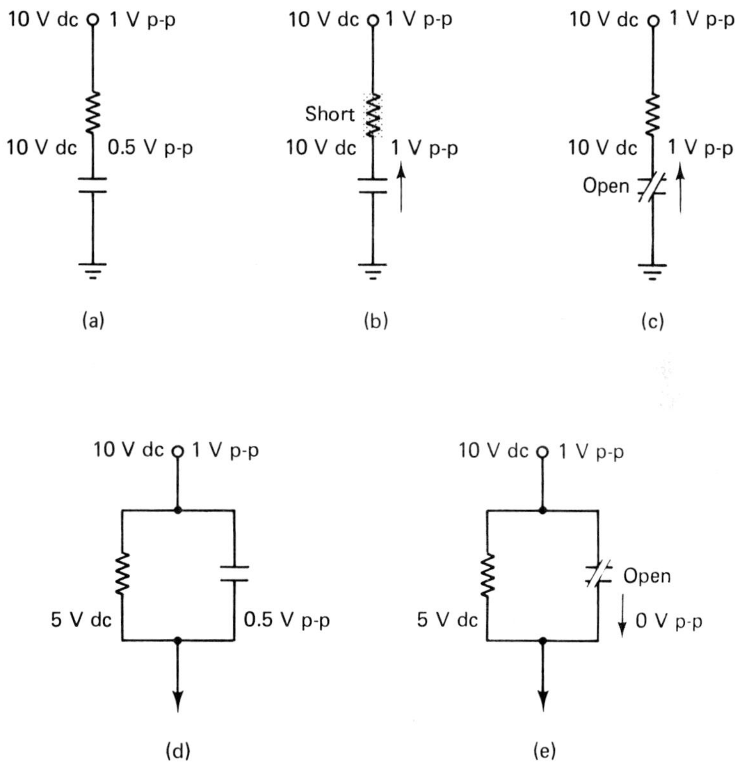

Figure 7-5 (a) Normal ac and dc voltages. (b) A short pulls up the ac voltage only. (c) An open pushes up the ac voltage only. (d) Normal circuit. (e) An open drops the ac voltage only.

However, neither will tell you which of the two parts is bad. You will have to disconnect one of the part's leads to isolate the bad part.

There is one case that we did not cover. If the capacitor opens, only the low-side signal voltage will fall [see Fig. 7-5(e)]. In this case the scope will reveal the open capacitor; the voltmeter will not.

7-7-2 *RC* Coupling

Figure 7-6 shows a very common *RC* circuit. Its purpose is to couple (to pass) a signal voltage from one circuit to another without upsetting the dc voltages in either circuit. In Chapter 8 you will find that if these dc voltages are upset, the ac signal voltages in these circuits will also be upset.

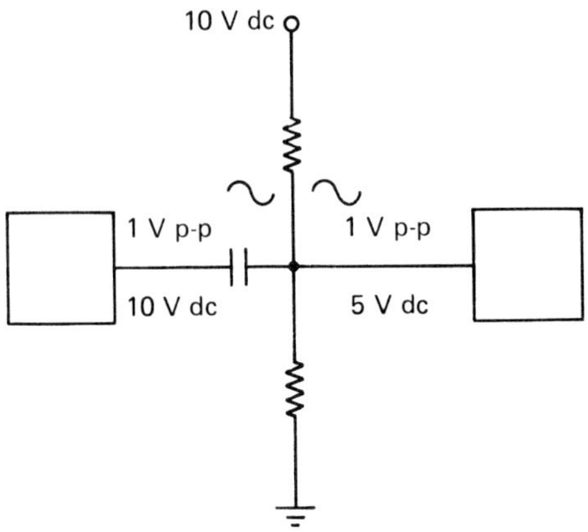

Figure 7-6 Commonly used circuit for coupling signals from one circuit to another. The capacitor blocks the dc voltages and maintains the desired dc voltages on both sides.

If the capacitor shorts completely, its dc input voltage will fall and its dc output voltage will rise. The ac signal voltage may change appreciably because of the changed dc voltages on both circuits.

If the coupling capacitor opens, the dc voltages do not change, but you lose the signal. This is easily picked up with a scope. You may have problems with a voltmeter, but there are ways to detect signals without the use of a scope. We take this up in Chapter 9 when we look at amplifier circuits and signal tracing. After finding a circuit with defective dc voltages, you can jump a suspected capacitor with a good one to see if operation can be restored.

7-7-3 Checking *RC* Filters

We met the *RC* filter in the power-supply and power-distribution circuits. The filter and bypass capacitors were large enough to bypass almost all of the ac voltage to ground. The result was almost ripple-free dc voltage.

In certain circuits only the higher frequencies are attenuated. In these cases the capacitors are not large enough to filter out the lower frequencies. Note the input and output waveforms in Fig. 7-7(a). A high-frequency wave is shown bypassed to ground

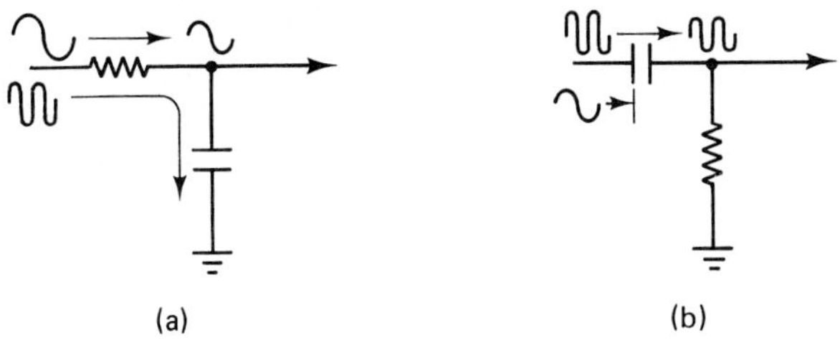

Figure 7-7 *RC* filters. The arrows show which frequencies pass through and which are stopped or bypassed to ground. (a) Low-pass filter. (b) High-pass filter.

by the capacitor. The low frequency passes through okay, although slightly attenuated. This circuit is called a low-pass *RC* filter. It is the similar to the one in Fig. 7-5(a), and checking it out is similar.

Figure 7-7(b) shows a high-pass filter. This time, the resistor is grounded. Its action is similar to the coupling capacitor circuit, except that this capacitor is smaller and can pass only the higher frequencies. Note that the low-frequency wave is blocked. This circuit is similar to the coupling circuit in Fig. 7-6. Checking it out is similar.

You will find these filter circuits in simple tone controls where the fixed resistors are replaced by variable ones. Many tone-control circuits are more complicated, but these simple filter-circuit examples should give you an idea of how they work. If you have a variable resistor in the circuit, you might use a scope and signal generator to see how the circuit responds to varying frequencies.

7-7-4 Estimating Normal Voltages in *RC* Circuits

In many of the cases so far, we relied on knowing what the normal voltages are in *RC* circuits. However, normal scope voltages are not always found in schematics. This may not be so bad if the *RC* filter parts are completely open or shorted, but for partial shorts and opens, things can be more difficult.

We already learned some ways of dealing with this: for instance, jumping capacitors and checking all the dc voltages in the suspected circuits. No method, however, beats knowledge and experience. Once you get to know the circuits you are working with, you will have a pretty good idea of how much signal voltages to expect at certain points.

> TIP 7-2
>
> Coupling capacitors usually do not drop much signal voltage. Filter capacitors also drop very little signal voltage (they bypass it to ground). Other capacitors are chosen to attenuate certain frequencies while passing others. Get to know the frequency range and the capacitance values used in various circuits.

Let us take an example. In audio equipment, 100 Hz is considered low, while 10 kHz is considered high. We can use some calculations if we want to. The equation for capacitive reactance tells us that a 16-μF capacitor will drop very little signal voltage at 100 Hz when connected to a 1-kΩ resistor in a low-pass filter. Experience tells us this, too. After a while, you will become familiar with the capacitor values used in various types of circuits.

The following list is based on the equation for capacitive reactance given in Section 7-4-1. It is made up of the frequencies and the capacitances that give us a reactance of 10, 100, and 1000 Ω. You can then compare these values with those in the circuit. Generally speaking, the reactance is considered insignificant if it is connected to a resistance that is 10 times larger.

FREQUENCY	**10 Ω**	**100Ω**	**1.0-KΩ**
100 Hz	160.0 μF	16.0 μF	1.7 μF
1 kHz	17.0 μF	1.7 μF	0.16 μF
10 kHz	1.6 μF	0.16 μF	0.016 μF
100 kHz	0.16 μF	0.016 μF	0.0016 μF
1.0 MHz	0.016 μF	0.0016 μF	160.0 pF
10.0 MHz	0.0016 μF	160.0 pF	16.0 pF
100.0 MHz	160.0 pF	16.0 pf	1.6 pf

This list gives you an idea of the reactance to expect for a given capacitance and frequency. It will give you an idea of how much ac will be built up, passed through, or bypassed by a capacitor at a given frequency.

Experience in working with actual equipment will provide you with a knowledge of their operating frequencies. The following list should give you an idea of the frequencies to expect in various types of equipment and circuits.

Audio	100 Hz to 20 kHz
Video	60 Hz to 6 MHz
TV circuits	60 Hz, 15.75 kHz, 3.58 MHz, 4.5 MHz
TV UHF circuits	470 to 890 MHz
Radio circuits	500 to 1600 kHz, 455 kHz
FM radio circuits	88 to 108 MHz, 10.7 MHz

7-7-5 Checking *RL* Circuits

Figure 7-8(a) shows an *RL* low-pass filter circuit. *L* stands for inductance. The coil opposes the high-frequency signals and passes more of the low-frequency signals. If the coil opens, you will have no signal at the output. The dc voltage is also lost. A voltmeter can pick this up, and you don't have to scope it.

If the coil shorts, the output will have the unwanted high frequencies. For very low resistance coils, the dc voltage will be practically unchanged, and a voltmeter may not detect the fault. A scope will detect it.

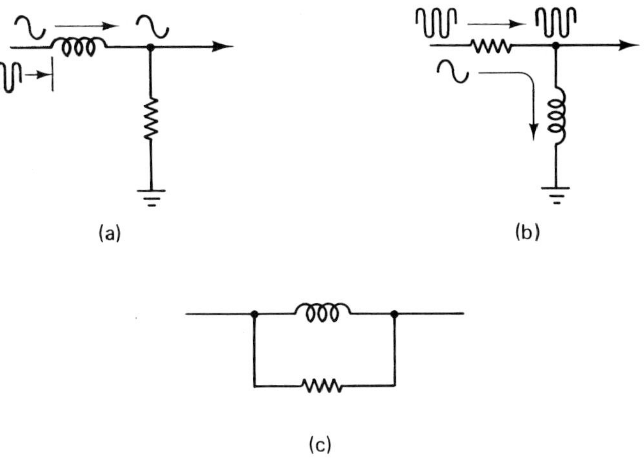

Figure 7-8 *RL* filters. (a) Low-pass filter. (b) High-pass *RL* filter. (c) Parallel *RL* low-pass filter.

Figure 7-8(b) shows an *RL* high-pass filter circuit. In this case the coil bypasses the low frequencies to ground and only the highs get through. The coil is holding the voltage down to practically 0 V. If the coil opens, the dc voltage above it will rise. A voltmeter will detect this. The scope will show a rise in the signal voltage, particularly a rise of the lower frequencies.

If the coil shorts, the signal voltage will be lost at the output. Neither the voltmeter nor the ohmmeter may detect the short. The scope is best in this case.

Figure 7-8(c) shows a parallel *RL* low-pass circuit. The coil bypasses the dc voltage across the resistor, so there is not much dc voltage drop across the circuit. If the resistor changes value, the dc voltages may not change much, hence a dc voltmeter may not do much good. The scope might detect the change if the original resistance value is not too high and the change in resistance value is large enough.

If the coil shorts, there may not be much change of dc voltage, but the ac voltage change may be substantial enough for the scope to detect it. If the coil opens, both the ac and dc voltage changes would be substantial, and either the scope or the voltmeter will detect the change.

7-7-6 Using the Scope to Check Capacitors and Coils

At times, practice is a lot easier than theory. Dc voltmeter and ohmmeter checks and part substitutions will cut through or bypass a lot of complicated theoretical considerations. Practically all faults can be found with simple dc checks.

This does not mean that you should neglect the scope—it is faster. You can make dc checks with the scope, too. Furthermore, the dc and ac voltages can often be seen simultaneously on the scope. In certain types of work where the dc voltages are about equal to the signal voltages, only the scope is needed.

In any case, if the multimeter cannot get the job done, use the scope. It is faster than jumping parts. The following tip may help. It sort of summarizes the basics up to this point.

> TIP 7-3
>
> When the dc checks fail to find the open capacitor or shorted coil, use a scope. It is faster than substituting capacitors or coils. An open capacitor will not pass the signal. A shorted coil will drop less signal voltage.

7-8 CHECKING *LC* CIRCUITS

Coil and capacitor or *LC* circuits are used to pass or block signals of various frequencies. They are used for signal selection and interference rejection. One way to see what these circuits can do is to look at their *frequency response curves.* These curves show what frequencies are passed and what frequencies are rejected.

7-8-1 *LC* Filters

Very efficient filter circuits are designed with coils and capacitors. We look at the simpler ones. In the low-pass filter in Fig. 7-9(a), the coil blocks most of the higher frequencies, and the capacitor bypasses what is left to ground. Note the arrow pointing down. On the other hand, the coil passes the lower frequencies. The curve shows how the signal voltage drops off as the frequency increases.

Figure 7-9(b) shows a high-pass filter. The capacitor blocks the low frequencies and the coil bypasses them.

Figure 7-9(c) is a series bandpass filter. In Section 7-6-1 we learned that this circuit has minimum reactance at the resonant frequency. The two frequencies f_1 and f_2 denote the band of frequencies that this filter passes. They show the *bandwidth* of this filter.

Figure 7-9(d) is a parallel bandpass filter. This circuit has the maximum reactance at only one frequency. It will pass this frequency and the frequencies close to it between f_1 and f_2. It will bypass all other frequencies.

7-8-2 Troubleshooting *LC* Filter Circuits

The series *LC* circuits above can be checked for shorted capacitors and open coils with a multimeter. On the other hand, parallel *LC* circuits, open capacitors, and shorted coils cannot easily be checked with the multimeter. They will have to be checked with a scope or by substitution.

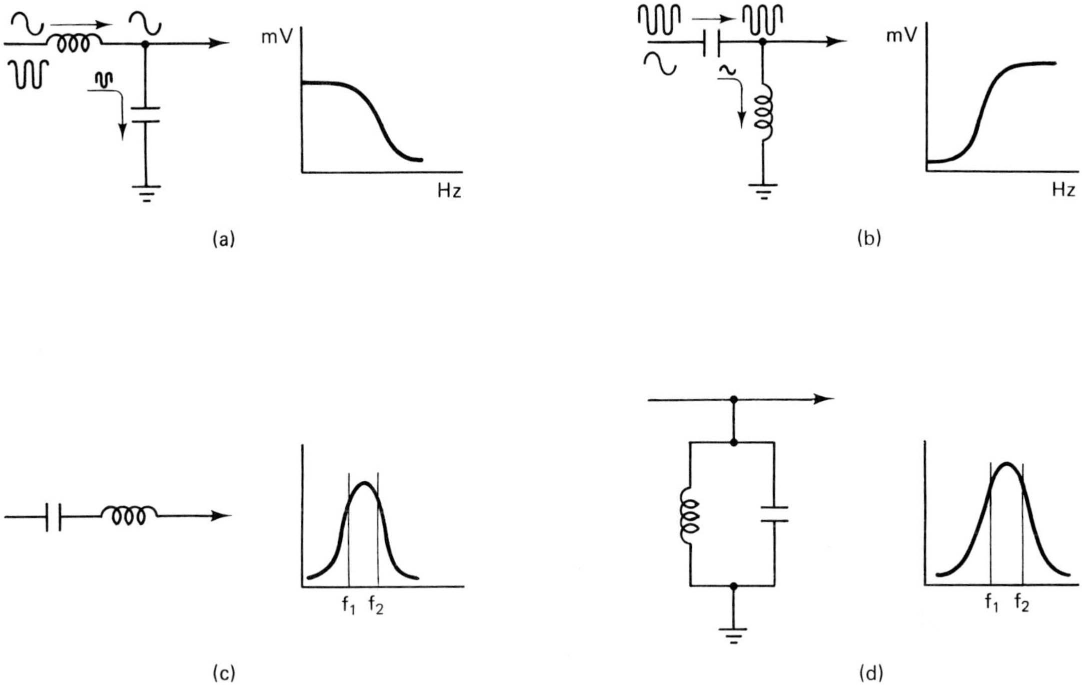

Figure 7-9 (a) Low-pass *LC* filter. (b) High-pass *LC* filter. (c) Series bandpass filter. (d) Parallel bandpass filter.

Partially defective parts may lower the output or change the frequency response of a filter. These faults can be detected with a scope and signal generator by checking the response at various frequencies. You can also learn to check the response by checking the symptoms at the output of the equipment.

In TV and video, a loss of high frequencies is shown by fuzziness in small objects and lines in the picture. A loss of low frequencies is indicated by a lack of darkness in dark scenes or large areas and a lack of brightness in bright scenes or large areas. In audio, a loss of low frequencies is indicated by a lack of bass in the sound. A loss of high frequencies is indicated by a lack of brilliance in the sound.

7-9 CHECKING DIODES IN AC SIGNAL CIRCUITS

We encountered diodes when working on power supplies. We found that the diode conducts only when it is forward biased—when the anode is more positive or less negative than the cathode. Let's see how this works in various circuits.

Figure 7-10(a) is similar to the half-wave power-supply rectifier. The input to the diode is at the anode. It differs in that it has a + 5-V dc level instead of 0 V. A 6-V p-p signal is riding on this 5-V dc level. On a scope you will see that the peaks of waveform are 3 V above and 3 V below the 5-V dc level. Actually, they are 8 V and 2 V above the ground or 0-V level. This is the anode voltage. This means that the diode will always be forward biased, since the cathode side is grounded through the

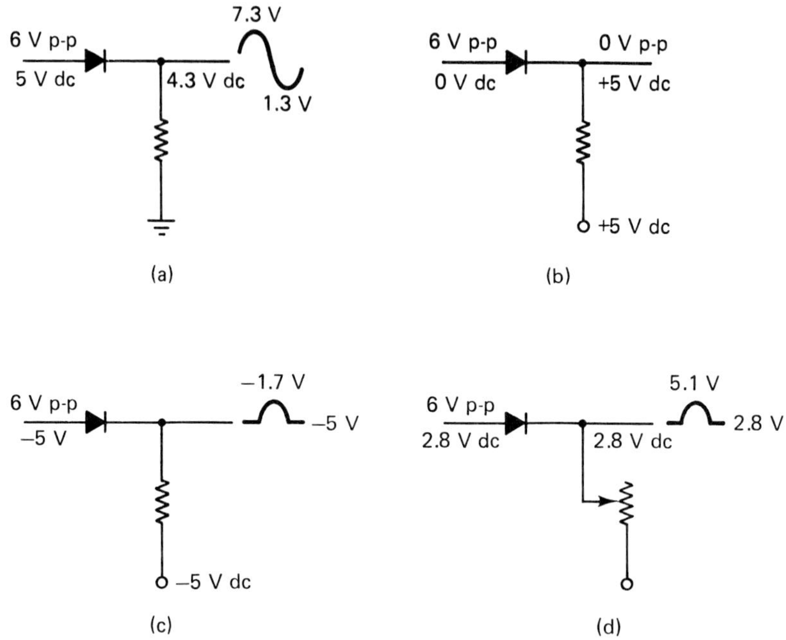

Figure 7-10 Diode switching and clipping circuits. Diodes can pass or stop a signal or clip off portions of a signal by varying the dc voltages on its terminals. Note the dc voltages on the anode and cathode in circuits (a) to (d).

resistor. The diode will always drop 0.7 V, which means that the output will always be 0.7 V less than the input. Hence the output will vary between 7.3 and 1.3 V and its dc level will be 4.3 V. You will still have a full-wave 6-V p-p signal at the output.

Figure 7-10(b) has +5 V dc on the cathode output and 0 V on the anode. None of the 6-V p-p (3-V peak) signal gets through. The anode is never more positive than the cathode.

Figure 7-10(c) has equal negative voltages on its input and output, each has −5 V dc. The input is 6 V p-p (3-V peak). It will vary between −2 V to −6 V. The diode will not pass any voltage that is more negative than −5 V. It will pass anything from about −5 to −2 V. There is also a 0.7-V drop across the diode when it does conduct. Therefore, you have a half-wave output, swinging between −5 V and −2.7 V. (*Note:* A diode does not begin to conduct until it has a small fraction of a volt across it. While it is not conducting, you will have -5 V on the anode output in this case.)

By varying the dc voltage level, you can determine how much signal gets through. For instance, Fig. 7-10(d) shows a 6-V p-p or 3-V peak signal on the anode and 2.8 V dc on the cathode. This dc voltage prevents any signal voltage below 2.8 V from getting through. The peak input voltage is 5 to 8 V. The 0.7-V drop across the diode will give you a 5.1-V peak at the output.

These circuits can be checked out with voltmeters, ohmmeters, and scopes. A dc voltmeter will not detect an open diode that is supposed to be reverse biased or have 0 V on both sides. Scopes will detect incorrect waveforms.

7-10 PULSE CIRCUITS

Besides smoothly fluctuating ac or sine-wave signals, there are pulsing signals—signals that abruptly change voltage during a cycle. Pulses do not react to *RC* circuits in the same way as sine waves. A sine-wave input to an *RC* circuit comes out as a sine wave at the output. There is no telling what the output would be when a pulse enters the same circuit unless we know something about pulse voltages.

Note: As mentioned before, we have another problem with pulse circuits. It is not easy to determine its dc level. For instance, a 10-V pulse signal riding on the zero-volt line can measure nearly zero dc volts on a meter if it is made up of widely separated spikes. You should be aware of this when making dc measurements in pulse circuits. Even with a scope, you must know where the 0-V level is by setting it before you read the pulse. Fortunately, the schematics give dc levels in pulse circuits as read from a specified voltmeter.

7-10-1 The Square Wave in *RC* Circuits

The circuit in Fig. 7-11(a) appears to be a low-pass *RC* filter, and it would be if it were used in a sine-wave circuit. In this case it is used for reshaping a square wave; it is called a *differentiator.*

When a positive-going edge of a square wave is applied to the circuit, the current rushes though the empty capacitor. The capacitor has minimum opposition, and the resistance below it has maximum voltage. However, as the capacitor charges up, less and less current gets through. Its opposition gradually increases, and the resistor voltage gradually goes down. A similar thing happens when the input square wave goes negative. This is shown in the figure by the similarity of the positive and negative spikes.

Figure 7-11(b) shows what happens if the capacitor is very small. The capacitor will charge up faster, and its current will decrease to zero much faster. As a consequence, the resistor voltage drops off more rapidly from its initial peak. This is shown by the sharp spikes in the ''differentiated'' waveform.

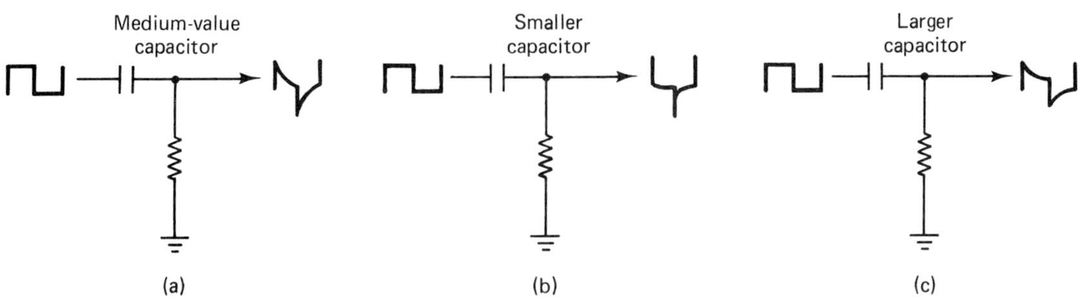

Figure 7-11 *RC* circuits can be used to change the shape of a pulse signal. (a) Differentiator circuit. (b) A differentiator with a very small capacitor for the same pulse. (c) A differentiator with a larger capacitor for the same pulse.

If the resistor or capacitor is bigger, as in Fig. 7-11(c), the capacitor will not have enough time to charge up to the full voltage. Its voltage drop will not reach maximum; therefore, the resistor voltage will not reach minimum. If the capacitor is large enough, you will have very little difference between the input and the output.

This circuit will work differently for varous types of pulse waves. We cannot look at all of them, but the square-wave example is basic for understanding them. There are other types of *RC* circuits for processing pulses. The integrator will be taken up in Chapter 13 when we look at video-display circuits.

7-10-2 Square Waves in *RL* Circuits

In the investigations below, we use such terms as ''slow'' and ''gradual.'' Actually everything is happening very rapidly, and we are only slowing down the action so that we can see it. Coils cannot hold their charges for very long.

Figure 7-12(a) shows a coil connected to a grounded resistor. When the input voltage suddenly rises, the coil has maximum opposition to the current. The voltage across the resistor is zero. During the time the input voltage is held unchanging at maximum, the current builds up and the resistor voltage builds up. When the input voltage shoots down, the coil again opposes this change, and the current can only decrease gradually to zero.

In Figure 7-12(b) the coil is *smaller,* and the current increases and decreases faster, and we have a square wave with the slightly rounded corners. If the coil is very large or the resistance very small, the coil acts almost like a power-supply filter [see Fig. 7-12(c)].

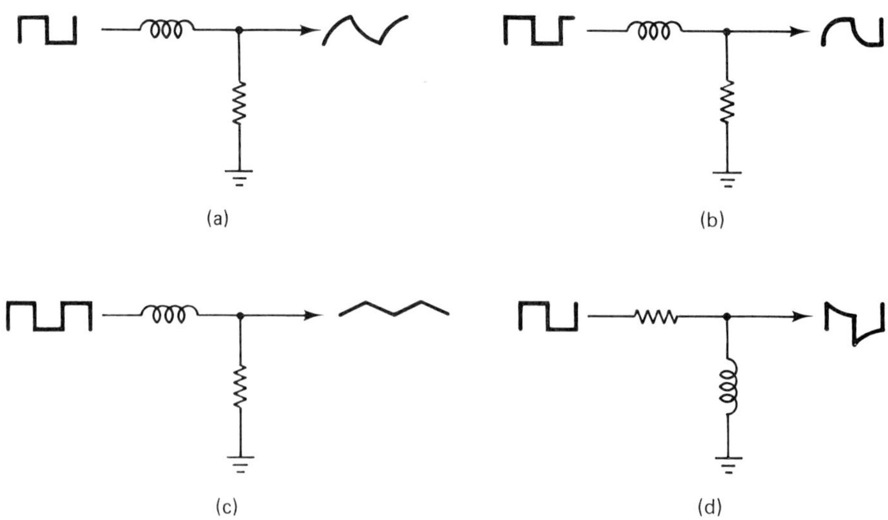

Figure 7-12 (a) An *RL* circuit opposes the sharp change of voltage. (b) A smaller coil has less effect. (c) A very large coil is almost like a filter. (d) An *RL* differentiator.

Figure 7-12(d) shows an *RL* circuit with the coil grounded. There is no delay through the resistor. When the voltage on the coil tends to shoot up, the opposition is high, so the voltage does shoot up. When the input voltage reaches the top of the square and remains steady, there is no opposition and the voltage on the resistor gradually falls. When the input voltage goes negative, the coil opposition is high and the negative voltage builds up rapidly (shoots downward). When the input voltage reaches bottom and remains steady, the opposition decreases and the voltage rises gradually to zero. The larger the coil and the smaller the resistor, the sharper will be the peaks. This circuit is called an *RL differentiator.*

If the coil is smaller and the resistance larger, the current will not have a chance to decrease or increase all the way, and the output will look more like a square wave with sloping tops and bottoms.

7-10-3 Checking Frequency Response with a Square Wave

A square-wave generator can be used to check the frequency response of a circuit. If a circuit can input a square wave and output an undistorted square wave, you can assume that it has good frequency response. The mathematical analysis of a square wave is quite complex. It makes sense, however, that a circuit with good high-frequency response would reproduce the sharp corners of a square wave, and a circuit with good low-frequency response would reproduce the flat, unchanging tops and bottoms of the square wave.

To check an audio amplifier, feed it a 1-kHz square wave and scope the output. The output square wave in Fig. 7-13(a) indicates good high- and low-frequency

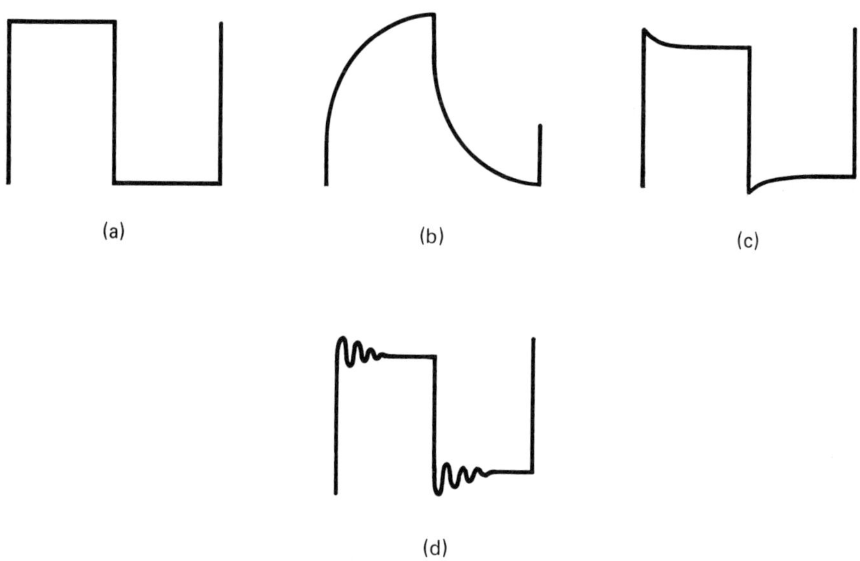

Figure 7-13 Use of a square waveform to show the frequency response of a circuit. (a) Good response. (b) Poor high-frequency response. (c) Poor low-frequency response.

response. Figure 7-13(b) shows poor high-frequency response and Fig. 7-13(c) shows poor low-frequency response. Figure 7-13(d) shows that a circuit has been shocked into oscillation by the abrupt change of voltage. This fault can be seen as extra lines on the vertical edges of objects in video displays.

7-11 DIODES IN PULSE CIRCUITS

The diode has many uses in pulse circuits. It can be used to alter the shape of a waveform and to switch voltages on and off. Let's look into these applications.

7-11-1 The Diode Clipper Circuit, Biasing

A diode can be used to clip off a pulse. In Fig. 7-14(a), when the pulses are swinging above and below the ground voltage, only the positive part of the pulse can go through the diode.

In Fig. 7-14(b), only the negative part of the signal can get through. Notice that in neither case is the entire signal clipped; there is always the drop across the diode (0.7 V, for silicon diodes).

You can also tie the diode to ground in order to clip a pulse signal as in Fig. 7-14(c). In this case the positive part of the pulse is shorted out.

In Fig. 7-14(d), the negative part of the signal is shorted out. When the line or cathode goes negative, the ground or anode goes positive. This forward biases the diode and it conducts (shorts) and pulls the line voltage down to almost zero.

In these basic clipper circuits the diode clipped off about half the waveform. If you want to clip off less, you can put a voltage on the diode. For instance, if you put −2 V on the diode as in Fig. 7-14(e), its input voltage will reach −2 V before it is clipped.

The dc voltage that controlled the clipping (switch off) point of the diode is called the *bias* voltage.

Diode clippers may be checked with dc meters, because their average voltage is above zero. However, you may not get much of an average if the pulses are very narrow and far between. Here again the scope is your best bet. If you do not have a scope, you can check the diode in-circuit with an ohmmeter. One trick is to check all the diodes on the board when you are stuck. There usually are not that many of them, and they are easy to spot and check.

7-11-2 The Diode Clamper Circuit

You might say that our input signals to the previous circuits were "riding" on a zero-volt dc voltage. That is, the voltage swung equally above and below zero volts. In the last example, we came across a signal riding on other than zero volts. It had a −2-V bias on it. Sometimes diodes and capacitors are used to maintain a dc level for the signal to ride on.

Figure 7-15(a) shows a clamping circuit. It allows the entire signal to ride above ground or 0-V level (0.7 V to be exact). When the input signal initially goes positive,

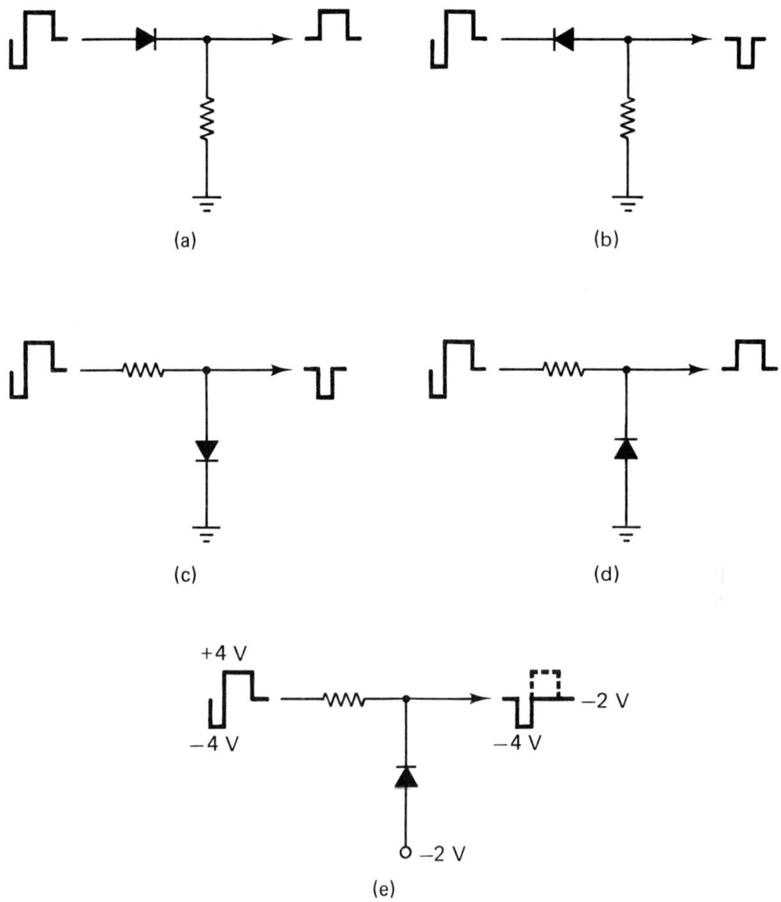

Figure 7-14 Diode clipper circuit. (a) The negative-going pulse is switched off. (b) The positive-going pulse is switched off. (c) The positive-going pulse is shorted out. (d) The negative-going pulse is shorted out. (e) The −2-V bias determines the clipping point. Everything above −2V is clipped.

the diode is switched off and its voltage rises to the +2.0-V peak. When the input signal goes negative, the diode conducts and the signal is shorted or "clamped" to 0 V (neglecting the 0.7-V drop, for simplicity). When the input goes positive, it goes up 4.0 V and so does the output. So we have a 4.0-V p-p signal riding on 0 V.

Figure 7-15(b) shows the same clamping circuit with extra bias on it. This bias is fed from an external source. It eliminates the effect of the 0.7-V drop across the diode, and the entire signal will pass through.

Clamping circuits can also be checked with dc voltage and resistance readings. The reason for clamping the dc bias voltage is usually to obtain a whole unclipped signal. This is best checked with a scope. Check for the dc level as well as the amplitude (height) of the wave.

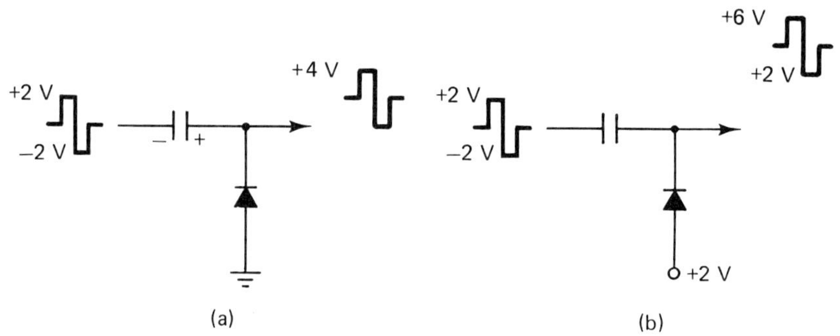

Figure 7-15 Diode clamper. (a) The diode clamps the negative going pulse to nearly 0 V (0.7 V, actually). (b) The negative-going pulse is clamped to +2 V.

7-12 AC SIGNAL PATHS THROUGH COUPLING CIRCUITS

Tracing signals from circuit to circuit to find out where the signal is lost, weakened, or deformed is an important part of troubleshooing. Figure 7-16 shows the signal path through several circuit blocks.

Intersecting this path are dc paths that maintain the dc voltage levels required by each circuit block. You can tell these are dc paths because they terminate at the supply line and ground. There are also paths to ground or the supply line that help maintain the signal level at a particular frequency or to determine its waveshape. They also maintain the proper dc level at the circuit inputs. There are four main types of coupling between circuits.

1. The path between CKT 1 and CKT 2 is a direct path. This method of coupling two circuits is called *direct coupling*. One problem with this coupling method is that the dc voltages in the two blocks are mutually dependent. We have a two-source (perhaps more than two inside the blocks) bridging circuit here (see Section 4-4).

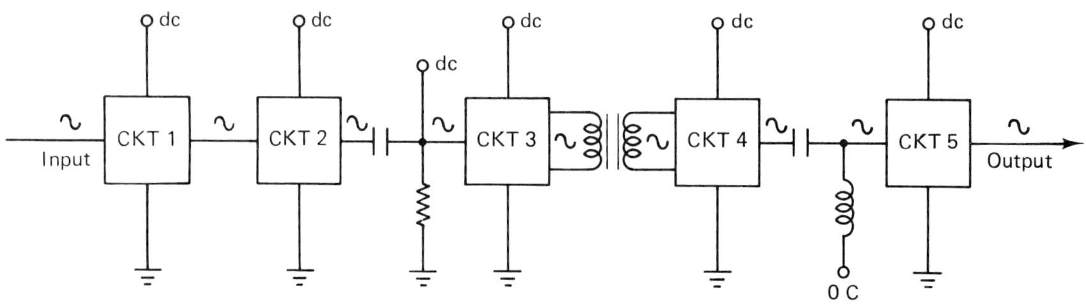

Figure 7-16 Methods of transferring signals from circuit to circuit. From left to right: direct coupling, capacitive coupling, transformer coupling, and impedance coupling.

2. The path between CKT 2 and CKT 3 is called *capacitive coupling*. This method isolates the dc voltages in each circuit and keeps them from disturbing each other. Each circuit requires its own dc operating voltages.
3. The path between CKT 3 and CKT 4 is called *transformer coupling*. This method also isolates the dc voltage of each circuit. It has other features, which are taken up in Chapters 8 and 9.
4. The path between CKT 4 and CKT 5 is called *impedance coupling*. The capacitor used with coil blocks the dc voltage.

Each of these four methods may simultaneously serve other purposes, such as filtering and waveshaping, but we are mainly interested in tracing the signal from circuit to circuit. You may see signals filtered through coils in the path between circuits. Diodes may also be found in these paths. They can serve as switches that determine when a signal can be coupled into a circuit. They can also process the signal before it is fed into a circuit. The steady 0.7-V forward-bias voltage across the diode is sometimes used for regulating or stabilizing the other voltages in circuit blocks.

The transistors inside the circuit blocks also pass signals; that is, they have signal paths. We learn about these in Chapter 8.

WHAT YOU LEARNED TO DO

In this chapter you learned how to:

1. Read waveforms in various ac and pulse circuits.
2. Detect faults in various types of *RC*, *RL*, and *LC* circuits by using the scope.
3. Troubleshoot diode clippers and clampers with the scope.
4. Use a square-wave generator and scope to check the frequency response of an audio amplifier.
5. Trace signals through capacitors, coils, transformers, and diodes so that you will be able to learn how to trace them from circuit to circuit in the next two chapters.
6. Use dc measurements while you learn more about ac and scope measurements.

QUESTIONS

1. What three basic measurements can be made on an ac wave? Can you make these measurements with a multimeter?
2. What can you do with a scope that cannot be done directly with a multimeter? What can you do with a multimeter that you cannot do directly with a scope?
3. Name the three basic adjustments for measuring ac waveforms on a scope.

4. In a series ac circuit that contains only resistors, do all the voltage drops add up to the ac supply voltage?
5. What is the difference between inductive and capacitive reactance as far as frequency is concerned?
6. If you read the ac voltages across each part in a simple *RC* circuit, will these voltages add up to equal the voltage across the circuit?
7. Draw the following: a low-pass *RC* filter, a high-pass *RC* filter, a low-pass *RL* filter, and a high-pass *RL* filter.
8. Should a coupling capacitor drop much signal voltage? How about dc voltage?
9. If a small capacitor is in series with a high resistance, will it drop much voltage at a low frequency?
10. Draw the shape of the frequency response curve of a series *LC* bandpass filter. Locate the point on the curve that represents the resonant frequency. Do the same for a parallel *LC* bandpass filter.
11. If the signal voltage across a series *LC* circuit is less than the sum of the individual coil and capacitor voltages, is the circuit good or bad? If the capacitor is open or the coil is shorted, will a dc voltage check reveal anything? How can you check the circuit without a scope?
12. Draw an *RC* differentiator circuit.
13. Draw a diode clipper that cuts off substantially less than half the input signal.
14. What is the difference between a clipper and a clamper?
15. Draw the frequency response curve of an audio amplifier with a poor low-frequency response.
16. Draw a waveform that shows a poor low-frequency response to a square-wave test. Draw a poor high-frequency square-wave response.

PRACTICAL TROUBLESHOOTING EXPERIENCE

You need an oscilloscope and a multifunction signal generator that can deliver 100 Hz to 10 kHz at 5 V p-p (1.768 V rms) for this job. This type of generator delivers square waves as well as sine waves. Preferably it should have a meter on its front panel. You will also need a low-voltage power supply for the last job.

1. Connect together the low-pass filter shown in Fig. 7-7(a). Use a 0.01-μF capacitor and 10-kΩ resistor. Send various sine-wave frequencies through the filter from a signal generator. Observe the results with a scope. Does the output vary inversely or directly with the frequency? Try capacitance values of 0.1 μF and 0.001 μF. Use the same 10-kΩ resistor. Does the output voltage vary with the capacitance?
2. Repeat instruction 1 with a square-wave input. Which values of capacitance, voltage, and frequency give you the sharpest spikes? Which values of voltage, capacitance, and frequency give you the least distortion?

3. Connect up the circuit shown in Fig. 7-7(b). Use a 1-kΩ resistor for the input and output. Vary the sine-wave frequency and try various values of capacitance. Does the output vary inversely with values of capacitance? Does it vary inversely with the frequency?
4. Do the same as instruction 3 with a square-wave signal. Which values of capacitance, voltage, and frequency give you the least distortion?
5. Connect the low-pass filter shown in Fig. 7-8(a). Use a small many-layered coil that has a nonmetallic core. This should give you a coil within the millihenries range. The resistor should have about 10kΩ. Send a signal through the filter and vary the frequency. What results do you get?
6. Do the same with a square-wave signal. What are the results?
7. Connect the circuit shown in Fig. 7-8(b). Vary the input sine-wave frequency. What are the results?
8. Do the same as above with a square-wave signal. What are your results? What frequency gives you the least-distorted waveform?
9. Wire up the circuit shown in Fig. 7-14(e). Use a signal diode and a 1-kΩ resistor. Send various sine-wave and square-wave signals of various frequencies through the clipper. Vary the dc voltage. Were you able to adjust the scope and read the results?

8

Troubleshooting Basic Transistor Circuits

EXPERIENCE REQUIRED

Before you can learn to troubleshoot transistor circuits, you should know:

1. How shorts and opens affect voltages in basic series and parallel circuits.
2. The difference between forward- and reverse-biased diodes.
3. How to check diodes and transistors with an ohmmeter.
4. How to check diode circuits with a voltmeter.
5. That dc voltages can be measured on signal lines.
6. How signals are depicted in waveforms.
7. How signals are passed, bypassed, and blocked by capacitors and coils.
8. That electrostatic charges can be built up on your body when various plastic materials such as cups, packages, cabinets, carpets, and so on, are touched, rubbed, or handled.

8-1 TRANSISTOR VOLTAGES

Now that we have learned to troubleshoot just about all the parts around it, we are ready to start working on the transistor itself. It is the most important part of electronic circuits. It is used for processing signal and control voltages. Transistors have more than two terminals, and each terminal requires a definite dc voltage level. This is why checking dc voltages is such an important part of electronic troubleshooting.

8-1-1 Transistor DC and Signal Voltages

In Chapter 7 we saw how signals were transferred from circuit to circuit (Fig. 7-16). In this chapter we look into these circuits. Just about all of them have transistors. Transistors require at least two terminals for the signal, one for input and one for output. Transistors also require at least two terminals for dc power. They have only three terminals. This means that the same terminal can have both signal and dc voltages. The signal voltage is ''riding'' on the dc voltage. We can use a meter to check the dc voltages and a scope for the signal voltages; or better yet, a scope to check both. The scope may be best; but as you gain experience, you will ''see'' the signal in your head when you read the dc voltage on a meter.

8-1-2 Transistor Inputs and Outputs

Transistors are good places to look for signals because nearly all system signals of one sort or other pass through them. Figure 8-1(a) shows the signal paths going through

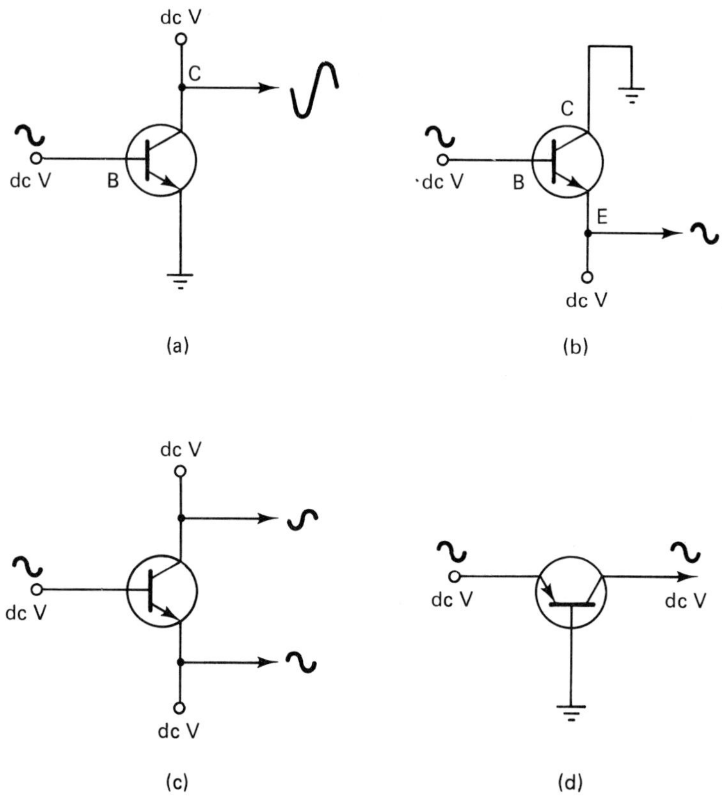

Figure 8-1 Signal path through a transistor. (a) From base to collector. (b) From base to emitter. (c) From base to collector and emitter. (d) From emitter to collector.

a transistor circuit. The signal goes into and out of the two terminals called base (*B*) and collector (*C*), respectively. Sometimes the signal goes through the emitter (*E*). The emitter has the arrow. If we know the inputs and outputs, we know which way the signals travel from circuit to circuit. This makes it easier to track down the trouble. The difficulty lies in the fact that transistors are not always connected in the same way. Let us look at the inputs and outputs of transistor circuits.

1. The base lead, *B*, is usually the input. In a schematic, the signal path usually goes from left to right, so you'll usually find the base on the left as shown in Fig. 8-1(a).
2. The output signal voltage is usually at the collector *C*, but not always. Sometimes it is at emitter *E*. In any case, if it is grounded or bypassed by a capacitor to ground, it cannot be an output. Of course, you can also identify the output by the fact that a signal path leads from it to the input of another circuit or device.
3. An emitter output is usually on the bottom side, as shown in Fig. 8-1(b). It usually outputs to the right. Look for a signal path leading from it to the input of another circuit or device.
4. Occasionally, there are two outputs, as shown in Fig. 8-1(c), one from the collector and one from the emitter. This is sometimes called a phase-splitter circuit. Both terminals are above "signal ground"; that is, they are not tied directly to ground or bypassed to ground.
5. In certain applications you might find an emitter input with a collector output. The base is grounded or bypassed to ground [see Fig. 8-1(d)].
6. There are special circuits in which a portion of the output of a transistor is fed back to its input. We will learn about these in Chapter 9.
7. The base is never the output.

> TIP 8-1
>
> When tracing signals, the transistor base terminal is usually the input, and the collector or the emitter is usually the output. Occasionally, the emitter is the input. The base is never the output. In schematics the signal usually goes from left to right.

8-1-3 Major and Minor DC Paths in Transistors

Besides signal paths, transistors have dc paths. In the early chapters we learned how to trace dc voltages in resistor, capacitor, coil, and diode circuits. Except for transformers, these parts had only one dc path. The transistor in Fig. 8-2(a) has two dc

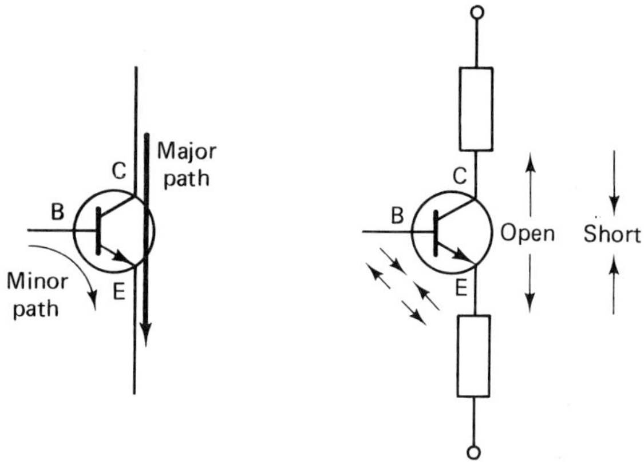

(a) (b)

Figure 8-2 (a) Major dc path through a transistor. (b) The arrows indicate what happens to the dc voltages on the terminals when the paths between them open or short.

paths. One is a major dc path from the collector to the emitter, the *CE* path. The other is a minor dc path from the base to the emitter, the *BE* path.

Like the transformer, there is a path for the input circuit and a path for the output circuit. The major path in the transistor is in the output circuit. It normally has a big dc voltage drop across it. (The transformer dc path normally has very little dc drop.)

These paths can short or open. You can apply the basic rules to them. When they short, the voltages above them fall and the voltages below them rise. When they open, the voltages above them rise, and the voltages below them fall [see Fig. 8-2(b)]. This information in itself may be enough to enable you to track down a defective transistor.

8-1-4 The Basic Transistor Action

The transistor shown in Fig. 8-3(a) has a resistor R_C connected in series with its *CE* path. The supply voltage is 20 V and it is dropped to 10 V at the collector. The *BE* path has a series resistor R_B that drops a 5-V dc input voltage to 0.7 V dc at the base.

The *BE* path can be treated as a forward-biased diode. It drops 0.7 V as a diode would, and when the input voltage is increased, the base voltage will not rise more than a few tenths of a volt.

The *CE* path can be treated as a variable resistor that is controlled by the current through *BE* [see Fig. 8-3(b)]. When you increase the input voltage, the *CE* resistance decreases and the voltage at *C* tends to be pulled down. The current increases. When maximum possible current is flowing, the transistor is said to be *saturated.* The *CE* path seems to be shorted.

Just the opposite happens when you decrease the input voltage; the voltage at *C* tends to be pushed up. When no current is flowing, the transistor is said to be *cut off.* The *CE* path seems to be open.

Furthermore, it takes only a small change of *BE* current to produce a large change of *CE* current. This is your basic amplifying action. (See Section 8-2-1 for more on this.)

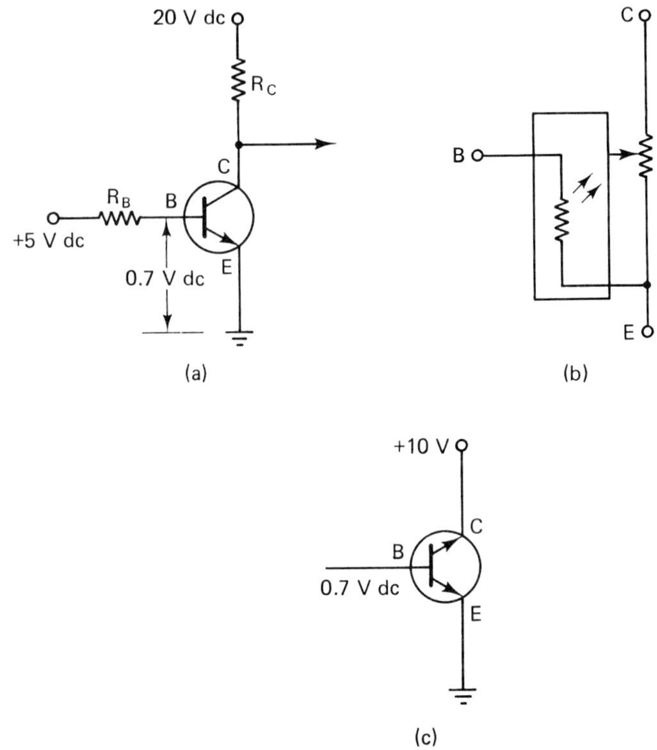

Figure 8-3 (a) Typical dc voltages on the transistor terminals. (b) The equivalent internal circuitry of a transistor. Changing the voltage on *BE* changes the resistance of *CE*. (c) The *BC* and *BE* terminals have back-to-back diodes connected to them.

Note: In some troubleshooting situations, you may want to know the ratio of the input current to the output current of a transistor. This value is listed in the transistor manuals in the h_{FE} column. To find the output current for a transistor, multiply its input current by its h_{FE}.

Usually, there is no *BC* path, and it can be treated as a reverse-biased diode. In this case, unless something goes wrong, we can ignore this path [see Fig. 8-3(c)]. Looking at the diagram, you may wonder how the current can go from *C* to *E* if it must go against the arrow *BC*. We'll find out a little more about this in Section 8-2; it has to do with the way a transistor is constructed and the type of materials used.

8-1-5 The Two Transistor Types: NPN and PNP

You'll find two types of transistors, PNP and NPN. The PNP type, shown in Fig. 8-4(a), has the arrow **P**ointing i**N.** The NPN type, shown in Fig. 8-4(b), has the arrow

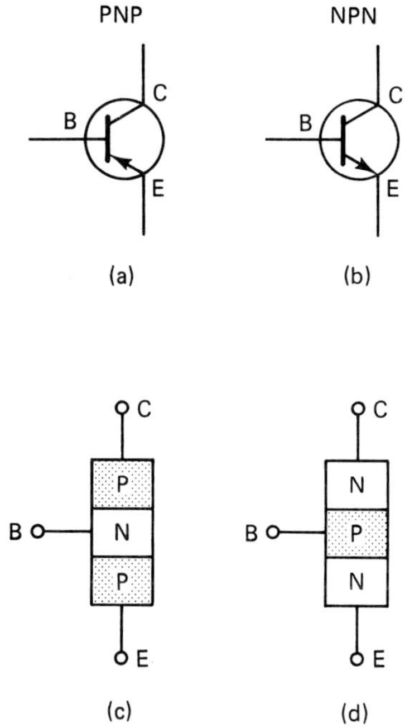

Figure 8-4 Two types of commonly used transistors: (a) PNP and (b) NPN. The transistor layers: (c) PNP and (d) NPN.

Not **P**ointing i**N.** The transistors are made with three layers of special semiconductor materials. Figure 8-4(c) shows a PNP type. It has an N-type material sandwiched between two layers of P-type material. The NPN type, shown in Fig. 8-4(d), has the P layer sandwiched between two N layers. The illustrations show what types of materials are connected, respectively, to the base, collector, and emitter.

The following information applies mostly to NPN transistors. Since they usually work with positive power supplies, they are the most commonly used. Except for the polarity of voltages, the PNP works in exactly the same way. We'll get back to the PNP type a little later.

8-1-6 A Basic Transistor Amplifier Circuit

A basic transistor amplifier circuit is shown in Fig. 8-5. Not all amplifiers are exactly like it; but for learning purposes, this one can be used as a basis for comparison. The input signal is at the base and the larger output signal is at the collector. The waveform is amplified with a minimum of distortion. This is one of the characteristics that makes this a class A amplifier.

As troubleshooters, we are interested primarily in the input and output signals and the dc operating voltages, as well as the resistors that maintain these voltages. The normal dc transistor voltages for amplifiers are often given in the schematics. However, you should remember the following points. For maximum undistorted operation, the collector resistor, R_C, drops the voltage down to about one-half the supply-line voltage. The emitter resistor, R_E, helps establish the emitter voltage. The emitter

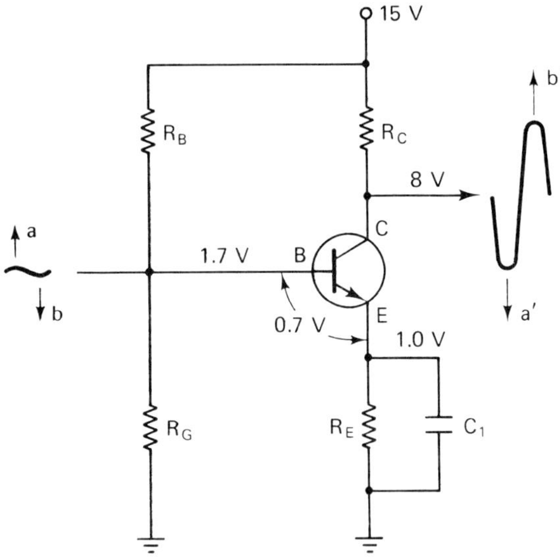

Figure 8-5 Basic transistor amplifier. Note the amplification and inversion of the waveform at the collector, *C*.

capacitor, C_1, bypasses the ac signal at the emitter. Otherwise, this signal will be subtracted from the signal at the collector.

The base voltage is determined mainly by R_B and R_G, the supply and ground resistors. These two resistors should drop voltages somewhat in proportion to their resistances; the transistor *BE* path resistance is too high to affect these drops very much. A 0.7-V dc drop is maintained across *BE*.

> TIP 8-2
>
> In the basic transistor amplifier (class A), the collector voltage for maximum undistorted operation is about one-half the supply voltage. The emitter voltage is much less. The collector and emitter resistors divide their voltage drops in proportion to their respective resistances. The base voltage is about 0.7 V higher than the emitter voltage.

The input to this circuit is at the base terminal. The input signal places a fluctuating voltage on the base. Let us slow down the action and look at the voltage as it goes high and as it goes low.

1. When the base voltage goes up, the collector current increases; that is, the *CE* resistance and the collector voltage both go down. Note the direction of the ar-

rows at *a* and *a'*. We see that the collector voltage is *inverted.* Note also that amplification also takes place.

There is no up-and-down action at the emitter, since the signal there is bypassed by C_1. However, if a defect in the base circuit raises the dc base voltage, the emitter dc voltage will also rise. The capacitor does not bypass dc.

2. When the base voltage goes down, the collector voltage goes up. Note the arrows at *b* and *b'*. The voltage change is inverted and amplified. The ac signal at the emitter is bypassed. The dc emitter voltage follows the dc base voltage and goes down. This can happen if a defect in the base circuit brings down the base voltage.

Tip 8-3

In a basic transistor amplifier circuit: When the base voltage goes up, the collector voltage goes down. When the base voltage goes down, the collector voltage goes up. In all other cases (in Section 8-1-7), the output voltage follows the input voltage.

The above is true for both NPN and PNP circuits and for both positive voltages and negative voltages.

The voltage changes covered above are fundamental to electronics. During normal circuit operation, they usually change very rapidly and cannot be read with a dc voltmeter. However, when something goes wrong, the dc voltage may be stuck high or low and enable you to use the voltmeter. This has important consequences. For instance, if the base voltage sticks high, the collector voltage will stick low.

8-1-7 The Emitter-Follower and Common-Base Amplifiers

There are three ways in which transistor circuits are wired up. We already looked at one of them, our basic amplifier in which the emitter terminal is common to both the input and output circuits. It is also called the common-emitter amplifier.

A second type is called the common-collector amplifier or emitter follower [see Fig. 8-6(a)]. The output is taken at the emitter, so the bypass capacitor, *C*, has been moved over to the collector. For ac signals, the collector terminal is common to the input and output circuits. The input is at the base, and the output is at the emitter. The output voltage follows the base voltage changes. Note the input and output waveforms in the illustration. This circuit does not amplify voltage; it amplifies current.

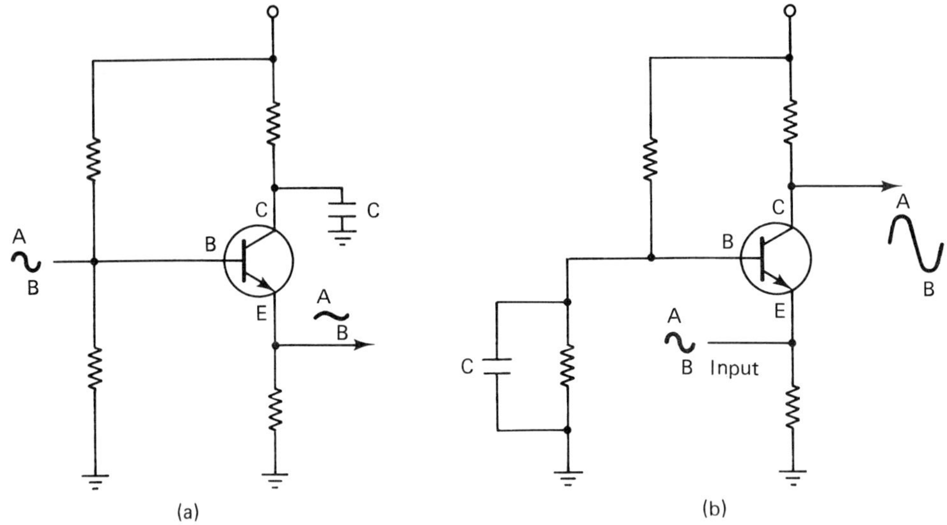

Figure 8-6 (a) Basic common-collector amplifier or emitter follower. (b) Basic common-base or grounded-base amplifier.

The third type is called the common-base amplifier [See Fig. 8-6(b)]. The base is common to the input and output circuits. The bypass capacitor, *C*, has been moved over to the base. The input is at the emitter, and the output is at the collector. The signal can be amplified, but it is not inverted.

> *Note:* In no case is the base the output, nor is the collector ever the input for signal circuits. (The collector can input dc for voltage regulation. See Chapter 10.)

8-1-8 PNP Transistor Circuits

PNP-type transistors in positive power-supply circuits may have their outputs at the emitter with the collector grounded [See Fig. 8-7(a)]. The same rules apply. The emitter always follows the base, and there is no voltage amplification.

In some cases the collector is the output; it divides the signal voltage with a ground resistor [see Fig. 8-7(b)]. The signal is inverted and can be amplified.

PNP transistors are often used with negative power supplies [see Fig. 8-7(c)]. The collector is connected to the negative supply side, and the emitter is connected to the ground side. You will find them connected in all three possible configurations: common-emitter, common-collector, and common-base. The same rules apply. The signal is inverted at the collector only. The emitter voltage follows the base voltage.

8-1-9 Looking for Hi–Lo–Lo's and Lo–Hi–Hi's

One method of troubleshooting a circuit board is to move your voltmeter probe quickly from transistor to transistor looking for one that does not seem to have normal

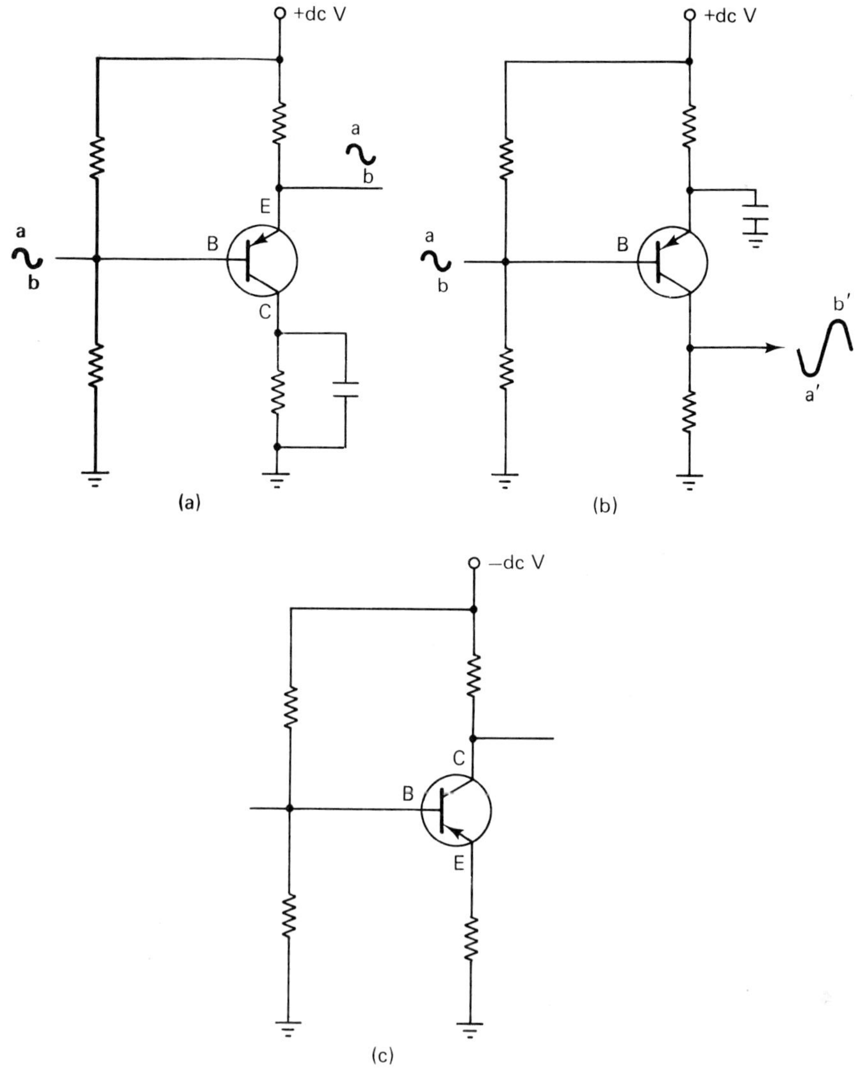

Figure 8-7 Basic PNP transistor circuits. (a) Common-collector amplifier or emitter follower. (b) Common-emitter amplifier. Note the amplification and inversion at the collector. (c) PNP transistor with a negative supply voltage.

voltages. If you have to look back and forth between board and the schematic, it is going to take a long time. You might save time by just seeing whether each transistor has one high voltage and two lows, Hi–Lo–Lo, or two highs and one low, Lo–Hi–Hi (see Fig. 8-8). The former is for NPNs and the latter for PNPs. If you do not find these voltage patterns, something could be be wrong and worth looking into. This is not a foolproof method; some special-purpose circuits do not have these voltage patterns. However, it may be worth a try. You can always make more extensive checks.

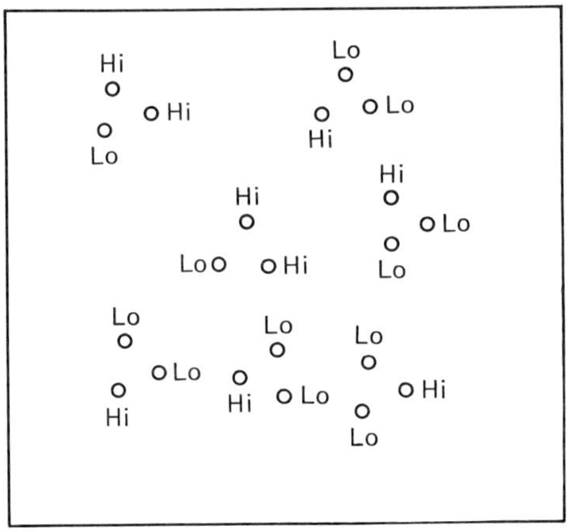

Figure 8-8 Transistors on a PC board can be checked quickly for hi–lo voltages. Follow up on the ones that do not show the typical hi–lo–lo or lo–hi–hi voltages.

8-2 TRANSISTOR DEFECTS

As troubleshooters, we note that the internal paths between terminals *BC*, *BE*, and *CE* have special characteristics. First, they have resistances. Second, they exhibit transistor action. Third, they act like diodes. Each of these characteristics determines what happens to the transistor voltages when something goes wrong.

All this is summarized in Tip 8-4 later in this section. It is a very important tip. The discussion that precedes it may help you to understand the tip a little better.

8-2-1 Transistor Junctions

Figure 8-9(a) shows the construction of an NPN transistor. We show an older type because it is easier to understand. The layer of P material is connected to the base *B*. The two buttonlike sections are connected to the collector *C* and the emitter *E*, so we have two junction diodes facing in opposite directions, as noted by the diodes in Fig. 8-9(b). It looks as if collector current cannot possibly flow through both junctions, especially since there is a reverse bias on the *BC* junction. But note the thin space in the middle of the *B* section. This seems to form a gap for current to flow through from *C* to *E*. It does—under certain conditions.

These conditions are met when forward bias is placed on the *BE* junction. According to researchers, a small current flowing through the *BE* junction overcomes

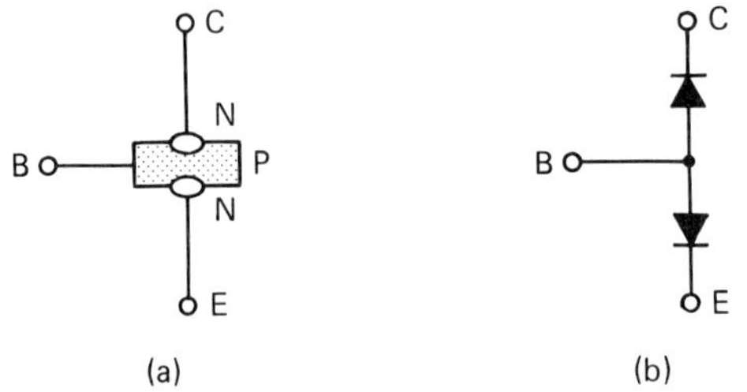

Figure 8-9 Transistor construction. (a) Note the narrow gap between the *C* and *E* buttons. This permits conduction through the reverse-biased *BC* junction when current flows from *B* to *E*. (b) The arrows pointing in opposite directions makes it seemingly impossible for current to flow from *C* to *E*.

the reverse bias on the *CB* junction and allows the *CE* current to flow through. The amount of *CE* current that flows is controlled by the amount of *BE* current. This is your basic transistor action.

Remember, the *BE* junction must be forward biased to allow the *CE* path to conduct. Zero or negative bias on *BE* will cut off the *CE* current. Also recall that the forward-biased *BE* diode voltage drop hovers around 0.7 V.

8-2-2 *BC* Junction Defects

In the basic amplifier, the *BC* junction is normally reverse biased, so the junction resistance is extremely high and hardly any current leaks through from *C* to *B*. The *BE* junction normally has 0.7 V. This allows some current to flow through *CE*. This is the situation when all the basic-amplifier terminals have normal voltages. In the following investigation, let V_B, V_C, and V_E represent the voltages on the respective terminals.

1. An open *BC* junction pushes V_C up and V_E down, as shown in Fig. 8-10(a). However, V_B may still be pulling V_E up, so V_E may not fall very much. (If the input resistors have very high values, V_E may fall drastically.) The *BE* voltage drop may be slightly above normal and hardly noticeable.

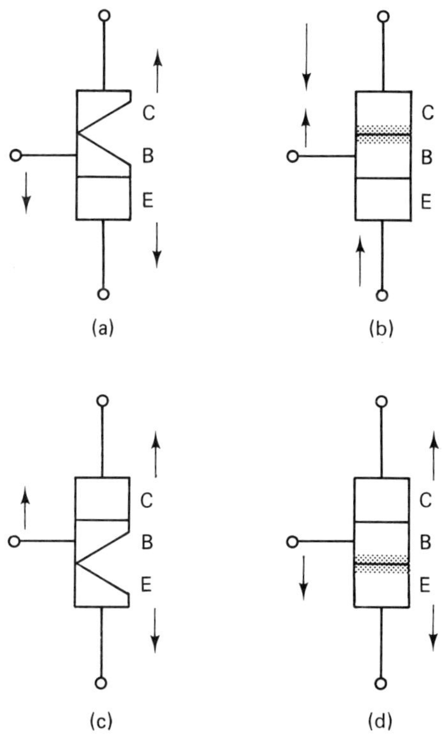

Figure 8-10 Transistor defects. (a) Open *BC* junction. (b) Shorted *BC* junction. (c) Open *BE* junction. (d) Shorted *BE* junction. The arrows indicate the direction of voltage changes.

2. A short in the reverse-biased *BC* junction in Fig. 8-10(b) will forward bias the *BE* junction, and a large current will flow through *CE*. In other words, path *CE* will seem to be shorted. Voltage V_C is pulled down and V_E is pulled up. Voltage V_B is pulled up by the shorted *BC* junction.

8-2-3 *BE* Junction Defects

The *BE* junction controls the *CE* path current or resistance. Therefore, it cannot be treated simply as a diode.

1. If the *BE* junction opens, the *CE* current, which must pass through the *BE* junction, is blocked, causing V_C to rise. The voltage V_E, which is below, falls. The voltage V_B, which is above it, also rises. The *BE* diode drop will increase to 0.9 V or more [see Fig. 8-10(c)].
2. If the *BE* junction shorts, its control function is lost [see Fig. 8-10(d)]. The reverse bias on *BC* regains control and blocks the *CE* current. This pushes up V_C and lets V_E fall. Voltage V_B falls with V_E. The *BE* diode drop will be far below 0.7 V, say 0.3 V or less for silicon transistors.

8-2-4 *CE* Path Defects

The first step in troubleshooting a transistor is to look at the voltages on its terminals. Usually, the collecter voltage is checked first. This is because the collector voltage, V_C, is usually the output and has a high value, and any fault that is transferred to it from the other terminals is easily seen. It can be treated as a resistance path that is subject to shorts and opens. However, remember that it is subject to transistor action. That is, a change in the *BE* circuit can cause a *CE* resistance change. In the briefing below, we are concerned only with internal defects. We assume that the external circuits are okay.

1. If resistance path *CE* opens, V_C will rise and V_E will fall. The voltage at V_B will depend on where the open is. If the *BC* junction opens, V_B will fall with V_E, as in Fig. 8-10(a). If junction *BE* opens V_B will rise, as in Fig. 8-10(b). If both open, V_B will rise.
2. If resistance path *CE* shorts, V_C will fall and V_E will rise, and the two voltages will be nearly equal. Voltage V_B will also be nearly equal to the other voltages. The *BE* drop will be less than normal if both junctions are shorted. It may be normal or slightly above normal if the *BC* junction is shorted. A shorted *BE* junction would open the *CE* path.

8-2-5 Internal Opens at the *B*, *C*, and *E* Terminals

In addition to the defects within the internal transistor paths and junctions themselves, we have internal opens at the terminals. Cracks in the respective *B*, *C*, and *E* layers may partially have the same effects.

1. When terminal *B* opens internally, the *BE–CE* transistor action is lost. Resistance *CE* virtually opens, the voltage at *C* rises, and the voltage at *E* falls. The voltage at *B* rises because of the open below it. The *BE* voltage drop rises to 0.9 V or more.
2. When terminal C opens internally, the voltage above it at *C* rises and the voltage below it at *E* falls, pulling down the voltage at *B*. The *B* and *E* voltages will be determined by the external circuit. The *BE* voltage drop will be slightly more than 0.7 V if noticeable.
3. When the *E* terminal opens internally, both voltages above it at *B* and *C* will rise. The voltage below it at *E* will fall. The *BE* drop will be 0.9 V or more.

8-2-6 Summary of Transistor Defects

Let's make a table to summarize the transistor defects. L stands for lower-than-normal voltage and H for higher. The *BE* voltages are given as 0.7 V + for slightly above normal. The abnormal *BE* voltages that definitely indicate a transistor defect are given by 0.9 V and 0.3 V. The math symbols > and < mean *greater than* and *less than*, respectively.

TRANSISTOR DEFECT TABLE

B	C	E	BE	Defect
L	H	L	0.7 V+	*BC* open
H	L	H	0.7 V+	*BC* short
H	H	L	>0.9 V	*BE* open
L	H	L	<0.5 V	*BE* short
H	H	L	>0.9 V	*B* terminal internally open
L	H	L	0.7 V+	*C* terminal internally open
H	H	L	>0.9 V	*E* terminal internally open

The following practical tip has been extracted from the information given above. It summarizes for the troubleshooter the contents of this entire section. *Don't forget*—we are always assuming silicon transistors.

TIP 8-4

The voltage drop across the base and emitter terminals in a basic transistor circuit is normally about 0.7 V.

Less than about 0.5 V difference between the forward-biased *BE* terminals indicates a shorted transistor. More than 0.9 V difference indicates an open transistor.

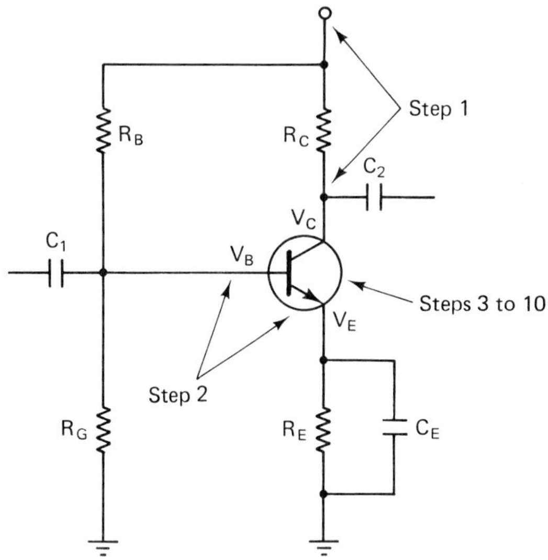

Figure 8-11 Troubleshooting steps for the basic transistor amplifier. Start with the collector and supply voltages. Then check the base and emitter voltages.

8-3 TRANSISTOR CIRCUIT DEFECTS

We have learned the possible voltage changes that can be caused by a bad transistor. Note that some of these changes can be caused by defects in the external circuit around the transistor. We will have to look at both the external circuit and the transistor to determine the actual cause.

8-3-1 What Happens When Each Part Goes Bad

We can now investigate the action of defective circuit parts around a transistor. This is usually the sequence followed in actual troubleshooting. The transistors are the most likely to go out, so you look at them first. Then you look for the defective parts. Let's find out what these defective parts can do to the voltages in a transistor circuit. We're looking at resistors R_C, R_E, R_B, and R_G and capacitors C_1, C_2, and C_E in the basic transistor circuit shown in Fig. 8-11. We are not overlooking the circuits that send or receive signals to or from this amplifer, or the PC board or the wiring and connections around the transistor. The terminal voltages are represented by V_B, V_C, and V_E.

1. An open-collector resistor, R_C, should drop the voltages in the path below it, and it does. Voltage V_C falls. V_E falls and brings V_B down with it. It should be noted here that the resistance of the *BE* junction decreases, so that its voltage drop decreases only a little, if it is noticeable at all.
2. A collector-to-ground short in the board or wiring reduces voltages V_C and V_E below it. *BC* will be forward biased, so V_B will be pulled down by both V_C and V_E. All voltages will be lower than normal. A shorted output coupling capacitor, C_2, will also bring down the collector voltage.

3. A collector-to-supply short tends to increase the voltages below it. However, as a net result of transistor action, only V_C increases significantly. The other voltages do not change significantly.
4. An open-emitter resistor, R_E, should increase all the voltages above it, and it does. The *BE* drop may go negative if you measure it from the emitter terminal. If measured from ground, V_E rises to a higher level than V_B.
5. An emitter-to-ground short pulls down V_E and V_B. The *BE* voltage drop tends to increase, but this is offset by transistor action, which tends to pull the emitter voltage back up. As a net result, the *BE* drop remains unchanged.
6. If the base resistor, R_B, opens up, voltage V_B falls, V_C rises, and V_E falls. Voltage drop *BE* decreases significantly and may go negative.
7. A base-to-ground short raises voltage V_C and lowers voltage V_E. The *BE* drop decreases.
8. A base-to-supply short, possibly in the sending circuit block, raises V_B, V_C falls; and V_E rises. The *BE* drop increases a little. The short to supply can also occur when the coupling capacitor, C_1, shorts.
9. If R_G opens, the voltage V_B rises and V_C will fall.

8-3-2 Summary of Basic Transistor Circuit Defects

Do not try to memorize everything. A basic knowlege of shorts and opens and a mental picture of the voltage paths and junctions in the transistor and Tip 8-4 should be enough. Much of the information above will be covered automatically as you follow the practical tips and methods in this chapter.

For instance, it does not take a lot of detailed knowledge to know that the transistor is the most likely part to go bad. As a matter of routine, you check it first. You check it with a voltmeter and then with an ohmmeter. You can also use an in-circuit transistor checker. After that, you check the parts.

If you are stuck, the following table may help. It lists high and low readings (H and L) for each transistor terminal, under the headings *B*, *C*, and *E*. The base–emitter voltage drop is listed under *BE*. The numbers 0.7 V+ and 0.7 V− stand for increased or decreased voltage drops. N stands for no discernible change for the troubleshooter. (−) indicates that the base voltage is lower than the emitter voltage, or reverse biased.

The *BE* voltage drop is typically about 0.7 V. An increased drop, due to a bad circuit part, would be about 0.75 V. A very high voltage would be about 0.8 V. More than 0.9 V would indicate a bad transistor.

8-4 TROUBLESHOOTING A BASIC TRANSISTOR AMPLIFIER

The major steps involved in troubleshooting the basic transistor circuit are: First, check the collector voltage. Second, check the *BE* drop to see whether the fault is in the transistor or the circuit. Third, check the circuit parts if you cannot find anything wrong with the transistor.

BASIC AMPLIFIER CIRCUIT FAULT TABLE

	Transistor lead voltages				
	B	C	E	*BE*	Cause
1.	L	L	L	0.7 V−	Open collector resistor
2.	L	L	L	L	Short, collector to ground, C_2 short
3.	N	H	N	N	Short, collector to supply
4.	H	H	H	0.7 V−	Open emitter resistor
5.	L	N	L	N	Short, emitter to ground, C_E short
6.	L	H	L	(−)	Open base-to-supply resistor
7.	L	H	L	(−)	Short, base to ground
8.	H	L	H	0.7 V+	Short, base to supply, C_I short
9.	H	L	H	0.7 V+	Open base-to-ground resistor

8-4-1 A Troubleshooting Procedure

The following example should give you an idea of how to check the basic transistor amplifier. With one or two modifications, it can be used to check other types of transistor circuits. The first step is to determine whether a suspected transistor circuit is defective. Refer again to Fig. 8-11. The coupling capacitors C_1 and C_2 at the base and collector serve to isolate our amplifier from the outside circuits. This simplifies things. In Chapter 9 we learn to troubleshoot amplifiers with no coupling capacitors. When working on actual circuits, all transistor voltage readings should be recorded.

We assume that you are working with a schematic in which the good voltages are given. Tip 8-2 also shows what the normal voltages should be. Tip 8-3 shows how good transistors respond to changed dc voltages. Tip 8-4 shows voltages that indicate bad transistors. The specifics for finding the bad parts are given in the preceding section, in the table in particular. We are focusing on the procedural aspect of troubleshooting transistors.

Step 1. Measure the dc collector voltage. This reading might let you know whether there is something wrong with the circuit (see Tip 8-2). If this voltage is okay, the trouble is probably in another circuit, but check the base and emitter anyway, since they are right next to the collector.

If the collector voltage is bad, measure the supply point voltage if it is easily located. If not easily done, go ahead with the base and emitter measurements first. If they do not reveal anything, find the supply-voltage point and measure it.

Step 2. Check the dc base and emitter voltages. Use the lowest meter range possible.

If the collector and emitter are responding normally to a changed (or defective) dc base voltage (as in Tip 8-3), the transistor is probably okay. A fault in the base circuit could change its voltage. If the collector and emitter are not responding normally, the transistor is probably bad.

At this point you have a pretty good idea whether the trouble is in the transistor or in the circuit. You may even have an idea of which circuit part is bad. Before you start pulling things out, however, you might want to confirm it. It is now worth your time to make more intensive checks.

Step 3. Make an in-circuit transistor check with an ohmmeter or transistor checker. If you find nothing definite with these checks or if a transistor checker is not on hand, note the following:

If there is zero *BE* voltage drop and the base-to-ground voltage is above normal, the transistor is bad (see Tip 8-4).

If there is more than 0.9 V between any two leads that are forward biased, the transistor is bad (see Tip 8-4).

The following tips should be helpful.

> TIP 8-5
>
> To check a low-power transistor, short the base to the emitter with a screwdriver, jumper, or test prod. (As you measure the collector with the red probe, short the *BE* terminals with the black probe.) If the collector voltage goes all the way up to the supply voltage, the transistor is okay and there is no dead short across the collector resistor.

> TIP 8-6
>
> To check a transistor, jump the base supply resistor with one of approximately equal value. If the collector voltage goes down, the transistor is okay.

If you are cutting a transistor lead to disconnect it, cut it so that it can be resoldered cleanly without overheating anything. It may be better, in many cases, to unsolder the lead on a printed-circuit board. Use a solder sucker or solder wick. You need not remove the lead from the hole. Move it so that it does not touch the edges of the hole. You can then touch the lead carefully with the test prod. To make sure that

TIP 8-7

To verify a transistor check in a basic amplifier circuit, disconnect the base lead first. Check all the transistor voltage readings with the base disconnected. The collector voltage should read high. The base voltage should be zero. The emitter voltage should equal zero on the lowest scale. If you do not find these voltages, the transistor is bad.

It is a good idea to recheck the transistor with an ohmmeter or transistor checker now. You will get a better reading this time.

you are getting a true reading, deliberately short the prod against the side of the hole so that you can compare the good reading with the shorted reading.

TIP 8-8

If an amplifier has a resistor from base to ground as well as from base to supply, the voltages should divide more or less in proportion to the specified resistances. If the base voltage is bad, disconnect the base terminal. If the voltage at the junction of the resistors returns to near normal, the transistor is bad.

At this point let us assume that you have determined that the transistor is okay and the trouble must be in the circuit. You can now make resistance checks on the parts or look over the voltages you have recorded. If you have difficulty, compare them to the chart. The resistance checks can be tried first since the multimeter is already on hand.

Step 4. Take an in-circuit resistance reading of the collector, emitter, and base resistors. Take the higher of two readings obtained after reversing the ohmmeter leads if you are using other than a low-voltage ohmmeter.

Step 5. If necessary, bring out the fault chart (you might have a copy pasted on the side of your meter) and compare it with the actual readings.

In addition to the resistors in the circuit, we have the capacitors. A shorted emitter capacitor will pull down the emitter and bring down the base and collector voltages with it. If the coupling capacitors short, they introduce bad voltages into the circuit.

At this point you probably have a very good idea as to which part is bad. It is time to pull things out.

TIP 8-9

If the base voltage in a basic amplifier is too high, the input coupling capacitor could be shorted. This could also bring down the collector voltage.

If the emitter voltage is shorted, it can pull down the base and collector voltages along with it. The collector voltage change may be much greater than the *BE* change.

Step 6. Remove the transistor from circuit. Check all resistors in-circuit with an ohmmeter. Check the voltage on the low side of the collector resistor. It should read high. If it reads low, there is a short between it and ground.

You already checked the base terminal. Check the emitter terminal voltage. It should have zero volts unless it is connected to another circuit.

Step 7. Replace the transistor with a known good one and check the operation.

Step 8. Replace other parts, if necessary, and check the operation.

Step 9. If you have a new trouble, check your soldering. Check whether your transistor leads are inserted in the right holes. It is very easy to become disoriented when you check out the leads looking at the transistor down side up, then turn it top side up for insertion. Recheck the voltages.

Step 10. If you still have the original trouble, do not give up; it happens to all of us.

Go back to Step 1 and start all over; you will be able to do it much faster this time. But make sure that you obtain good readings.

Record all your readings if you already have not done so. Compare your new readings with the old readings if you have them.

Could the trouble be in another circuit? Is your base voltage being thrown off by a fault in another circuit? Is your collector voltage being thrown off by another circuit? Did you check the ground connections?

These steps are for the basic amplifier. Steps for other amplifier types may require fewer steps or more. However, note that even though the signal paths differ for different types of amplifiers, the dc paths remain very much alike [see Fig. 8-6(a) and (b) and Fig. 8-7(a) and (b)].

8-5 VARIATIONS IN THE BASIC TRANSISTOR CIRCUIT

If you understand the reasons behind the basic transistor amplifier checks, you should have no trouble with various other types of amplifiers. As mentioned before, the dc voltage paths through these circuits are pretty nearly the same.

8-5-1 Amplifiers with No Base-to-Ground Resistors

Figure 8-12(a) shows an amplifier with no base-to-ground resistor. In this case a higher-value base-supply resistor R_B is needed. The current flow is very low through this resistor. The voltage is also determined by the emitter resistor. In fact, the voltages across the collector and emitter resistors are nearly in proportion to their resistances.

Troubleshooting this circuit is practically the same as troubleshooting the basic amplifier that has a base-to-ground resistor. All the tips apply. The only difference is that there is no ground resistor that can open up to raise the base voltage. If the base voltage rises above normal and if there is an input coupling capacitor, the transistor is more likely to be bad. Check the transistor first, then check the coupling capacitor. In the next chapter we find out how to check the circuit if there is no coupling capacitor.

8-5-2 Amplifiers with No Emitter Resistors

You will find many circuits with the emitter-to-ground resistor missing, as in Fig. 8-12(b). In these circuits you may have difficulty checking the base voltage. It is said that transistors respond to current changes rather than to voltage changes. In the basic amplifier the current changes were associated with substantial base and emitter voltage changes caused by the emitter resistor. Without the emitter resistor, the base voltage changes are going to be very small. Large changes in the collector voltage can result from small or no changes at the base.

However, the *BE* path does have some resistance, and some voltage does build up at the base. The circuit can be checked out in the same way as the basic amplifier. The emitter voltage is zero volts under all circumstances. The collector voltage should rise when the base goes down and fall when the base goes up.

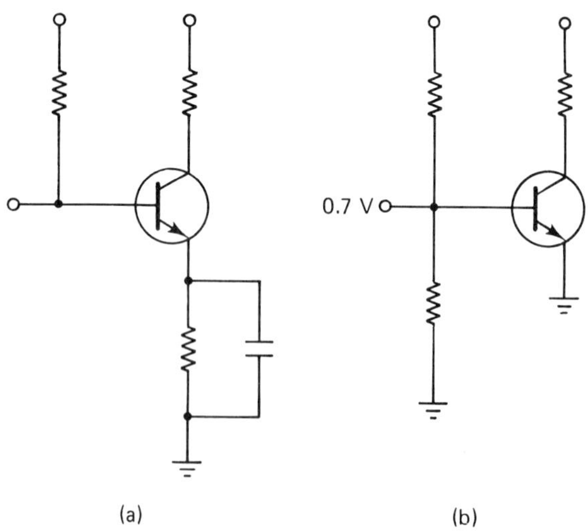

Figure 8-12 (a) Amplifier without a base-to-ground resistor. (b) Amplifier without an emitter resistor. Troubleshooting these variations is basically the same as for the basic amplifier shown in Fig. 8-11.

1. You can ground the base to check the transistor. See if the collector voltage rises when you ground the base.
2. If you have more than 0.9 V at the base, the transistor is bad.
3. If you have 0 V at the base in this circuit, the collector voltage should be above normal. If so, the base resistor could be open. Check the transistor in-circuit with an ohmmeter. If the transistor checks good, the resistor is almost sure to be open.
4. If collector voltage is not above normal when the base is 0, the transistor could be bad. Check the transistor in-circuit with an ohmmeter. If the transistor checks good, the trouble could be in the collector circuit.

8-6 AMPLIFIER DISTORTION

The basic amplifier should not have much distortion. That is, the output waveform should look pretty much like the input waveform, though amplified and inverted. This means that as the input voltage changes, the output voltage should change proportionately. For instance, if the amplifier gain is 10, for every tenth of a volt change at the input, the output voltage should change a full volt—no more, no less. But suppose that when you increase the voltage from 1.9 V to 2.0 V, and all you get is a change from, say, 19 V to 19.5 V, an increase of 0.5 V instead of a full 1.0 V. The gain drops off at this higher input voltage. We now say that the change is *nonlinear.* The amplifier does not maintain its normal gain at the higher input levels.

Perfect linearity is imposssible to achieve in the basic amplifier. The transistor gain falls off when its *BE* voltage and current is too high. We say that the transistor *saturates* at the high input levels. The transistor gain also falls off when the *BE* voltage and current is too low. We say that it *cuts off* when this happens.

Figure 8-13(a) shows what can happen when an amplifier with a 10-V peak-to-peak sine-wave output is overdriven. The ideal output is shown with dashed lines, and the actual output is shown with solid lines. Note how the gain drops off at the highest and lowest parts of the voltage swings. But notice that the actual wave follows the ideal wave pretty closely near the middle of the voltage swings, when the voltage swings are smaller.

Figure 8-13(b) shows what happens when you feed a much smaller signal to the base. The output is now reduced to 5 V peak-to-peak, as shown by the solid line. But look at what happens to the actual output wave; it is almost perfect—no distortion. It does not reach the saturation and cutoff levels as shown by the distorted wave in dashed lines.

In this case distortion was caused by *overdriving* the input. Both the top and the bottom of the wave are flattened out. This type of distortion can also occur when the collector voltage is too low. They say that the collector is more easily driven into saturation and cutoff when the collector voltage is too low.

In some amplifiers, distortion occurs when the base voltage is too low, or as we say, "incorrectly biased." Suppose that you have a very low *BE* voltage drop that is more or less fixed at 0.5 V by a circuit defect, and you have a 1.0-V p-p signal input. The upward swing will be amplified okay, but the downward swing will cut off the

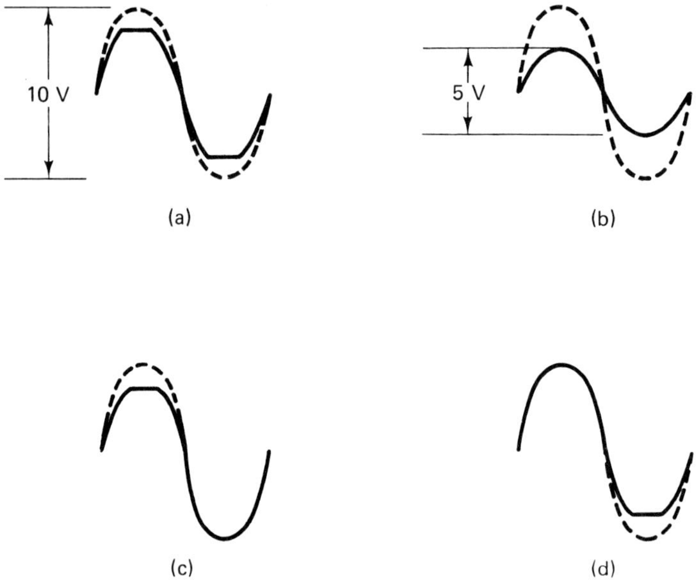

Figure 8-13 Amplifier distortion. (a) The solid line shows the distorted wave. This can be caused when the input wave is too big. (b) A smaller input to the same amplifier shows less distortion. (c) Insufficient dc voltage across *BE* will cause the collector voltage to flatten out (cut off) at the top. (d) Excessive *BE* voltage causes the collector wave to flatten out (saturate) at the bottom.

transistor and put maximum voltage on the collector for a large part of the collector upswing. (Recall that the collector voltage is inverted.) Figure 8-13(c) shows a sine-wave input and output for the amplifier with bias fixed too low. Note that the top part of the waveform is flattened out.

Distortion also occurs when a *BE* voltage is fixed too high by a circuit defect. In this case the collector saturates on the upswing of the input. This causes the bottom part of the collector waveform to flatten out [see Fig. 8-13(d)].

The information above shows why many technicians rely so much on the voltmeter. The dc voltages enable them to ''see'' the waveform without using a scope. Generally speaking, if you have a bad dc voltage on a transistor, you are going to have either distortion or abnormal gain.

So far, we looked at the external faults around the transistor. Do not forget the transistor itself.

The symptoms for distortion in various types of equipment will be taken up in later chapters. Knowing the symptoms would allow you to ''see'' the waveforms without a scope.

8-7 TROUBLESHOOTING TRANSFORMER-COUPLED AMPLIFIERS

Some amplifiers use transformers rather than resistors in their input and output circuits (see Fig. 8-14). The base and emitter circuits of this amplifier still need the same

TIP 8-10

In the basic amplifier, distortion occurs when:

a. The input signal is too big and overdrives the amplifier. This results in a waveform that is flattened out at the top and bottom.
b. The collector voltage is too low and is more easily saturated and cut off. This results in a waveform that is flattened out at the top and bottom.
c. The *BE* voltage is fixed too low by a circuit defect and tends to cut off the transistor. This results in a output waveform that is flattened at the top.
d. The *BE* voltage is fixed too high by a circuit and tends to saturate the transistor. This results in a waveform that is flattened at the bottom.

TIP 8-11

Distortion can be caused by a defective transistor.

resistors to maintain the correct voltage. We already know how to check these resistors. Let's look at the transformers.

If the input-transformer secondary opens, the dc base voltage will rise. If it shorts, the dc base voltage will change very little, if at all. Recall that coils drop very little dc voltage. Use a scope to check for signal at the primary and secondary. Jumping a capacitor across the transformer from primary to secondary may restore a signal to some extent. If it does, the transformer is probably bad.

The output-transformer primary winding at the collector has little or no dc voltage drop, so the collector has practically the full supply voltage on it. Some schematics may not show the collector voltage; they expect you to know that the collector voltage equals the supply voltage.

If the output-transformer primary opens, the collector will lose its dc voltage. If it shorts, there will be no evidence of anything wrong at the collector if you check it with a dc meter. Check it with a scope. If there is no signal, check for it at the emitter with capacitor disconnected. If you have a signal here, the transformer is shorted.

A short in the secondary circuit might also kill, or subtantially weaken, the signal. In this case, disconnect one of the secondary leads of the transformer to free it of the short. If you regain the signal at the collector, the transformer secondary circuit is bad.

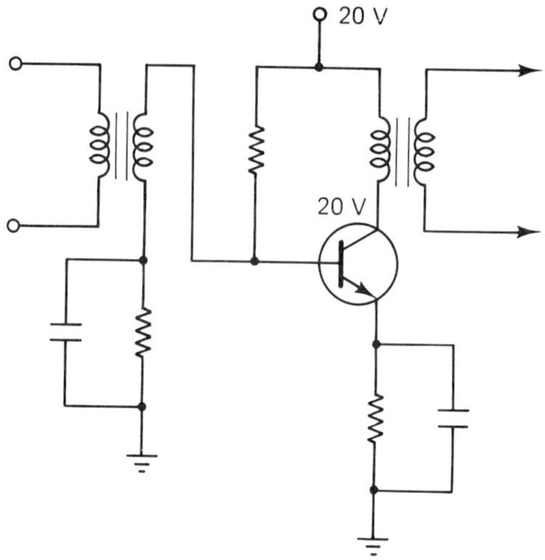

Figure 8-14 Transformer-coupled amplifier. There should be very little dc voltage drop across the individual windings. Any changes in the dc voltages in the source and load circuits should not affect the dc voltages in this circuit.

If neither a scope nor a substitute transformer is available, insert a coil in series with the transformer. If you now obtain a signal of sorts at the collector, the transformer is shorted. Preferably, the coil should have about half the size and weight of the suspected transformer. Actually, even a resistor having about five times the resistance and wattage of the emitter resistor would do.

In addition to checking for a signal, if you cannot get the test resistor to drop any voltage, the transistor could be open. Recheck the base and emitter voltages before you draw any conclusions (see Tip 8-3).

TIP 8-12

An open output-transformer primary will cause a loss of dc collector voltage. A shorted primary will cause a loss of signal but no dc voltage loss at the collector. A scope will detect the loss. A signal at the emitter with its bypass capacitor disconnected indicates a bad transformer. Open a secondary lead to see if the secondary circuit is causing a loss of signal at the collector.

You can substitute the primary with another transformer, a coil about half the size and weight of the original transformer, or even a resistor with about five times the resistance and wattage of the emitter resistor will do. Now, if you get a signal at the collector, the transformer is bad. You can also short the *BE* terminals; the collector voltage should rise. A 1-W resistor should handle low-power transistors. A 10-W resistor should handle most power transistors.

8-8 TROUBLESHOOTING THE EMITTER FOLLOWER

In Fig. 8-15 we have an emitter follower. The collector is connected directly to the supply line. The emitter resistor value is high enough to take about half the supply voltage. The dc voltage at the collector does not change, nor should there be a signal there. The emitter should have the signal. The main dc checks are made in the base and emitter circuits. For these circuits, the troubleshooting steps are similar to those taken for the basic amplifier. The dc wiring is almost the same as for the basic amplifier. There might even be a dropping resistor in the collector circuit. However, it would be bypassed by the collector capacitor and would be considered a supply-line resistor. Check its voltage drop if you have this resistor.

8-9 TROUBLESHOOTING BASIC TRANSISTOR SWITCHES

Besides the basic transistor amplifier that is widely used in radio and TV transmission, the transistor switch is finding ever wider use in all fields of electronics, especially in computer circuits.

8-9-1 The Transistor Switch

Figure 8-16(a) shows a transistor switch in series with the load. When the base voltage goes high (0.7 V), the collector goes down to nearly 0 V (about 0.2 V). The collector is said to be saturated. The low collector voltage turns on the load, since it now has nearly the full supply voltage on it.

When the base voltage goes low, the collector voltage goes up to nearly the full supply voltage. The load now has nearly 0 V, and is turned off. (These voltages will vary with the circuit parts and the transistor used.)

Figure 8-16(b) shows a transistor switch that shunts the load. In this case the load is bypassed to shut it off. This is accomplished by putting 0.7 V on the base, thus driving the collector down to nearly 0 V. To turn on the load, the input voltage must go low, below 0.7 V, to bring the collector voltage up into the operating range of the load.

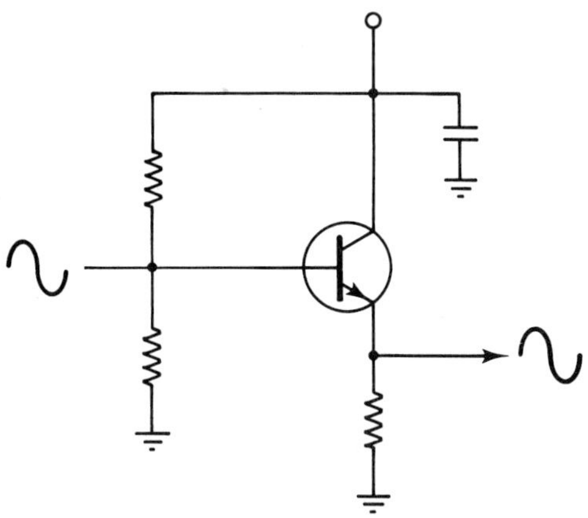

Figure 8-15 Troubleshooting the emitter follower is about the same as for the basic amplifier. The dc voltage checks for the base and emitter are the same. The collector has the supply voltage. It is bypassed for signal.

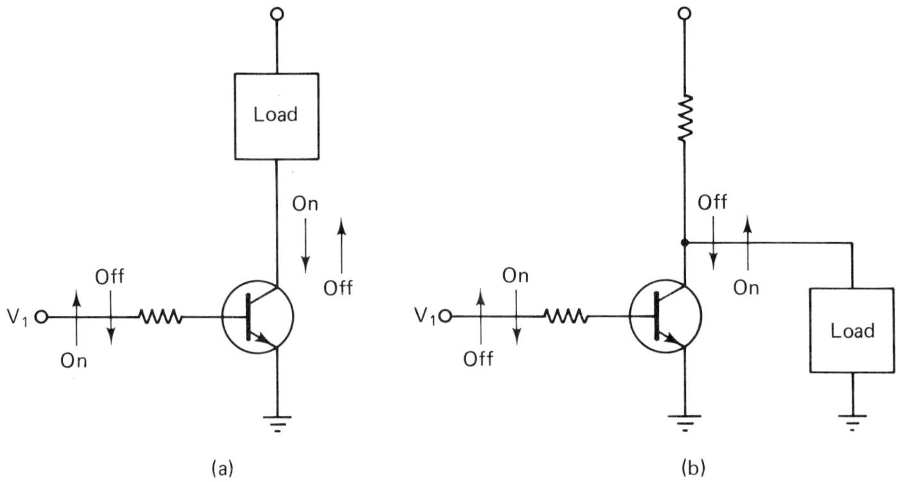

Figure 8-16 Basic transistor switches: (a) series load and (b) shunt load.

8-9-2 Checking a Series Transistor Switch

To troubleshoot a series transistor switch, you can follow the suggestions below, more or less in the sequence given [refer to Fig. 8-16(a)].

Check 1. The first step might be to check the load by grounding the collector. If this does not turn on the load, measure the supply voltage. If the supply voltage is good, the load is bad.

Check 2. If the load is good, measure the base, collector, and input voltage V_1. If this last voltage is okay and high, the voltage at the base should be about 0.7 V, and the collector should have nearly 0 V, perhaps about 0.2 V. Otherwise the transistor is probably bad, or the base resistor value could be too high.

Check 3. If the input voltage is okay and low, the base should be low, say about 0.1 to 0.4 V, and the collector should have the maximum supply voltage on it. Otherwise, the transistor is probably bad.

Check 4. If the switch cannot be turned off, try grounding the base. If this does not turn it off, the transistor is bad.

Check 5. If the switch cannot be turned on and V_1 is high, jump the base resistor with one of equal value. If this does not turn on the load, the transistor may be bad. If it does turn on, the resistor could be bad.

Check 6. Check the base resistor with an ohmmeter.

8-9-3 Checking a Shunt Transistor Switch

There are three major parts to check out when troubleshooting a shunt transistor switch: the load, the collector resistor, R_C, and the transistor. The following checks are helpful.

Check 1. If the collector voltage goes higher than normal, the load could be open or the collector resistor could be shorted. In this case, if the collector voltage can be driven down by raising the input voltage, the load is open.

Check 2. If the collector voltage cannot be raised, either the collector resistor R_C is open or the transistor-load parallel circuit is shorted. Check both resistances.

Check 3. If the resistance readings do not show anything definite, disconnect the load or the transistor. The load may be easier to disconnect. If the load is disconnected and you can raise the collector voltage, the load is shorted.

Check 4. If the transistor is easier to disconnect, do so. If the load can now be operated with the power turned on, the transistor is shorted.

Check 5. If the load cannot be turned off, either the transistor or the input circuit is bad. Check the input and collector circuit voltages; the off-and-on voltages should be the opposite of those you encountered in the series transistor switch.

Check 6. If the input voltage at the base resistor is good, you can jump the base resistor with one of equal value. If this does not turn the load off, the transistor is bad.

8-9-4 A Pulse-Circuit Switch

Sometimes a transistor switch is turned on and off repeatedly with a pulsing voltage. The transistor may not be able to react fast enough to the fast-changing pulse voltages. The transistor switch in Figure 8-17 has a speedup capacitor across the input resistor. It bypasses the resistor and the voltage at the base reaches maximum or minimum peak much faster.

This circuit is checked out like the basic transistor switch. In addition, a scope check on the base should show sharp peaks on the leading downgoing edges of the waveform, as shown. A below-normal dc drop across the resistor would indicate a shorted capacitor. If a scope is not available, jump the capacitor with a good one to see whether it is open.

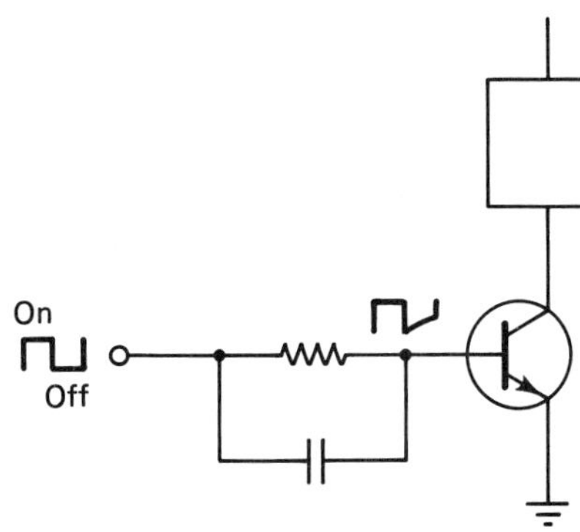

Figure 8-17 The capacitor in this transistor switch speeds up the voltage at the base.

8-10 CHECKING THE FIELD-EFFECT TRANSISTOR

The transistors we have worked on up to now are called bipolar transistors. The *field-effect transitor* (FET) is another type you will find occasionally in high-frequency, low-noise, low-distortion applications. There are several types of FETs. We look first at the junction (JFET) type.

Figure 8-18 shows a basic N-channel JFET amplifier circuit. It is similar to the basic NPN transistor amplifier we already worked on. The *drain D*, which has a high positive voltage, replaces the collector, the *gate G* replaces the base, and the *source S* replaces the emitter.

The operation is similar, too. The gate is the input that controls the drain-to-source voltage. The output at the drain is amplified and inverted. The source voltage "follows" the gate voltage.

There are two big differences between bipolar transistors and FETs. First, the FET has a very high input resistance. Second, unlike the bipolar transistor, a zero-volt drop between the gate and the source (the *GS* voltage) will cause maximum current to flow through the major path from drain to source (the *DS* path). To keep this current within normal limits, a negative voltage is placed on the gate. This restriction of current by the negative voltage is called the *field effect*. This negative voltage is measured from the +5.0 V at the source terminal. Measured from ground, the gate has 0 V. Except for the polarity, this reminds us of the *BE* voltage of the NPN transistor.

To check this circuit it would be helpful to know that the gate forms a reverse-biased diode with the source and the drain (see Chapter 5). Note the direction of the diode arrow in the illustration. You have hi–lo resistance readings between the gate and other two terminals. When checked out-of-circuit the high reading should be infinite. In-circuit, it should equal the value of the gate resistor, R_G, plus R_S. Also take the drain-to-source resistance reading. It should measure lo–lo.

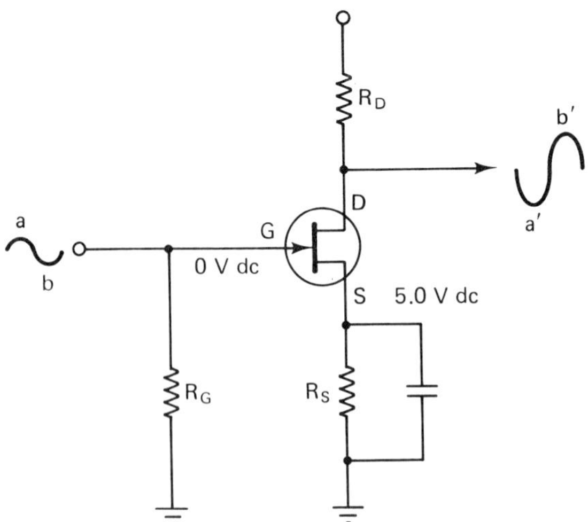

Figure 8-18 Typical JFET amplifier. Note the 5-V dc gate-to-source voltage and the gate 0-V dc voltage.

The gate resistor, R_G, has a very high resistance, in the megohm range. This is desirable, since it does not load down the source circuit. With an analog meter, you have to take the higher of two readings, reversing the ohmmeter readings.

The drain resistor, R_D, is in series with the very low drain-to-source resistance and R_S. The resistance across it should not measure more than its specified value.

A milliameter check will reveal whether the drain-to-source transistor path is open. Connect the meter between source and ground. ''No current'' indicates an open drain-to-source path.

Connect the meter between drain and ground. No current should flow if the drain resistor is open. If it is partially open, divide the supply voltage by the current ($R = V/I$).

If source resistor, R_S, has very low resitance compared with the rest of the circuit, you should get a fairly accurate reading with an ohmmeter. Wait for a steady reading before drawing any conclusions.

TIP 8-13

To check a basic N-channel FET amplifier, temporarily short the source terminal to ground. The drain voltage should fall.

As you can see, the FET basic amplifier is pretty much the same as the NPN amplifier. There are also versions that are similar to the common-emitter and the common-collector amplifier. They are the common-source and the common-drain amplifiers respectively.

The JFET we just looked at is called the N-channel JFET. There is also a P-channel JFET that corresponds to the PNP bipolar transistor. Note the direction of the arrow in Fig. 8-19(a). This type has the drain connected to the negative-supply side or the source connected to the positive supply side.

There is also a MOSFET type, which has infinite resistance, both ways, between gate and the other two terminals [see Fig. 8-19(b)]. It works in very much the same way as the JFET, except that no current flows until a positive voltage is placed on the gate. This type is called an *enhancement mode* MOSFET. There is also a *depletion–enhancement mode* [see Fig. 8-19(c)]. It is similar to the FET in that current flows at zero gate voltage.

Care should be taken when working on MOSFETs that are identified as non-gate protected. They can be damaged by electrostatic charges that are always present to varying degrees in the work area. Before working on a MOSFET circuit, you should touch ground to discharge yourself. Use grounded tools to stay discharged.

MOSFETs can also be damaged by sudden surge voltages (transients) that occur when a unit is inserted or pulled out of a circuit. Turn off the power before inserting, pulling out, or cutting out a unit in a MOSFET circuit.

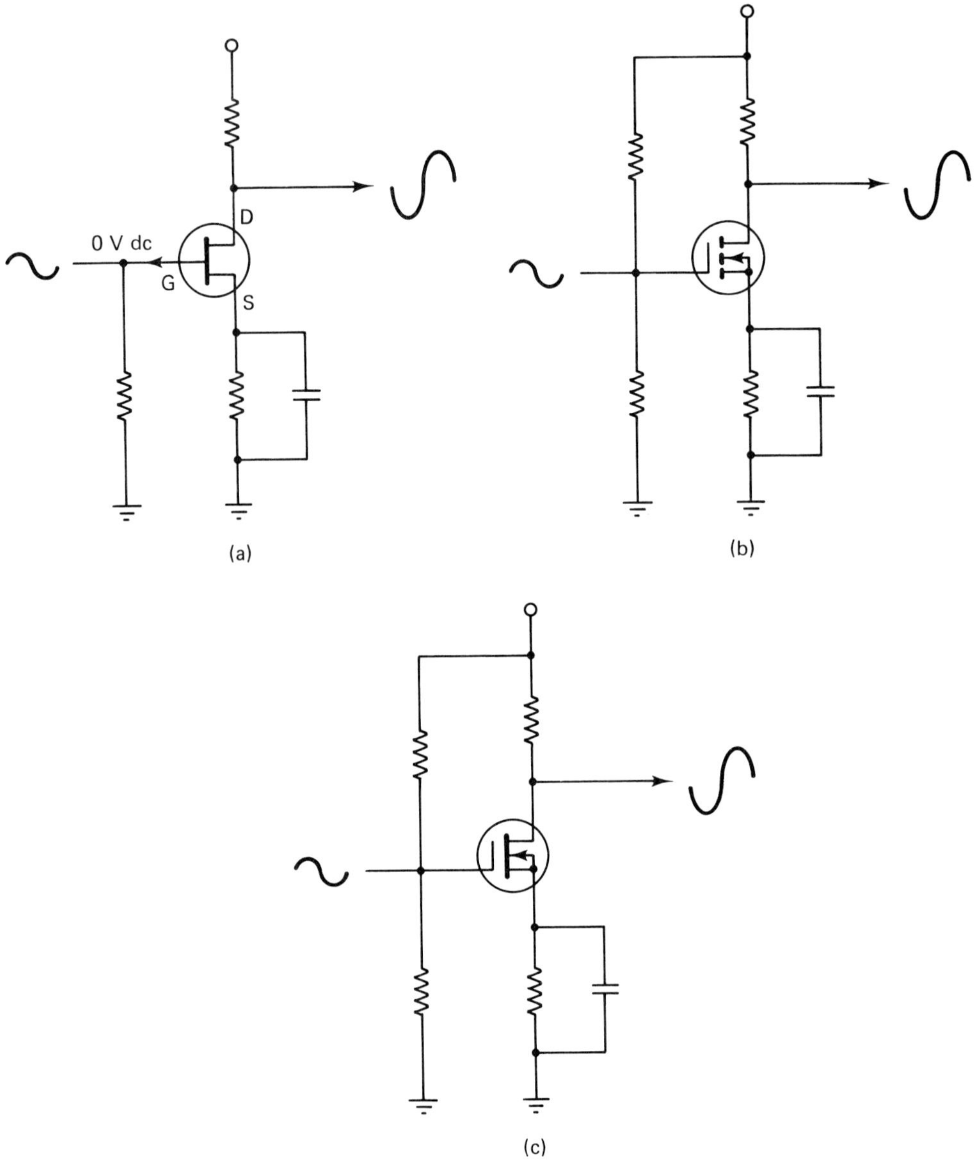

Figure 8-19 Field-effect transistors: (a) P-channel JFET; (b) Enhancement mode MOSFET; (c) depletion–enhancement mode MOSFET.

8-11 STUDYING A VIDEO OUTPUT AMPLIFIER SCHEMATIC

Figure 8-20 shows the schematic of an RCA PC-board circuit that is mounted on a color display tube socket. There are three transistor amplifiers, one for each basic color. They are very much like the basic amplifier you saw in this chapter. There are some added parts, but we need not go into the reason for them here. The main thing

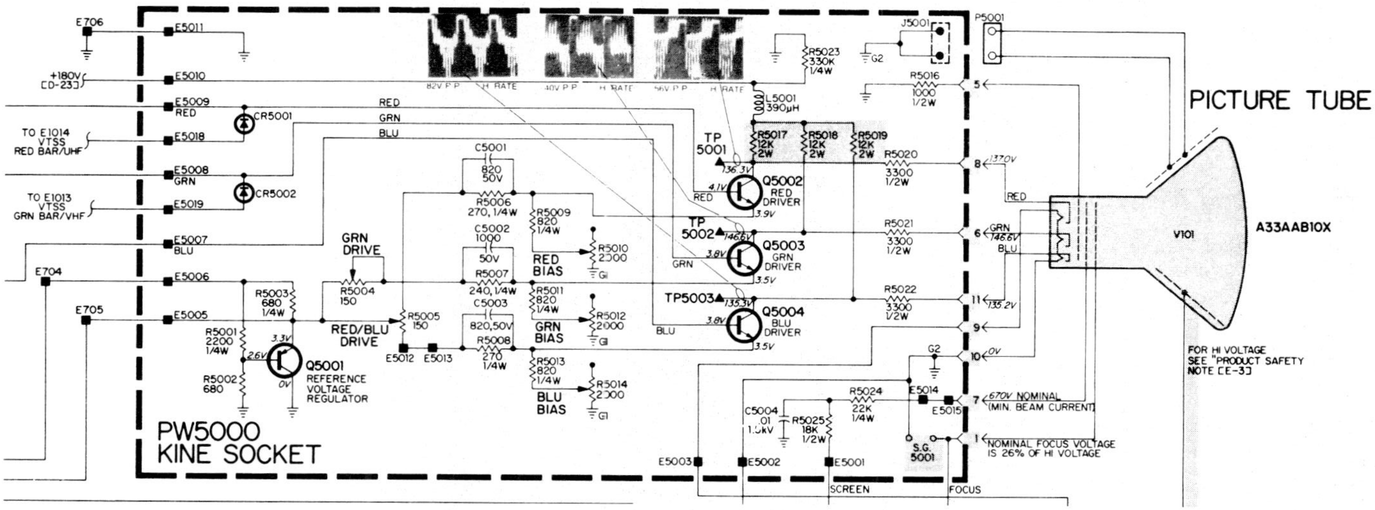

Figure 8-20 Actual RCA schematic shows basic transistor amplifier circuits. (Reprinted courtesy of Thomson Consumer Electronics)

now is to learn to locate the basic circuit parts in a schematic. If you can do this, you should be able to troubleshoot the circuits.

Let's just look at the Q_{5002} transistor amplifier on top. The other circuits are nearly the same. R_{5017} is the collector resistor. R_{5009}, R_{5010}, R_{5006}, and the Q_{5001} transistor circuit together make up the emitter resistance. The supply-to-base resistance is on the preceding page. It need not necessarily be a resistor; it could be a transistor or an IC circuit. R_{727} is the base-to-ground resistor.

Now that we have found the basic parts that are on the board, we can use the information in this chapter to troubleshoot it. Briefly, we can go through some of the possible faults and some of the possible causes.

If the collector voltage of Q_{5002} is too high, the transistor could be open. Other possible causes are an open in the base supply circuit outside the board; an open in the emitter circuit, including Q_{5001}; an excessively high voltage at connections E_{5006} or E_{5005}; an open in the base supply outside the board; a short in the base circuit connections to ground; or a short from a high-voltage point on the PC board or socket connections to the collector itself.

If the collector voltage is too low, we could have an open R_{5017} collector resistor; an open R_{727} base-to-ground resistor outside the board; a shorted base-to-supply circuit, also outside the board; an excessively high voltage on the E_{5018} connector; a shorted Q_{5001} in the emitter circuit; or an excessively low voltage at the E_{5005} connector.

SUMMARY OF SELECTED TIPS

Tip 8-2. In the basic transistor amplifier (class A), the collector voltage for maximum undistorted operation is about one-half the supply voltage. The emitter voltage is much less. The collector and emitter resistors divide their voltage drops in proportion to their respective resistances. The base voltage is about 0.7 V higher than the emitter voltage.

Tip 8-3. In a basic transistor amplifier circuit: When the base voltage goes up, the collector voltage goes down. When the base voltage goes down, the collector voltage goes up. In all other cases (in Section 8-1-7), the output voltage follows the input voltage.

The above is true for both NPN and PNP circuits and for both positive voltages and negative voltages.

Tip 8-4. The voltage drop across the base and emitter terminals in a basic transistor circuit is normally about 0.7 V.

Less than about 0.5 V difference between the forward-biased *BE* terminals indicates a shorted transistor. More than 0.9 V difference indicates an open transistor.

Tip 8-5. To check a low-power transistor, short the base to the emitter with a screwdriver, jumper, or test prod. (As you measure the collector with the red probe, short the *BE* terminals with the black probe.) If the collector voltage goes all the way up to the supply voltage, the transistor is okay and there is no dead short across the collector resistor.

Tip 8-6. To check a transistor, jump thc base supply resistor with one of approximately equal value. If the collector voltage goes down, the transistor is okay.

Tip 8-7. To verify a transistor check in a basic amplifier circuit, disconnect the base lead. Check all the transistor voltage readings with the base disconnected. The collector voltage should read high. The base voltage should be zero. The emitter voltage should equal zero on the lowest scale. If you do not find these voltages, the transistor is bad.

It is a good idea to recheck the transistor with an ohmmeter or transistor checker now. You will get a better reading this time.

Tip 8-9. If the base voltage in a basic amplifier is too high, the input coupling capacitor could be shorted. This could also bring down the collector voltage.

If the emitter voltage is shorted, it can pull down the base and collector voltages along with it. The collector voltage change may be much greater than the *BE* change.

WHAT YOU LEARNED TO DO

In this chapter you learned how to troubleshoot basic transistor circuits by reading the terminal voltages. You learned how to:

1. Identify the input and output terminals of a transistor.
2. Trace the major and minor dc voltage paths inside the transistors.
3. Trace the signal paths through a transistor.
4. Recognize faulty dc voltages on the terminals of a good transistor.
5. Check the *BE* voltage, and learned what it does to the emitter and collector voltages.
6. Troubleshoot a transistor path as you would a resistor or a diode, or as a transistor junction.
7. Quick-check transistors using the hi–lo–lo, lo–hi–hi methods.
8. Follow a procedure for troubleshooting the basic transistor amplifier.
9. Troubleshoot an emitter-follower circuit.
10. Troubleshoot a basic transistor switch circuit.
11. Troubleshoot a basic FET amplifier.

QUESTIONS

1. Is it possible with a scope to read both dc and signal voltages on the same line? Is it possible with a voltmeter to read the dc voltage on the same line?

2. The collector voltage of a basic amplifier reads above normal. What three parts in the circuit are most likely to be open or shorted? Which possibility will you tackle first? How would you eliminate each possibility? (Check out each possibility.)
3. The collector voltage of a basic amplifier reads below normal. What two parts in the circuit are most likely to be open or shorted? Which would you take on first? How would you eliminate each possibility?
4. The emitter voltage of a basic amplifier is found to be higher than normal. What two parts are most likely to be open or shorted? How would you eliminate each possibility?
5. If the emitter voltage is lower than normal, cite three parts that are most likely to go bad. Which of these is most likely to go bad? Which of these is least likely to go bad? How would you go about eliminating each possibility?
6. You find that the base voltage of a basic amplifier is too high. What two parts in the circuit are most likely to be bad? Which possibility would you check first? How would you eliminate each possibility?
7. Besides the parts in the circuit, what else can cause defective voltages to appear on the transistor terminals?
8. An amplifier has an emitter bypass capacitor. What happens to the dc voltages on each terminal when the capacitor shorts? When it opens? What happens to the signal voltages on each terminal when it opens? When it shorts?
9. What are the main differences between an NPN amplifier and a JFET N-channel ampifier?

PRACTICAL TROUBLESHOOTING EXPERIENCE

You will need a signal generator and scope as well as a multimeter for this job. You may need a power supply if the equipment has a shorted power-output transistor or defective power supply.

Obtain a discarded board or chassis that has small transistor amplifiers. Check the power supply. If it is a simple type and is not working, you can repair it. Otherwise, use a power supply. You may have to disconnect the power-output transistors or circuits and the main power-supply line if they draw excessive current (see Chapter 6).

Things will go a lot faster if you have a schematic. All the procedures below are useful ones that you will be using at one time or another in actual troubleshooting.

> *Note:* It is a good idea to go through the steps below on a schematic before working on the actual chassis or board. Any similar schematic will do. In fact, if you cannot find an old chassis, just going through the steps on a schematic will be very helpful.

1. Find a basic NPN amplifier. Check the transistor in-circuit with an ohmmeter. If it is not good, come back to it later. Find another circuit with a good transistor in it.
2. If you do not have a schematic for this circuit, draw one. Label the parts: R_B, R_G, R_C, and R_E.
3. If the power supply is not working, find the dropping resistor to a low-voltage supply line that you want to work on and disconnect it. This will isolate the shorted power suppply and the power-output circuit.

   Connect an external power supply to the low-voltage line. Bring up the voltage slowly. Monitor the collector voltage and bring it up slowly to normal. If a part starts smoking, cut the voltage in half and look for the shorted part that is in series with it. The bad part should have a drastic loss of voltage across it.

   If it is a transistor, check its base and emitter voltages (refer to Tips 8-3 and 8-4). If the voltages look bad, turn off the power and make resistance checks. Remove the base lead and try Tip 8-7.

   If the transistor is protected by input and output capacitors, check them and replace the transistor if you want to. If there are no coupling capacitors, the problem may be beyond our capabilities at this point—cut out the burned part.
4. If nothing is smoking, check out various transistors using the hi–lo–lo and lo–hi–hi approach of Section 8-1-9.
5. Check out various transistors using the selected tips listed near the end of this chapter. Try out as many tips as are applicable. Do not forget that defective dc voltages can be found in a good transistor circuit if it is not protected by coupling capacitors. However, the tips will indicate whether or not a particular transistor is bad.

9

Troubleshooting Basic Amplifier Stages

EXPERIENCE REQUIRED

Before you can make use of the troubleshooting information in this chapter, you should know something about the following:

1. How to trace signal and dc paths through a single transistor amplifier circuit.
2. That the base is never an output, the collector inverts the base voltage, and the emitter follows the base voltage.
3. That coupling capacitors and transformers (other than autotransformers) block dc voltages and pass ac signal voltage.
4. That voltages in a good transistor circuit can be thrown off by a defect in another circuit.
5. That diodes pass conventional current and voltages only in the direction of the arrow (when forward biased) and that they rectify ac voltages.
6. How to draw solder out of PC-board connections with a solder sucker or wick.
7. How to apply the basic rules for shorts and opens in a transistor circuit.

Knowing how to use the signal tracer would be helpful. Its use is described briefly in this chapter.

9-1 BASIC SIGNAL TRACING

A first-time look at the insides of an electronic system may reveal a huge conglomeration of chips, resistors, transistors, capacitors, and coils massed together to produce a desired output. To the beginner this can be a formidable sight. To make things easier, break the system down into smaller sections. Divide the system into sections that have particular jobs or functions. The bad section can be identified by looking at the output symptoms to see what functions are missing or defective. In later chapters we identify the symptoms. Chapter 1 gave you the overall approach. In this chapter we trace the signals through the circuits.

9-1-1 Tracing Signals from Section to Section

Figure 9-1 shows signals passing from section to section. These signals can be traced with a scope or signal tracer. The signal tracer is a test instrument with a probe that picks up the signal at various points in the system. It amplifies the signal and feeds it through a built-in speaker. Thus you can listen for the signal at the output and input of a suspected section.

While you can hear low-frequency nonaudio signals as tones or buzz with a signal tracer, you can do better with a scope. The scope will pick up distorted audio signals and give you clues as to what is causing them. In either case, if you have a good input and a bad output, the section is bad. This, basically, is signal tracing.

The illustration shows that the signal is coming from a signal generator at the system input. The generator is used when the radio or TV-station signal, or any other normal signal source, is missing or not suitable for testing purposes. Most often, however, the transmitting station signal is sufficient.

A very popular signal generator of sorts for use in low-voltage low-power systems is a test prod, screwdriver, or the tip of your finger. This ''test instrument'' picks up 60-Hz ac signals that are always present in the surrounding space and transfers them to the circuit.

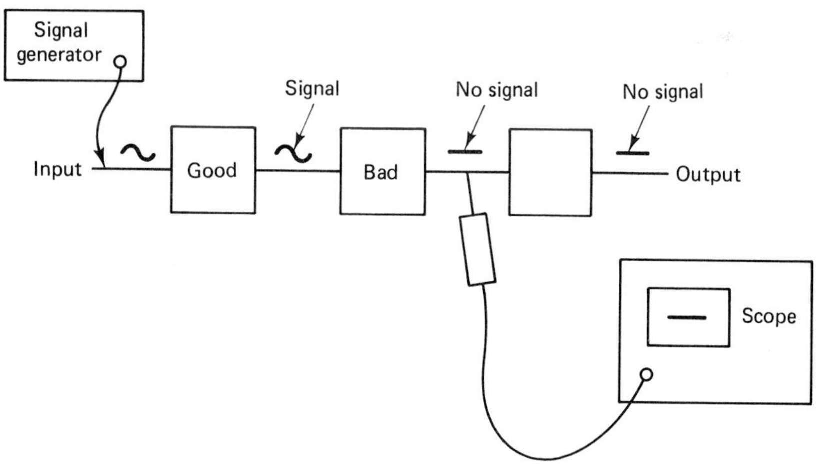

Figure 9-1 Tracing a signal path through several sections with a signal generator and oscilloscope. The bad section has a good input signal, but no output.

Caution: If you are working on a power amplifier, turn down the volume control if you are going to touch a point near the input. If you have no volume or level control after the point you are about to touch, do not touch it. You might blow out an output transistor or device. Also: Hands-off high-voltage circuits.

> TIP 9-1
>
> If an amplifier has no output, turn the volume or level control down low and touch the base of a transistor with your finger. If you get a response at the system output, the circuits after that point are probably okay. You can also touch the base with a screwdriver or a test prod for this purpose. **Caution:** You can blow a speaker if the volume control is not turned down low.

9-1-2 Tracing Signals: Its Limitations

Within each section you may see signal paths going from transistor circuit to transistor circuit. These circuits are called *stages.* Each stage changes the waveform in one way or another, so we can check it with a scope or signal tracer. We can then trace the signal from stage to stage until we find the stage that is not doing what it is supposed to do. This is the defective stage.

This sounds easy. Unfortunately, the method works only if the stages are isolated from each other with coupling capacitors or transformers. If the the stages are not isolated, the bad stage will transmit bad dc voltages to the other stages. These stages will be disabled and you will have to search back until you find a signal. In later chapters we work on certain types of feedback control circuits that are impossible to check by signal tracing. We have to rely on dc voltage checks.

9-2 TROUBLESHOOTING DIRECT-COUPLED AMPLIFIERS

In Chapter 8 we learned to check out an amplifier stage that had coupling capacitors. Now we are going to work on stages that have no coupling capacitors. They are called *direct-coupled amplifier* stages. They're affected by dc voltage changes in other stages. We have to keep this in mind when working on them.

9-2-1 Where's the Bad DC Voltage Coming From?

Let's examine a three-stage amplifier (see Fig. 9-2). To keep things simple we left out the bypass capacitors. They have no effect on dc voltages, anyway. We have Q_1, an NPN amplifier, feeding Q_2, a PNP amplifier, which in turn feeds Q_3, an NPN am-

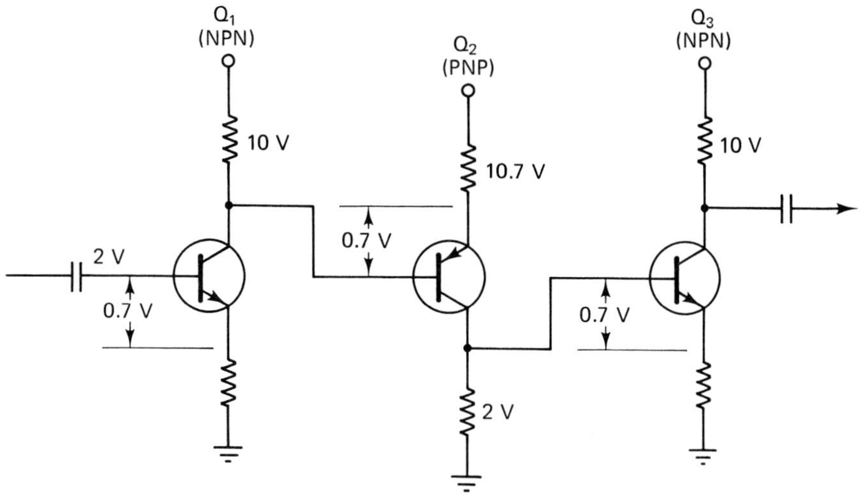

Figure 9-2 Three-stage direct-coupled amplifier. Note how the NPN and PNP voltages match-up.

plifier. This arrangement eliminates the need for coupling capacitors, as explained below.

Recall that the emitter of a PNP amplifier can be connected to the high-side voltage, and the collector can be connected to the ground side. The base voltage must therefore be high since it is only 0.7 V less than the high emitter voltage. This is why the base of the PNP transistor can be connected directly to the high 10-V collector of Q_1. Similarly, the low 2-V base of Q_3 can be connected directly to the required low collector voltage of the PNP transistor. No blocking capacitors are needed.

When checking these stages, we recall what happens when the base voltage is changed. The collector is inverted, and the emitter follows the base. This is in the forward direction. If it does this, the transistor is good. What happens in the backward direction? What happens to the input side when something goes wrong on the output side?

In the backward direction, the output terminals are hit first. That is, either the collector or emitter is affected first. Figure 9-3(a) shows an NPN basic amplifier (grounded emitter) that is fed back a higher-than-normal voltage on the collector. This voltage will tend to pull up the emitter and base voltages only a little. The important thing to note here is that the change in base voltage does not substantally affect the voltages in the preceding stage. Also note that the collector change is not the inverse of the base change. We can say, in this case, that the bad voltage is not passed back.

Now let us look at a PNP emitter follower. The one in Fig. 9-3(b) has a higher-than-normal voltage fed back to it. Note that both the collector and the base rise. The change in the base voltage is not enough to affect the voltages in the other stages. Also, the collector change is not the inverse of the base change.

Figure 9-3(c) shows a basic amplifier preceded by an emitter follower. In this case the voltages in the preceding stage are substantally affected by the bad voltage

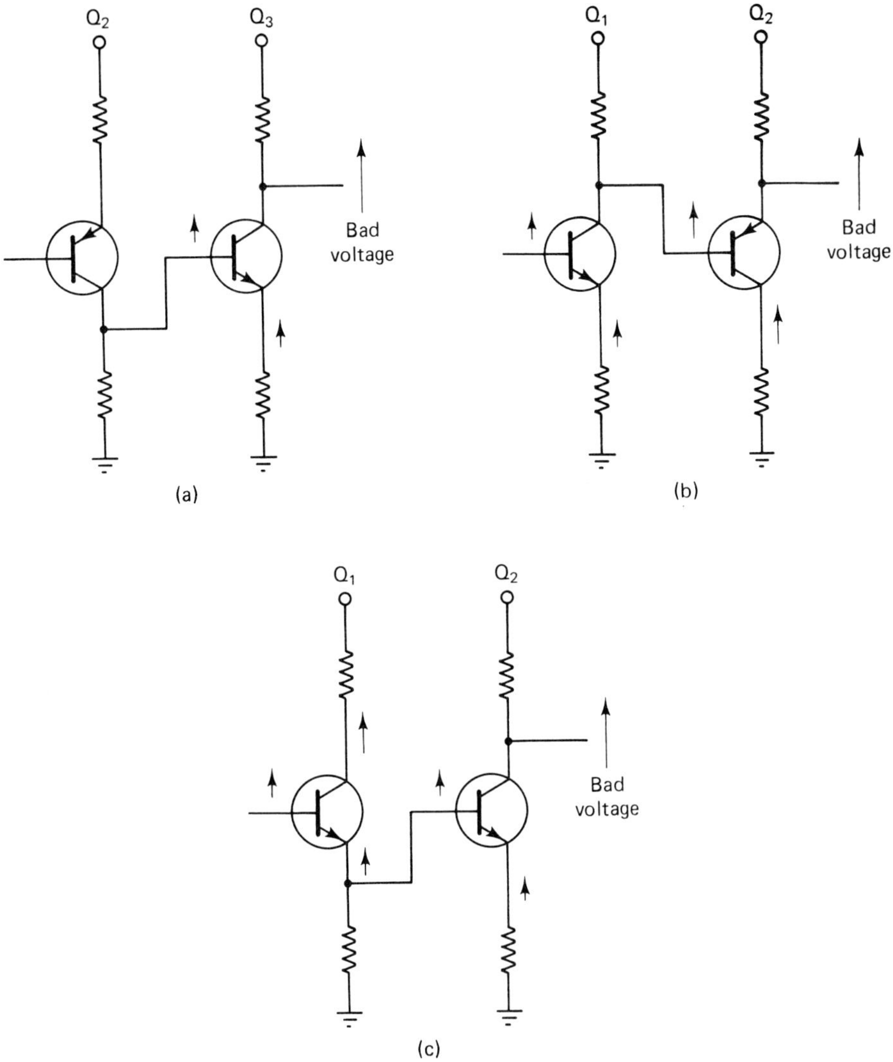

Figure 9-3 Defective voltages cannot be passed backwards significantly through more than one good stage, nor is the collector inverted in any case.

being fed back. However, note that the collector voltage change is not the inverse of the base change for both transistors.

Note that in the cases above, the collector voltage is not inverted when compared to the base voltage. This is what happens when the voltage is transferred backwards. In the forward direction, the collector voltage is the inverse of the base voltage.

Note also that a bad voltage is not substantially passed backward from the collector to either the base or the emitter terminals. Further investigation will show you that these backward voltage transfer characteristics are true for all types of amplifiers, NPN and PNP, as long as neither the collector nor the base is tied directly to ground or the supply line.

> **TIP 9-2**
>
> Defective dc voltages are transferred from stage to stage in the forward direction, with the collector voltage inverted (as long as neither base nor collector is tied directly to ground or supply line). This does not happen when the defective voltages are being transferred backwards.

Here is another tip that can help you trace bad voltages along the dc signal path. Looking at the cases above, where the preceding stage has a collector output, we see that the voltages in the preceding stage are not substantially altered.

> **TIP 9-3**
>
> A bad dc voltage cannot be substantially passed back from the collector to the emitter or base of a stage.

> **TIP 9-4**
>
> A bad voltage at the emitter may not appear to be transferred backward if the collector is directly tied to a regulated supply line. Look at the preceding or following stage.

We now have the information needed to help us track down bad dc voltages from stage to stage. Let's see how we can do this.

9-2-2 The Trouble Lies between Two Points

In Chapter 2 we learned that if you find a bad dc voltage between two good points in a series path, the trouble must lie somewhere in between. These two points were usually supply point and ground. Perhaps we can do something similar with signal paths.

Well, we know that dc voltages do not pass through coupling capacitors and transformers. Suppose that we have capacitors or transformers at two points in a signal path. If we find a bad dc voltage between these two points, we know that the trouble is not coming from some place outside these points. The trouble must lie between these two points.

> TIP 9-5
>
> If a bad dc voltage lies in a signal path between two coupling capacitors or transformers that are not autotransformers, the trouble lies between these two devices.

Figure 9-4 shows a four-stage dc signal path. It begins with an input coupling transformer and ends with an output coupling capacitor. This is a long path. Most transistor dc signal paths have fewer stages. Since there are so few, it does not really

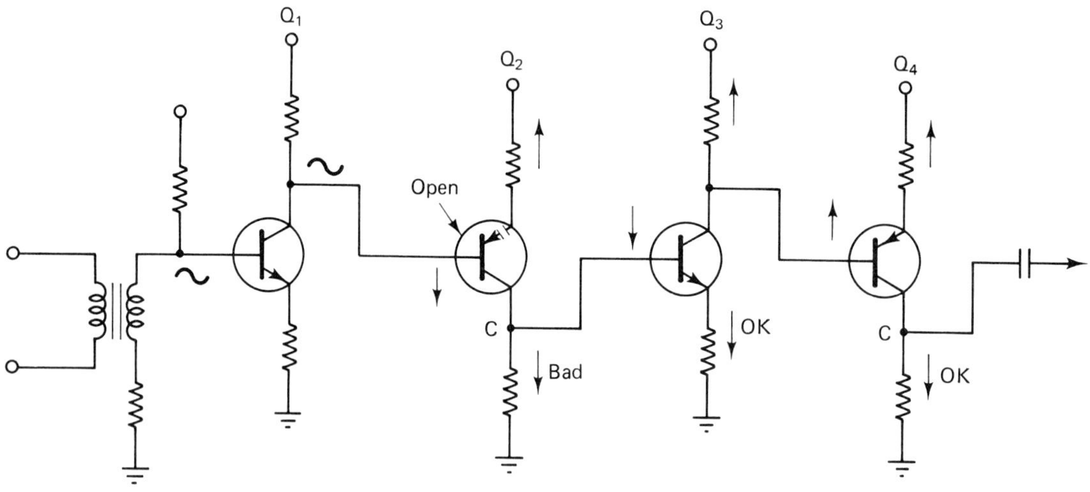

Figure 9-4 Troubleshooting a four-stage amplifier. The trouble is isolated between the coupling transformer and capacitor. Note how the bad voltage is transferred in the forward direction as expected.

matter where you start. Let's start at the end; it is easier to keep track of things that way.

Let's look at the collector voltage of Q_4. It is too high and is the inverse of the base voltage. Going over to the emitter, we see that it follows the base. Note the arrows in the illustration. The transistor is doing what it is supposed to be doing. It is passing forward the bad input voltage. Things look good, so far.

We go backward to the preceding stage, transistor Q_3. Here, too, the collector and emitter are doing what they should be doing to the base input, which is too low. This stage looks good, too.

Moving back again to the next preceding stage at Q_2, we finally find a transister that is not doing what it is supposed to do. The collector voltage is not inverted. Looks bad. Let's stop here. A closer check reveals that the *BE* voltage drop is more than 9.0 V. Q_2 is definitely bad (recall Tip 8-4). It has an open *BE* path.

Note that in this case Q_1 has a signal on it and looks okay; its dc voltages are okay. If Q_2 had a *BC* short, Q_1 will not look so good. In the latter case, check Q_2 first. Disconnect its base first, and Q_1 should have good voltages on it.

TIP 9-6

When checking dc signal paths, if a transistor passes a bad voltage forward; work backward, stage by stage, until you find a stage (the suspect stage) that does not pass a bad voltage forward. If the preceding stage has good voltages, the suspect stage is bad.

Otherwise, check the transistor in the suspected stage with an ohmmeter. If it checks bad, recheck it by disconnecting its base (Tip 8-7). If this produces good voltages in the preceding stage, the suspected stage is bad.

Here is another one for your bag of tricks:

TIP 9-7

When working on a dc signal path, you can short the *BE* terminals of a transistor at some point in the path. If the resulting voltage change does not pass forward to a second point down the line, something is wrong between the two points. Move forward and repeat the shorting until you find the bad stage.

9-3 DC VOLTAGES THAT SHOW THE EXISTENCE OF SIGNALS

We have a very important clue for checking the existence of signals in a multistage signal path. It does not require the use of a scope. Certain diode and transistor circuits develop dc voltages when signals pass through them. (We deal with these circuits in more detail in Chapters 11 and 12.) Right now we are interested in checking signal paths. As we learned in Chapter 6, a diode is capable of rectifying 60-Hz ac voltages to produce dc voltages.

Diodes in signal circuits can do the same thing. They rectify signals to produce pulsating dc voltages that may be completely or partially filtered. In any case they produce dc voltages only in the presence of signals. These dc voltages can be checked with a voltmeter.

If you can read a dc voltage on a signal-rectifying diode, the entire signal path up to it is probably okay [see Fig. 9-5(a)]. These diodes are called *detectors.* We'll come across them again in Chapter 11.

Besides detectors, certain transistor circuits can also produce signal-dependent dc voltages. Their *BE* junctions form diodes that rectify the signal voltages. In Fig. 9-5(b) the *BE* junction diode develops negative voltage at the base in the presence of signal. Note the negative-charge arrow. It is on the negative side of the capacitor. Compare this with the capacitor at the output of D_1 in Fig. 9-5(a). Resistor R has a

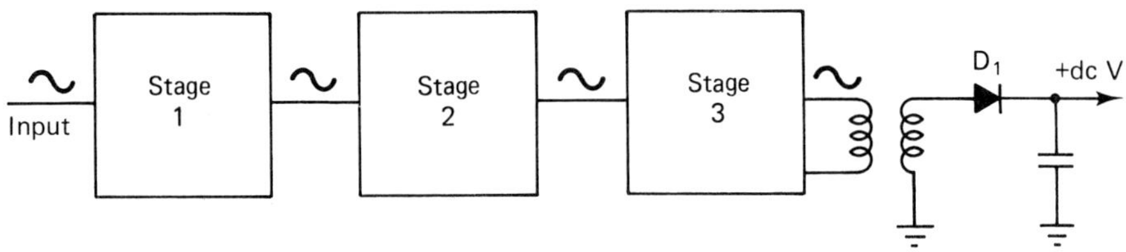

(a)

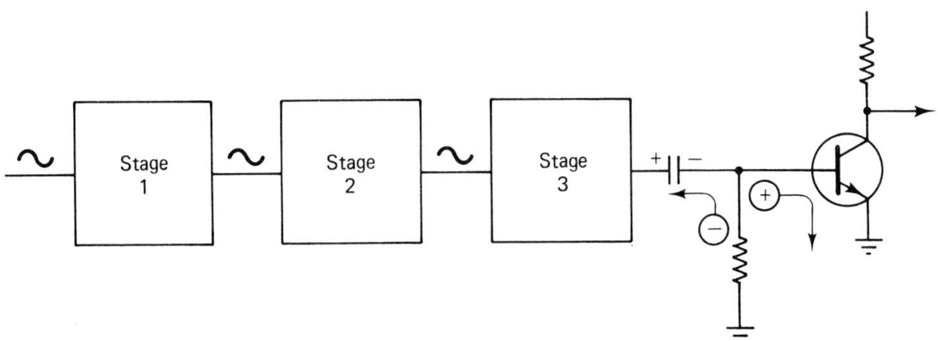

(b)

Figure 9-5 Signal-dependent dc voltages. (a) D_1 develops a dc voltage only when a signal is impressed on it. (b) This transistor develops a negative dc voltage only in the presence of signal.

high value; this limits the discharge current. This circuit will be reviewed in Chapter 12 when we take up "sync clippers" in display-tube circuits.

TIP 9-8

Certain signal-rectifying circuits such as detectors and sync clippers develop dc voltages in the presence of signals. To check for signal in these circuits, turn the signal on and off at a preceding stage. If the dc voltage at the rectifying circuit responds, the intervening stages are probably okay.

9-4 UNDESIRABLE SIGNALS IN BASIC AMPLIFIER STAGES

Besides nicely passing normal signals, multistage signal paths can be plagued with unwanted signals such as hum, interference, noise, and oscillations. We already learned how open power-supply filter capacitors can allow hum voltages to spread throughout the system. Let's find out about other types of hum and noise signals in ordinary amplifier stages.

9-4-1 Hum and Interference Problems

The 60-Hz ac power lines surround your equipment with fluctuating electromagnetic fields. You cannot escape them. They induce 60-Hz ac voltage or "hum" into the circuits. Anything conductive picks them up—even you. Just try touching your finger to an ungrounded scope probe. You will see a very large 60-Hz waveform. This should give you an idea of why hum pickup is such a big problem in amplifier stages. You can check for it by touching your scope probe to various points in the circuit. Adjust your scope to the lowest range possible for obtaining a readable wave. Later, you'll learn to jump capacitors to find the source of hum pickup.

The following tip is based on the rule that an open below raises the voltage above it.

TIP 9-9

When checking for hum or interference, look for open ground connections, among other things.

Before going into the amplifier stages, check all the external connections to the system. Check all the cables and connectors. Reverse the ac plugs if possible. Sometimes, reversing the ac plugs can solve the problem. If there are provisions for external connections to ground, try grounding them. Also try removing these ground connections. We will have more information on this in Chapter 10.

> **TIP 9-10**
>
> When checking for hum and other types of interference, check all the external connections first. If possible, reverse the ac plugs and reinsert them in other outlets. Remove or connect provisions for external grounding.

After checking the external connectons, look into the input amplifier stage; it is a primary suspect. The hum picked up here is amplified by the later stages.

> **TIP 9-11**
>
> Remember to check the input amplifier for hum and interference. The low-level interference voltages there will be amplified by the later stages.

The input circuits are carefully designed to handle low-level hum and other types of interference. You might find shielded wire connected to their inputs. The shield is grounded to short out the hum it picks up. If its ground connection opens, its hum voltage will rise and couple over to the inner wire. As you might guess, the shield and the inner wire form a capacitor.

Another cause of hum pickup is a weak transistor stage. The signal output of the stage is reduced, but not the hum pickup at the next stage. In technical language, they say that the *signal-to-noise ratio* is reduced. When you crank up the user gain control to increase your signal level, the noise or hum comes out strong.

The scope is the best instrument for tracing hum. Most signals have a higher frequency than the 60-Hz hum. If the scope is adjusted to read the signal voltage and is adjusted for 60 Hz (about 16.0 ms), the signal waves will be so crowded that only a straight bar across the screen can be seen [Fig. 9-6(a)]. A wavy bar would indicate excessive hum [see Fig. 9-6(b)]. If so, trace backward until you locate the stage where it is picked up. Readjust the scope voltage control at each stage. The last stage that had hum is the culprit.

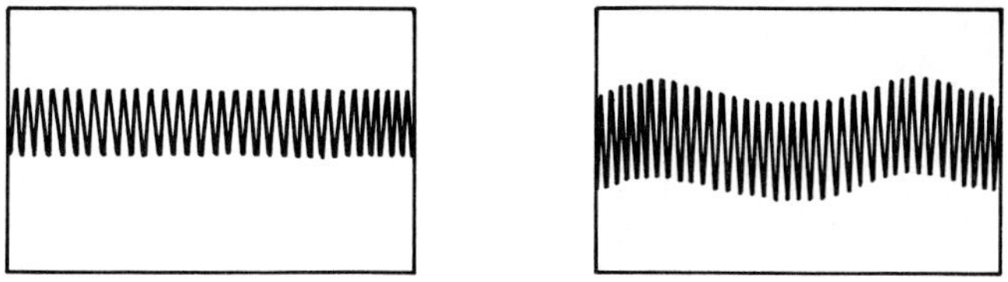

(a) (b)

Figure 9-6 You can check for 60-Hz hum with a scope. (a) Here you have good signal and no hum. (b) You have hum here.

TIP 9-12

When checking hum problems, look for incorrect voltages on the transistor terminals. Insufficient amplification will reduce the signal-to-noise ratio when the noise is picked up by a later stage.

You can also use a signal tracer to locate hum. Check the output first to hear the relation between hum and signal. Then work back to find the bad stage. As a finer check, note whether the hum or the signal is being amplified more as you move from stage to stage.

If no test instrument is available, the speaker or the video display tube of the system will monitor the hum for you. Hum is seen as a wide vertical bar across the display screen. You can then check out a stage by disabling it and noting whether the hum goes away. Do this by shorting the *BE* terminals. Try this on several stages until you find the hum source. This method is limited to stages with coupling capacitors or transformers.

You can use a bypass capacitor to short out a signal. This has the advantage of cutting out the interference without disabling the stages. If hum is lost at the output when you short out a point, the trouble lies before the shorted point. Check backwards. If the signal is not lost when a point is shorted, check forwards. Use a capacitor that is about 10 times bigger than the largest-value coupling capacitor in the signal path. The idea is that if the capacitance is big enough to couple a signal, a capacitance much larger than it is should be big enough to bypass it.

Another advantage of this method is that you can check other points besides the transistor *BE* terminals. A good point is the volume control; it may be easier to spot and short.

The controls themselves can serve as test instruments. If the hum doesn't go away when the control is adjusted for minimum output, the source of hum lies after the control.

There are other sources of interference besides hum. *Atmospheric noise* and electric motors are common sources of interference. The methods described for locating hum pickup can also be used to search out the location of noise interference

> **TIP 9-13**
>
> The scope is the best instrument for tracking down the source of hum or interference pickup. The signal tracer can also be used. If no instruments are available, disable a stage by shorting its *BE* terminals (the stage must be capacitive coupled), or bypass the hum with a capacitor. If hum still appears at the output, it must arise after the disabled stage.

pickup. The big difference here is that noise interference has a higher frequency than hum. It often has a random frequency that may be difficult to stop with a scope; but this does not prevent you from seeing it and knowing where it is present. If you can stop it on the scope, it often looks like grass. You can recognize the sound of noise from a speaker. It is a rushing sound. On a video-display tube it looks like snow or whirling grains of sand.

Another type of interference is from installations that radiate electromagnetic waves—radio, TV, or FM station interference, for instance. Amateur and CB transmitters are other sources. Nearby equipment such as computers and TV and radio receivers also send out interference signals. Ordinarily, these sources do not cause much trouble in equipment that has been properly installed and operated. However, bad transistor stages can sometimes pick up undesirable levels of interference.

The types of interference from transmitters, computers, radios, and TV sets show up on a video display as regular patterns, stationary or moving. On the scope you can see them as periodically repeating waveforms along with the regular signal. From the speaker, they come out as regular tones, sometimes harmonious.

There are special circuits that are designed to reduce this type of interference. We'll see them in Chapter 11. In this chapter we concentrate on what goes on in the basic amplifier stages.

9-4-2 Internally Generated Noise Problems

In the preceding section we were concerned with noise coming from the outside. Noise can also be generated within a signal path. Transistors, resistors, and bad connections are the primary sources of internally generated noise. Noise is often associated with heat: "thermal noise," as we call it. The hotter a part gets, the more noise it generates. The problem is aggravated when the parts and connections are defective.

Here again, you can use the signal-tracing methods described previously. The trouble is most pronounced when it is generated in the early stages. Special low-noise components may be used in these early circuits, but not always, and they can go bad, too.

The noise generated in the early stages may be too small to detect with a scope. Start from the output and work your way back as far as you can. Note how much you have to crank up your signal-tracer or scope output for each stage. Look for the first stage in the signal path that has a high noise-to-signal ratio. If the scope does not have enough gain to pick up the noise, you must use other methods.

You can use a capacitor to short out the noise at each stage. Try the later stages first, just to see if you can knock out the sound with the capacitor. This is a very effective method.

Try spraying a suspected transistor or resistor with coolant. If cooling reduces the noise output, the part may be bad and deserves further checking.

> TIP 9-14
>
> Internally generated noises can be tracked with a signal tracer or with a jumper capacitor that bypasses the noise to ground. Look for noisy transistors or resistors in the early stages. Cooling off a noisy part with spray may reduce the noise.

9-4-3 Oscillations in Basic Amplifier Stages

Unwanted signals can also be generated within the stages themselves. These signals are called *parasitic oscillations.* They are said to be ''regenerated'' when the output of one or more stages is fed back to the input. Oscillations in general are discussed in more detail in Chapter 11.

Oscillations cause unstable operation, loss of output, and extra power consumption. In audio equipment you hear squeals, motorboating, and extra rushing sounds. In equipment with metered circuits, extra current can be noted when oscillations occur. You can see them on a scope as waves that override the desired signal.

In basic amplifier stages, the interaction can take place through the supply line. Ordinarily, the power-supply filter and bypass capacitors posted along the supply distribution lines prevent unwanted signals from getting on the lines. The bypass capacitors are often called decoupling capacitors; they decouple signals in the supply line between stages.

When bypass capacitors open, they permit oscillations that can be heard in speakers as a motorboating sound. Sometimes these oscillations are inaudible but still harmful.

In high-frequency circuits, oscillations can result from the inadvertent coupling between various conductors and parts in the circuit. We'll have more on this in Chapter 11, but we should mention some measures taken at the factory to prevent oscillations. Factory placement of parts and leads can be very critical in some circuits. Be careful about moving leads or positioning parts when you replace them. If oscillations occur, move these parts or leads to see if you can kill the oscillations.

Note: In this section we found that oscillations were caused when unwanted signals from the output were fed back to the input. This type of

TIP 9-15

Unwanted oscillations, which result in motorboating and squeals, are often caused by open decoupling capacitors in the supply lines. Jumping these capacitors might cure the problem.

feedback is called *positive feedback*. The feedback signal was added to or aided the input signal. In the next section we deal with *negative feedback:* The feedback signal will oppose the input signal.

9-5 FEEDBACK CIRCUITS

Modern circuits are characterized by circuits that are self-regulated. Their output levels do not have to be controlled by the user. This is not only convenient; it is also more accurate and faster. Troubleshooting these circuits calls for special methods.

9-5-1 Voltage Feedback

Previously in this chapter we saw how signals were coupled from stage to stage. We should now investigate a form of coupling called *negative feedback*. Figure 9-7(a) shows a path running back from the output stage to the input stage. Its purpose is to feed back a correction voltage to counteract any distortion that may arise within the stages. We see the original input signal (1) and the distorted output signal (3). The distortion takes place in the second stage.

At the input in Fig. 9-7(b) we see the original wave (1) and the reduced feedback wave (4). All this takes place almost instantaneously. The two waves combine to produce wave 5 and amplified, produce wave 6. Note that wave 6 looks very nearly the opposite of the distorted output wave (3). When this oppositely distorted wave 6 is ''redistorted'' in the second stage, the result is an almost undistorted wave (7). Remember, all this is taking place almost instantaneously.

If there is no distortion within the stages, the entire signal will be reduced, but not distorted, by the small feedback signal. However, this is a small price to pay for the reduced distortion that results.

Feedback is used in all types of circuits. We introduce the basics here in preparation for its use in more advanced circuits. The main idea is that a sample correction voltage is fed back from the output to the input.

9-5-2 Emitter Feedback

Another way to feed back a signal is through the emitter. This is called *current feedback* rather than voltage feedback. If you remove the emitter bypass capacitor from

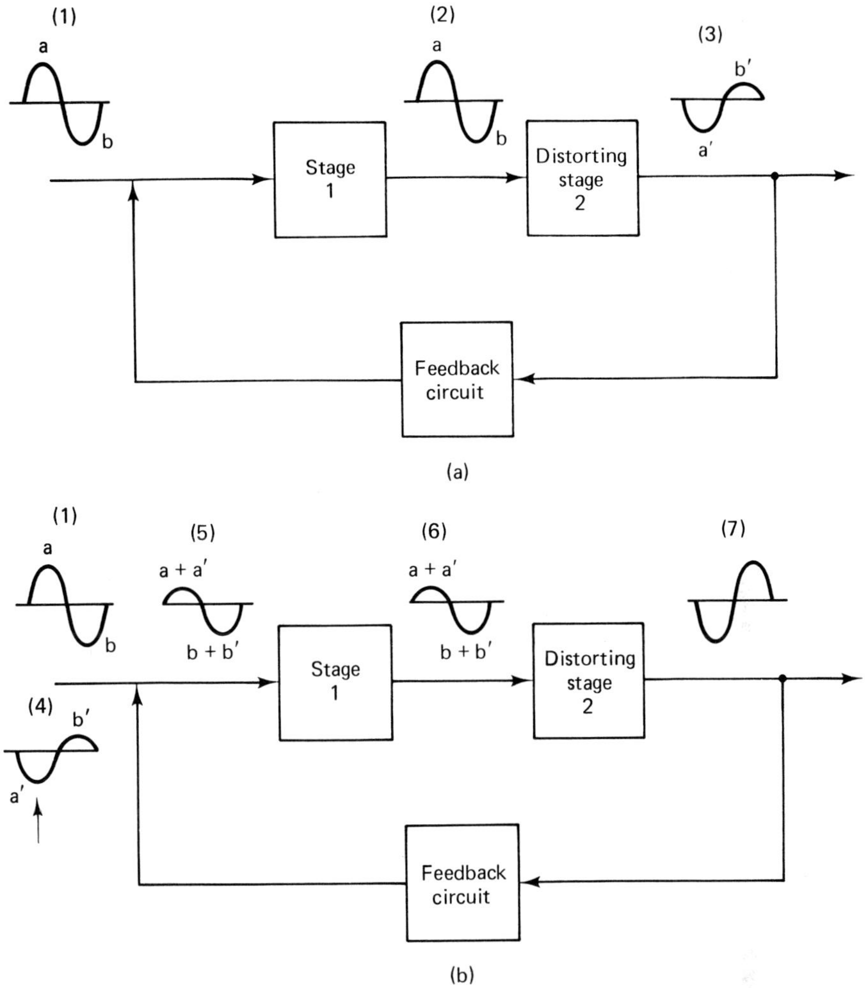

Figure 9-7 (a) Stage 2 distorts the signal by flattening it out at b' and stretching it out at a'. (b) Stage 2 stretches the flattened part and flattens the stretched part to produce a normal wave.

your basic amplifier, you will have a current-feedback circuit (see Fig. 9-8). At the input we see the input wave (1). A distorted wave (2) is shown at the output. It represents what would happen if there were no feedback. The arrow shows the amount and direction of the distortion. At the emitter we have the feedback wave (2′). Note the direction of the distortion. Now think of wave 2′ as being fed back through to the collector. (Any voltage will tend to be fed through.) This wave will not be inverted at the collector. (Inversion only takes place from base to collector.) When the distorted feedback wave from the emitter is redistorted, the original waveshape is restored. See wave 3, which is the "distorted" shape of wave 2′, but has the same shape as the input wave (1).

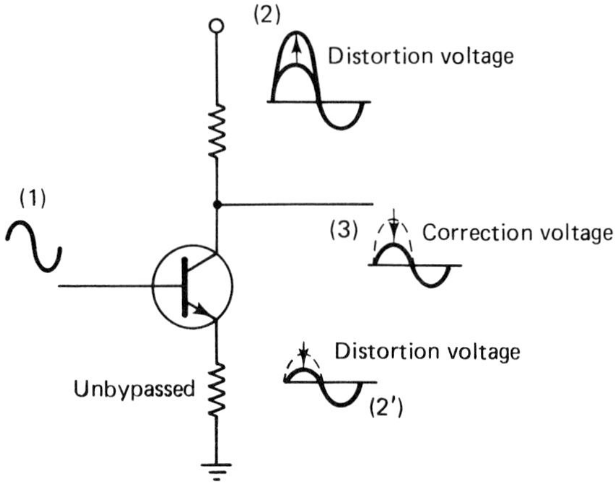

Figure 9-8 An unbypassed emitter resistor produces another form of feedback correction. The arrows represent the distortion and correction voltages.

> TIP 9-16
>
> When a dc coupled feedback loop goes bad, you may not be able to scope it. With a voltmeter, look for the stage that does not transfer the bad dc voltage in a normal manner from input to output. See Tip 9-6, Tip 9-7, and Fig. 1-6.

9-6 MORE DIRECT-COUPLED AMPLIFIERS

While we are on the subject of dc amplifiers, we should cover two widely used types that are now used in integrated-circuit chips (Chapter 13): the Darlington amplifier and the differential amplifier.

9-6-1 Checking a Darlington Amplifier

The Darlington amplifier in Fig. 9-9 has very high current gain and a high input impedance. It delivers a lot of power without loading down the source circuit. Both transistors are emitter followers, so the final output voltage follows the source signal voltage. There is no voltage gain, but you obtain a very high output current with a small input signal.

The Darlington pair is often enclosed in a single-transistor casing, and the whole unit may be called a Darlington transistor. It can be tested almost like an ordinary transistor. Take hi–lo readings on the terminals and compare them with a good Darlington transistor. You can also check the voltages. The input base-to-output emitter voltage should be about 1.4 V. Shorting these two terminals should lower the output voltage at the emitter.

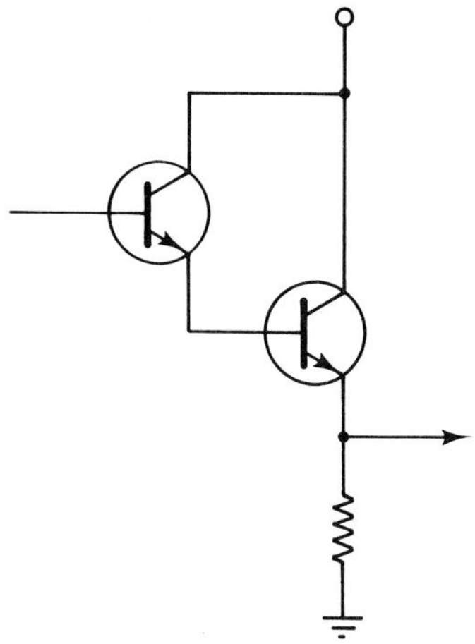

Figure 9-9 Darlington amplifier. You check it out as two emitter followers.

If the circuit consists of a discrete pair of transistors, you can check the voltages or resistances of each transistor separately. Shorting the base-to-emitter terminals of either transistor should lower the output voltage.

9-6-2 Checking a Differential Amplifier

Another direct-coupled amplifier is the two-transistor differential amplifier shown in Fig. 9-10. It has two inputs. Its output depends on the difference between the two input voltages. The inputs are at the bases, and the outputs are at the collectors. You'll find various combinations of inputs and outputs. For instance, you might find one base input and one collector output.

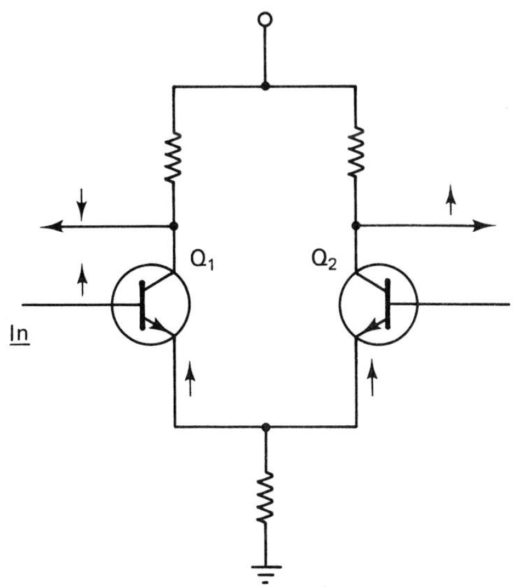

Figure 9-10 Differential amplifier. The Q_1 output is inverted, and the Q_2 output is noninverted.

To see how this amplifier works, let us make Q_1 the input and Q_2 the output. Look at the arrows. In this case, Q_1 is an emitter follower and Q_2 is a common-base amplifier. Note the direction of the arrow at the collector of Q_2. It is not inverted.

Now look at the the collector of Q_1. Its arrow is inverted. So we have two kinds of outputs, inverted and noninverted. If we make Q_2 the input, the roles will be reversed. This time, the collector of Q_1 is noninverted and that of Q_2 is inverted.

Let us try something else. Let us apply the same signal to both inputs, as in Fig. 9-11. Note that their voltages go up and down in unison. The solid arrows in the figure represent the output from the Q_1 input, and the dashed arrows represent the outputs from the Q_2 input. Notice that the collector voltages cancel each other. In a perfect differential amplifier, there will be zero output.

This canceling action is called *common-mode rejection.* This enables the amplifier to reject any noise or hum that is picked up simultaneously by both inputs.

TIP 9-17

To check a differential amplifier, short the two inputs together. You should get zero volts between the two collectors. Also short the inputs to ground, one at a time. Both output voltages will go up and down as you alternately ground the terminals. One output should go up when the other goes down.

Without going into the details, we can note that the differential amplifier can have high gain, high input impedance, and good stability. It is widely used in integrated-circuit chips called op amps (see Chapter 12).

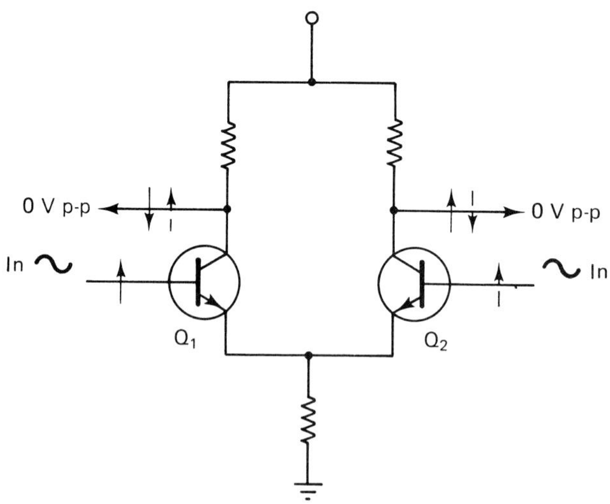

Figure 9-11 Two equal voltages at the inputs of this differential amplifier are canceled out at the outputs.

9-7 THE UNCERTAINTY OF VOLTAGE AND RESISTANCE READING

In going over the preceding circuits, we stressed voltage and resistance measurements. Sometimes we find that the voltages we read in the circuits are far different from those in the schematic. This happens even in good circuits. How, then, can you tell whether something is really wrong when you have an off-voltage reading? The answer lies in trying different approaches. This is one reason why we have more than one way for checking parts.

There are many reasons why voltage readings are so uncertain. The factory line voltage might be different from yours. The factory meter sensitivity and accuracy might also be different. The parts tolerances (usually 10%) may add up. The factory signal inputs may be different. The temperature conditions might differ. Sometimes at the factory itself, different circuits may be measured by different people on different days or in different rooms. Sometimes a draftsperson will inadvertently sneak a wrong voltage into the schematic.

You can get around these differences, on the one hand, by using the same type of meter, by adjusting your line voltage, by using the same input signal, and by adjusting the room temperature in hot weather. These conditions may be listed in a corner of the schematic. They list the sensitivity of their meter in ohms per volt. If your meter has a low ohms-per-volt rating, it might pull down the voltages too much. It really pulls down hard in circuits that have high resistances.

On the other hand, if you have no factory recommendations, take note of the line voltage, the supply voltages, and a few other voltages here and there. Compare them with the schematic voltages and adjust your readings accordingly. That is, if these voltages are 5% lower, add 5% to your readings. If they are higher, subtract.

One nice thing about old-fashioned nondigital meters is that you can readily see the angle or percentage by which the needle is off. However, the higher ranges are more accurate for reading voltages in high-resistance circuits, so you should take this into account. Also, if the voltages in the high-resistance circuits are further off, take this into account and adjust for it.

9-7-1 A Procedure for Reading Uncertain Voltages

If you are uncertain of your voltage readings, you might follow the steps below.

Step 1. Take more than one round of voltage readings. Take all the collector voltage readings first. Note whether the voltages are consistently higher than normal or lower than normal. Note which ones are definitely worse than the others.

Step 2. Do not forget to read the supply voltages, including the line voltage. Compare these voltages with the collector voltages.

Step 3. Measure the voltages on all the more suspicious transistors (the ones with the worst collector voltages).

Step 4. Measure the voltages on all the transistors in a dc signal path. The dc signal path, as you learned, is the path through several transistors between two blocking capacitors or transformers.

Step 5. Always record your voltages. Look for the most drastic departures from normal voltages and follow up on them. If all readings are consistently off by about the same percentage, but one of them is off by a different pecentage, it should be followed up.

9-7-2 Resistance Measurements

Resistance measurements are often uncertain, because the parts measured are not measured under actual working conditions. A fault may develop only when the working voltage is applied or after the part has warmed up. Also, resistance readings are made difficult by parts that are connected in parallel.

However, resistance measurements are easy to take and safe. Moreover, when a part checks bad with an ohmmeter, it is definitely bad. Only when it checks good is there uncertainty, and you have to follow up with other checks.

9-8 READING SCHEMATICS

Perhaps the biggest timesaver in the shop is a file cabinet full of schematics. It is possible to troubleshoot without them, but you will lose a lot of time. Many technicians will not or cannot work without them. As helpful as they are, schematics can be just as intimidating as the PC boards themselves. This is why we have to break them down into sections and stages, and their inputs and outputs, and track down the voltages.

9-8-1 The Major DC Voltage Path in a Stage

Each transistor stage in a schematic contains a major dc voltage path that can be traced from supply line to ground. Figure 9-12 shows a major voltage path that may not be very realistic, but it has nearly all the basic parts we have encountered so far. It starts from the 70-V supply point above R_2 and ends at ground just below R_3. The schematic also shows some minor paths that do not substantially affect the dc voltages in the major path. Starting at the supply point:

1. The major path goes through resistors with a voltage drop. (see R_1).
2. A minor path goes through R_2. This path has such a high resistance that it is virtually open. Any changes on the other side of R_2 has almost no effect on the voltages in the major path. We can neglect this path when we are working on the major path. (Things would be different if R_2 shorted, but this is not likely to happen.)
3. The dc path does not go through capacitors. If C_1 shorts, the voltage will be pulled down. If C_2 shorts, the major path will change if the other side does not have 20 V.
4. The path goes through coils with little or no voltage drop. Coil L_1 drops 0 V dc, practically speaking (see also the primary of T_1).

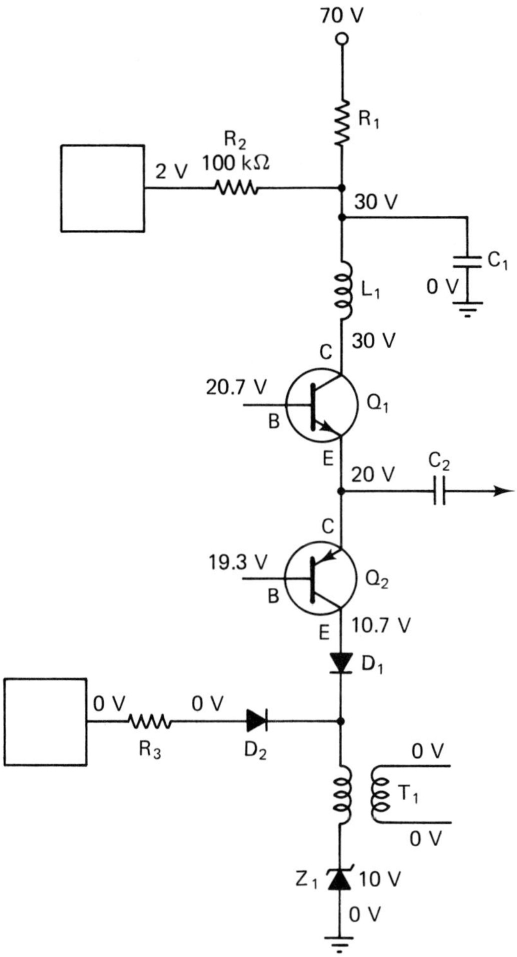

Figure 9-12 How to trace out a major voltage path from supply to ground. This path can be checked out pretty much like a simple series circuit.

5. There is a path through a transistor (see Q_1 or Q_2) between collector C and emitter E. This path may be turned off or on. In most cases there is a substantial dc voltage drop between C and E.
6. Normally, there is no dc path between collector C and base B. It is normally reverse biased (see Q_1 and Q_2).
7. There is a minor path between B and E. The BE voltage drop is never more than 0.9 V. It ranges from 0 to 0.7 V. In some circuits it can have a negative voltage drop.
8. The path goes through diodes when forward biased (see D_1). When forward biased it goes through with a small drop, about 0.7 V for silicon diodes.
9. It cannot go through a diode when it is reverse biased (see D_2). It is reverse biased or switched off.
10. The path cannot go through transformers from one winding to another if there is no dc connection between them (see T_1).

11. It goes through a reverse-biased zener diode and has the zener voltage across it. In this case the zener voltage is 10 V.

Once we trace out a major dc voltage path, we can apply the basic rules for shorts and opens, or decreased and increased resistances, to find the bad part in these paths.

9-8-2 Tracing Supply Lines and Simple Voltage Paths

One of the first things that a troubleshooter does when looking at a schematic is to look for the supply line connected to the suspected circuit. This is not always easy. You will see lines running all over the schematic, often from one page to the next, sometimes in parallel and sometimes not. You cannot always tell which is a supply line and which is not.

> TIP 9-18
>
> In cases where it is difficult to follow the power-supply lines in a schematic, trace over the main lines with a thin red pen and ruler. Use orange for a lower voltage line. For other voltages, use dashed lines. Reserve green for signal paths.

Many schematics make it easier by omitting all the supply lines. The sources of the supply lines and their various terminal points are labeled. The lines are understood to exist between these labeled points (see Fig. 9-13). The 20-V supply point is

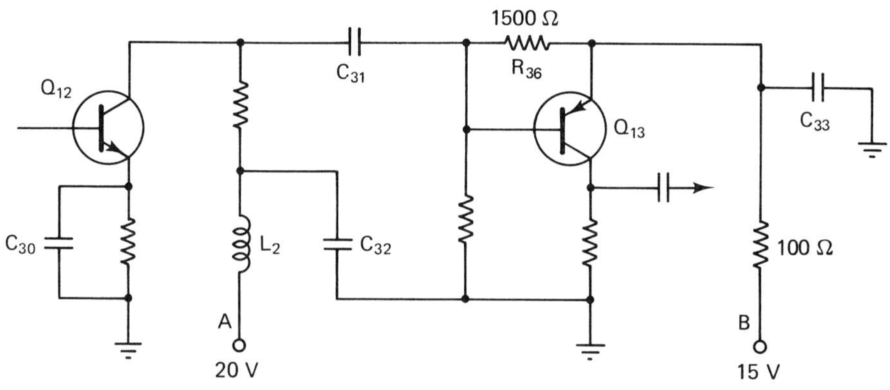

Figure 9-13 Two typical paths, *A* and *B*. The path of least resistance in *A* is through the coil L_2. The 1500-Ω resistor R_{36} helps isolate major path *B*.

understood to be connected to the 20-V source in the power-supply distribution system. Similarly, the 15-V supply point is understood to be connected to a 15-V source.

After locating the supply point, you trace the path leading from it to ground (see Fig. 9-13 again). It shows two simple voltage paths. Path *A* through Q_{12} is kept in line by blocking capacitor C_{31}. The coil L_2 does not drop any dc voltage. The dc path does not go through C_{30} or C_{32}. As noted before, the path goes through the transistor from collector to emitter in the direction of the arrow. Notice too that the path goes from the higher voltage to the lower voltage, from 20 V to 0 V.

Path B through Q_{13} is kept in bounds by capacitors C_{31} and C_{33}. R_{36} has such a high resistance that the major path does not go through it. The main reason that we can overlook this path for the moment is that when something goes wrong with it, a more drastic change will be produced in the major path. You look for this drastic change first. The path through *BE* is a minor one and can be ignored for the moment. The major path is between *C* and *E*.

Once you trace them out like this, it is not that hard to troubleshoot these major dc paths. It is like a simple series path. All we have to do is to apply the basic rules. A short below a point pulls its voltage down. A short above pulls it up. An open below lets it rise, and an open above lets it fall.

However, you will remember that good transistors apparently do "short" and "open" (saturate or cut off). Nevertheless, a bad voltage is a bad voltage, even if the parts are good. We learned earlier in the chapter how transistors transfer bad voltages from circuit to circuit.

9-8-3 Reading Schematics: More Details

Figure 9-14 shows how to deal with two commonly encountered minor paths. They split off from major paths *A* and *B*. Looking at major path *A*, we see that it goes through coil L_{10} and the *CE* path of Q_{15}, but not through R_{15} and path *BE* of Q_{15}. This

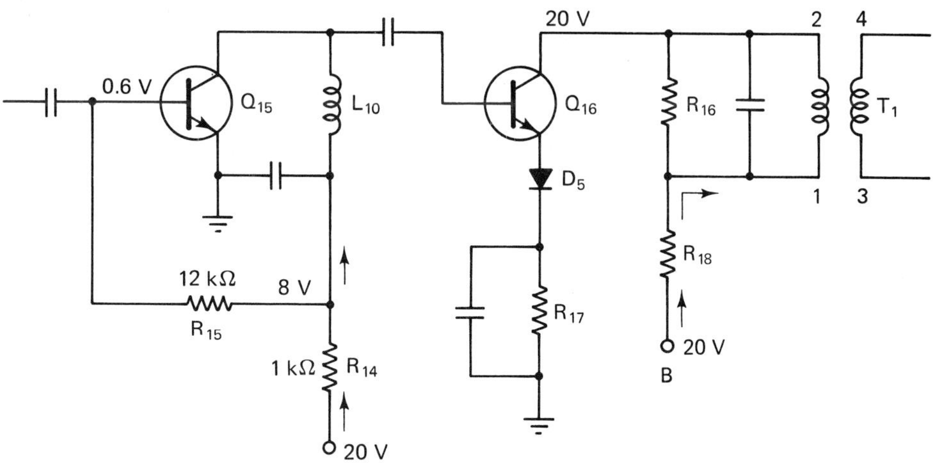

Figure 9-14 *A* and *B* show that the path of least resistance is through the coils.

is because the resistance of R_{15} and *BE* is very much greater than that of L_{10} and *CE*. The resistance of the minor path *BE* is always very much greater than that of the major transistor path *CE*. To sum it up, major voltage path *A* starts at the 20-V supply point and runs through R_{14}, L_{10}, Q_{15}, and ends at ground.

Going over to major path *B*, we see that it bypasses R_{16} and goes through a winding (1–2) of part T_1, a transformer. Coils have near-zero resistance and drop nearly zero volts. This small resistance shorts out the voltage drop across R_{16}, so we say that voltage path *B* starts from the 20-V supply point and goes through R_{18}, T_1, Q_{16}, D_5, and R_{17}.

Once we have traced out the major path, we can figure out what parts are responsible for the bad voltages in the path. Just apply the basic rules.

9-8-4 Reading Schematics: Multiple-Source Circuits

Figure 9-15(a) shows a multiple-source circuit with supply points at R_{51}, R_{54}, and R_{56}. Our major paths of interest are the ones that run through Q_{24} and circuit 2. Notice that there is a bridge running across from one path to the other. It runs from *E* of Q_{24} to the center of R_{55}.

This is a complex circuit. We will have to use a combination of approaches to troubleshoot it. Right now, we are trying to trace out the paths. At least we can recognize a bridge. Notice that it seems to be balanced. It has 1.6 V on one side and 1.7 V on the other. Let's redraw the circuit so that it looks more like a bridge [see Fig. 9-15(b)]. We tied the two 20-V supply points together on top. We give you some hints based on what you have learned up to now:

1. An open *BE* path in Q_{24} will raise the voltage at *C*. A shorted *BC* junction will bring it down. A defect in the *BE* junction, short or open, will raise the voltage at *C*. (*Optional:* Recall that the *BE* junction and current counteracted the reverse bias on *BC* when the transistor was good.)
2. An open R_{51} will bring voltage *C* down. A short across it, possibly in the PC board, will raise the voltage at *C*. An open R_{50} will raise all the voltages above it.
3. As for the bridge between *E* of Q_{24} and R_{55}, you can check the coil with an ohmmeter. It should have near-zero ohms.
4. R_{53} can be checked with an ohmmeter if it has a low resistance compared to the others. In any case it should not be more than its rated value.
5. With a scope, there should not be much signal loss across R_{53}, since it is supposed to pass a signal to circuit 2. This would be even more probable if R_{54}, R_{55}, and R_{56} are very large compared to it.
6. Do not forget that the most likely part to go bad is the transistor. Checking it out with a voltmeter will solve the problem in most cases.
7. Since there is only 1.6 V on emitter, you can short it to ground. If there is no effect on the other side of R_{53}, it is open.

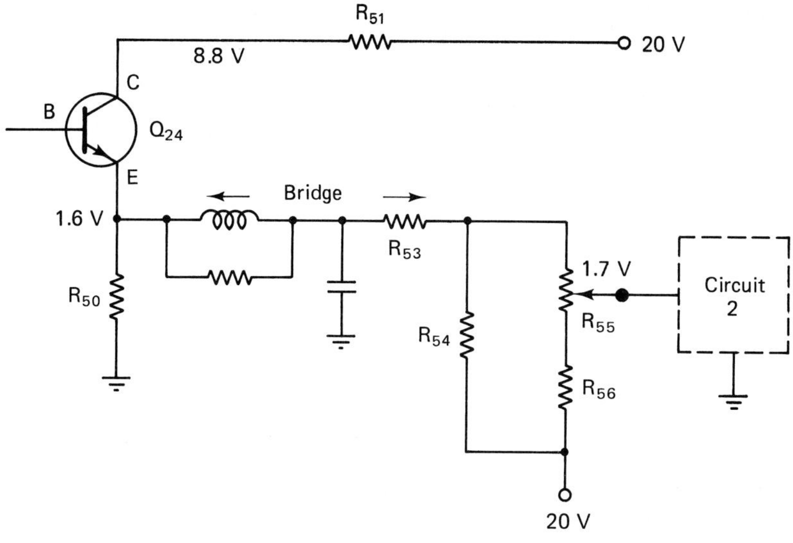

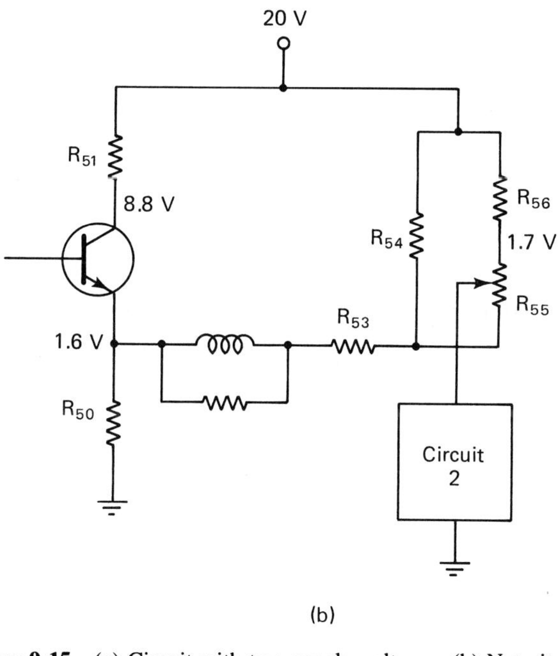

Figure 9-15 (a) Circuit with two supply voltages. (b) Note its similarity to a bridge circuit.

SUMMARY OF SELECTED TIPS

Tip 9-1. If an amplifier has no output, turn the volume or level control down low and touch the base of a transistor with your finger. If you get a response at the system output, the circuits after that point are probably okay. You can also touch the base with a screwdriver or a test prod for this purpose. **Caution:** You can blow a speaker if the volume control is not turned down low.

Tip 9-3. A bad dc voltage cannot be substantially passed back from the collector to the emitter or base of a stage.

Tip 9-5. If a bad dc voltage lies in a signal path between two coupling capacitors or transformers that are not autotransformers, the trouble lies between these two devices.

Tip 9-6. When checking dc signal paths, if a transistor passes a bad voltage forward, work backward, stage by stage, until you find a stage (the suspect stage) that does not pass a bad voltage forward. If the preceding stage has good voltages, the suspect stage is bad.

Otherwise, check the transistor in the suspect stage with an ohmmeter. If it checks bad, recheck it by disconnecting its base (Tip 8-7). If this produces good voltages in the preceding stage, the suspect stage is bad.

Tip 9-7. When working on a dc signal path, you can short the *BE* terminals of a transistor at some point in the path. If the resulting voltage change does not pass forward to a second point down the line, something is wrong between the two points. Move forward and repeat the shorting until you find the bad stage.

Tip 9-8. Certain signal-rectifying circuits such as detectors and sync clippers develop dc voltages in the presence of signals. To check for signal in these circuits, turn the signal on and off at a preceding stage. If the dc voltage at the rectifying circuit responds, the intervening stages are probably okay.

Tip 9-9. When checking for hum or interference, look for open ground connections, among other things.

Tip 9-13. The scope is the best instrument for tracking down the source of hum or interference pickup. The signal tracer can also be used. If no instruments are available, disable a stage by shorting its *BE* terminals (the stage must be capacitive coupled), or bypass the hum with a capacitor. If hum still appears at the output, it must arise after the disabled stage.

Tip 9-14. Internally generated noises can be tracked with a signal tracer or with a jumper capacitor that bypasses the noise to ground. Look for noisy transistors or resistors in the early stages. Cooling off a noisy part with spray may reduce the noise.

Tip 9-15. Unwanted oscillations, which result in motorboating and squeals, are often caused by open decoupling capacitors in the supply lines. Jumping these capacitors might cure the problem.

Tip 9-16. When a dc coupled feedback loop goes bad, you may not be able to scope it. With a voltmeter, look for the stage that does not transfer the bad voltage in a normal manner from input to output (see Tip 9-6, Tip 9-7, and Fig. 1-6).

WHAT YOU LEARNED TO DO

In this chapter you learned how to:

1. Signal-trace transistor stages with scope, signal tracer, and signal generator. You discovered that bad voltages may be transferred through several stages and make it impossible to trace signals in certain types of feedback control circuits that are found in later chapters.
2. Trace bad voltages from stage to stage to find the bad stage. You found out that dc troubles are confined between coupling transformers or coupling capacitors.
3. Check a feedback loop with a dc voltmeter when a scope cannot do the job.
4. Detect the presence of signals up to a diode detector or a transistor with signal developed bias by checking their dc voltages.
5. Trace undesirable signals to their source or pickup point in basic amplifier stages.
6. Deal with uncertain voltages that vary considerably from schematic voltages.
7. Break down schematics into sections and stages, and trace signal paths through them. You learned to trace dc power supply lines and dc voltage paths to ground.

QUESTIONS

1. What can happen in dc-coupled stages that might prevent you from signal tracing them?
2. If you find several dc-coupled stages with bad dc voltages, how can you tell which stages are good?
3. Can bad voltages be transferred back from collector to collector of good stages?
4. When checking circuits with a dc voltmeter, how can you tell whether a signal has reached a diode detector or transistor clipper stage?
5. Can interference voltages be traced like regular signals?
6. What defect might cause ordinary amplifiers to oscillate?
7. What are the advantages of a Darlington amplifier?
8. How can a differential amplifier be tested?
9. Can you give three reasons why a schematic voltage may differ from your own voltage reading in a good circuit?
10. What can be done to facilitate the tracing of supply lines in a schematic?

11. Why is it possible to treat a major voltage path as a simple series circuit even when there are minor paths connected to it?

PRACTICAL TROUBLESHOOTING EXPERIENCE

A. Reading Schematics

Obtain a schematic of a major piece of equipment. Trace over the main dc supply line (the one that goes to most of the circuits) in red. Then trace several major dc paths through the transistors from supply line to ground in blue. Trace several signal paths from input to output in green. Trace several dc signal paths from coupling point to coupling point in orange.

B. Tracing Signals

This job requires a PC board with several dc-coupled amplifier stages. You should also have the schematic for the board. The supply voltage for the stages should be about 20.0 V or less. You will need a power supply, multimeter, signal generator, and scope or signal tracer. If you do not have a scope, generator, or tracer, that is okay; just check the dc voltages as called for. Connect the power supply to ground and to the supply line to the stages.

> *Note:* If you cannot find a suitable board for this job, perform a "dry run" on a schematic. That is, locate in the schematic the circuits, parts, and test points indicated in the steps below. Imagine you are performing the checks that are called for and predict the outcome if a part is bad. Predict the outcome if the part is good.

1. Introduce a 500-Hz signal at the input stage of a series of dc-coupled stages. Vary the signal level and check the output of that stage with a scope or signal tracer. If you get a good signal, the stage is okay and so is your test setup. You can go on to step 2.

   If you cannot get an output signal from this stage, check the dc voltages. Don't forget the supply and ground points. Check all the ground points with the ohmmeter for continuity to the ground terminal of the supply. An open ground terminal will raise the voltages above it. If necessary, tack-solder ground wires between various ground points on the PC board ground terminals. Do the same, if necessary, to the supply points.

   If you have bad transistor voltages, check the transistor in-circuit with an ohmmeter. If it seems good, perhaps the following stage is bad. Check it to see if it is a good stage that is transferring a bad voltage forward.

   If it checks bad, or doubtful, remove the base lead and recheck it. Use both voltage and resistance checks. Check the other parts, too. Replace the defective transistor or part, or both.

If necessary, check the voltages in the following stages. Replace the transistor or other defective part.

2. Introduce a signal at the input stage and note whether you can get a signal at the output of the last stage. If you can get an output, go on to step 3.

   If you cannot get an output, check the collector voltages of each stage that has a collector output. Find the stage that is not transferring the bad voltage correctly from base to collector. If you find a bad stage, check its transistor. Make an in-circuit resistance check. Remove the base lead and check the transistor with both ohmmeter and voltmeter. Check the other parts. Replace the bad transistor or part.

3. If you can get a signal at the output, jump-short a collector resistor in an intermediate stage. Check the voltages in the following and preceding stages. Can a bad collector voltage be transferred forward? Can it be transferred backward?

4. Short an emitter resistor and see how far the bad voltage can be transferred forward and backward.

10

Regulated Power Supplies

EXPERIENCE REQUIRED

If you observe the safety rules and have done some work with basic power supplies, amplifiers, and feedback circuits, you should not have much trouble working with regulated power supplies. In the latter part of this chapter you will come across regulators containing advanced circuits that are taken up in later chapters. These circuits are contained in blocks, and only the input and output voltages are given. You should learn to work with circuit blocks because most circuits are now integrated into small chips with only inputs, outputs, control, and supply terminals. Chips are taken up in Chapter 14. You should recall, in particular, that the collector inverts the base voltage; in all other cases, the output follows the input.

10-1 POWER-SUPPLY REQUIREMENTS

The basic power supply that we studied in Chapter 6 was good enough for small radios and other equipment with noncritical requirements. However, many of today's sophisticated instruments require better regulation and higher efficiency. That is, they need steadier output and less power loss. Modern circuits can be very sensitive to variations in the power-supply voltage. They must be protected against sudden surges and power dropouts.

10-2 THE TRANSISTOR REGULATED SUPPLY

We are going to look at three voltage regulators. Each is an improvement over the previous version. The first uses a zener diode to control the main regulator transistor. The second uses a feedback transistor. The third uses an added protective circuit.

10-2-1 Checking a Zener-Diode-Controlled Transistor Regulator

Figure 10-1 shows a voltage-regulator circuit using a power transistor with a zener diode, *Z*, connected to its base. After initial turn-on, the zener maintains a steady voltage on the base of the transistor, 10 V dc in this case. When the power is first turned on, the base acquires a positive voltage causing the transistor to conduct, and a steady 0.7 V is established across the *BE* terminals.

If changes in the load cause the emitter voltage to rise, the *BE* voltage will decrease and the emitter voltage will fall, thus counteracting the original rise. Just the opposite happens if the output voltage falls. Thus the output voltage is regulated against changes in the load circuits.

The input voltage changes are also counteracted. If the input voltage rises due to changes in the ac line, the emitter output voltage will not change, for two reasons: Transistor action does not transfer voltage changes from collector to emitter, and the zener prevents any changes at the base that would otherwise be passed to the emitter.

A Troubleshooting Procedure. Let's look at some checks that we can make on this circuit. The ac input circuit is assumed to be okay. In the regulator circuit there are only a few parts to check, so you could check them one by one. However, it is usually best to check the input and output voltages of the regulator first. It should drop some voltage; otherwise, it would not be able to counteract an input voltage that falls below the desired output voltage.

1. If the input voltage is fairly close to normal and the output is low, check the base voltage.

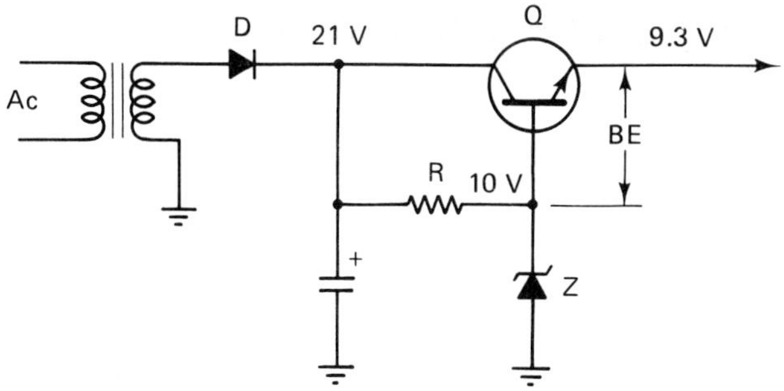

Figure 10-1 Zener-diode-controlled transistor voltage regulator. The 10-V zener voltage on the base of *Q* holds the emitter to 9.3 V. Normal transistor action prevents changes at the collector from affecting the emitter.

a. If the base voltage is low, the problem could be a shorted load and shorted transistor *BE* junction, a shorted zener, or an open or a greatly increased-value resistor. (In the latter two cases, the lowered base voltage would be followed by a lower emitter voltage.)
b. If the base voltage is okay, the transistor *BE* junction is open and the load may be partially shorted. (The zener diode will maintain the base voltage, while the open *BE* junction will lower the emitter voltage.)
c. If the base voltage is high, the zener diode is open.

2. If both the input and output voltages are below normal and equal, the transistor could be shorted and the load could be partially shorted (see Chapter 11 for shorted load circuits).
3. If the output voltage is higher than normal, the transistor *CE* path is shorted. In a rare case, the base resistor could be shorted.
4. If the input voltage is low and the output voltage lower still, the input filter capacitor could be open.

You can follow up on each of the steps above with the checks given below.

1. If *R* opens partially, the trouble may not show up; or if the resistance increase is large enough, there may be borderline operation. In the latter case, a fall in the line voltage may result in a fall in the output.
2. You can tell if *R* is open when the output is low because a shorted load would blow the fuse, rectifier, or transistor. An open *BC* junction in the transistor would also lower the output voltage, but this would put too much strain on the *BE* junction through *R* and damage it, too. At any rate, check the resistor with an ohmmeter. A higher-than-normal resistance definitely indicates an open.
3. If the zener holds up 10 V, it is very likely to be okay. If you can get a hi–lo resistance reading, it is even more likely to be okay. If you have more than 10 V, the zener is definitely bad.
4. The transistor itself can be checked with dc voltage and resistance measurements (see Sections 5-2-10 and 8-3-2). You should have about 0.7 V *BE* voltage. More than 0.9 V or less than 0.3 V indicates a bad transistor provided that the collector voltage is higher than the base voltage and the base voltage is higher than the emitter voltage.
5. The transistor can also be checked by shorting the *BE* terminals momentarily. This should bring the emitter voltage down momentarily.
6. An open filter capacitor will increase the ripple voltage at the input and the dc input voltage will fall. Since this circuit has a high input voltage, 21 V, the lower voltage should not affect the zener. Also, the filter capacitor may not be very large because the regulator does a lot of the filtering. If the input voltage does fall below the zener voltage as the result of an open capacitor, the transistor base voltage will ripple and the emitter output will follow suit. Jumping the filter capacitor with a good one will cure these faults (recall Tip 6-6, open input capacitor).

7. If the filter capacitor shorts, the rectifier or fuse will blow. With the rectifier or fuse blown and not yet replaced, use an ohmmeter to check the capacitor for a short. If you find a drastic short across the capacitor, it is probably bad, but check the zener and the transistor before replacing it. If you find a partial short across the capacitor, you should definitely check the other parts.

10-2-2 Checking a Feedback-Transistor Voltage Regulator

Figure 10-2 shows a widely used transistor regulator. The dc input voltage to the regulator is at the collector of Q_1, a power transistor. The output voltage at the emitter is controlled by changes at the base. The base voltage is controlled by a feedback path consisting of R_2, R_3, and Q_2.

The downgoing arrow at the output shows that the voltage is too low at this instant. This error voltage is fed back and inverted by Q_2 and fed to the base of the regulator, Q_2. Note the direction of the arrow at this point. The dashed arrow that follows it at the emitter is the correction voltage. It cancels the error voltage and pulls the output voltage back up to normal.

The zener diode anchors the emitter of the feedback transistor Q_2 to a steady voltage. Without this anchor, the emitter voltage of Q_2 will follow the base, and the collector will tend to follow the emitter and keep Q_2 from amplifying properly.

A Troubleshooting Procedure. The variable resistor, R_3, controls the operating voltage of the regulator so that its output voltage can be adjusted over a wide range. Assuming that the ac input circuit is okay, the following steps should lead you to the defective part.

1. See if you can vary the output voltage above and below the specified normal voltage.

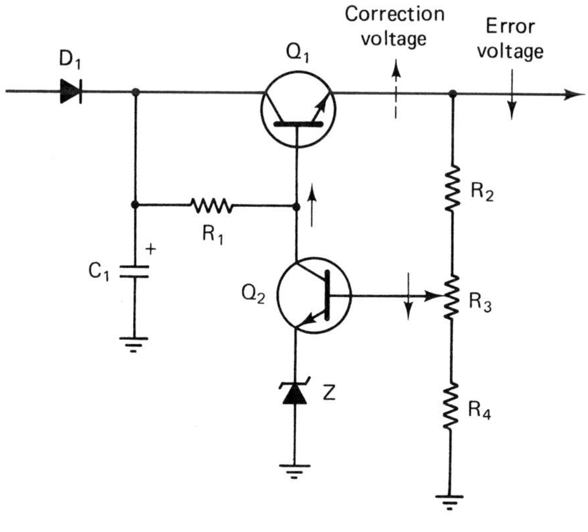

Figure 10-2 Transistor feedback voltage regulator. This circuit can be checked out like two directly coupled amplifiers.

2. If both the output and input are too low, check for an open input filter capacitor. A shorted capacitor would blow an input fuse (not shown) or rectifier. A partially shorted one might also blow these parts, or its effect would be counteracted by the regulator. A shorted R_1 is not likely to occur.
3. In all other cases, check the transistor voltages and see if you can vary them by turning R_3. A digital meter will show base voltage changes even though the zener diode keeps the changes very small.
   a. The base voltage of Q_2 is held in place by the zener diode Z, and is also held within limits by R_2 and R_4. It should be 0.7 V higher then the zener voltage. If you can increase the base voltage more than 1 V, the zener diode or the *BE* transistor junction is open.

      If the collector does not respond to even a small change in the base voltage of Q_2, the latter is bad. Shorting the *BE* terminals of Q_2 should increase its collector voltage and also the output voltage.
   b. If you can vary the base of Q_1 but not its emitter, it is probably bad. Check its *BE* voltage drop. Recall that for a good transistor it should not be less than 0.5 V or more than 0.9 V. Shorting the *BE* terminals of Q_1 to test it should lower the output voltage.
4. The checks may also indicate a possible defect in the feedback resistors. Check the resistance from the base of Q_2 to ground; you should be able to vary it by adjusting R_3.

   An open or partially open R_4 would cause the output voltage to fall. An open R_2 would cause the output voltage to rise. Also, you may not be able to bring up the base voltage to equal the zener voltage or slightly more.

   An open in R_3 will depend on what part of it is open. An abrupt change of the output voltage would indicate that R_3 is open. Absolutely no change at all would indicate an open center terminal connection—internal or external.
5. R_1 and Z can be checked as in the preceding section.

The following tip contains checks that are really helpful when voltage and resistance checks cannot be made in shorted voltage regulator and load circuits. They can be applied to other types of voltage regulators.

> TIP 10-1
>
> A voltage regulator and load circuit can be checked by applying power to its input through a series lamp. If the lamp glows brightly, there is a short. Turn off the power and disconnect the load circuit. Reapply power. If the lamp still glows bright, the regulator is shorted. Otherwise, check the voltages in the regulator and load circuits. After repairing the regulator, check the load circuit under reduced voltage.

Using a Variable DC Power Supply to Check a Voltage Regulator and Load Circuit. One way to check almost any type of voltage-regulated power supply and load circuit would be to insert a variable dc power supply at various points in the circuit. It should have meters for monitoring voltage and current. It should be capable of delivering as much current and voltage as the power supply being tested. Assuming that the dc input circuit is okay, we can make the following checks. Always bring up the voltage from 0 V for each test, and always monitor the current and keep it within limits.

1. Apply the normal voltage to the input of the regulator. The dc input current should not be too far off normal. It should be somewhat less than the power-supply output fuse rating—if there is one. If there is no fuse, keep the current under the current rating of regulator as listed in a transistor manual. (The actual current drawn includes the Q_2 path, but staying below the Q_1 rating alone provides a margin of safety.)
2. If the normal current is substantially exceeded, you have a short. Disconnect the load.
   a. If the current drops to far below normal, the load circuit is shorted.
   b. If the current does not drop substantially, the regulator circuit or capacitor C_1 is shorted.
3. If the current is very low with the load still in the circuit, you could have an open. Check the output voltage.
   a. If the input voltage is about normal and the output voltage is low, either Q_1 or R_1 could be open. R_4 could be open. (However, the feedback transistor, Q_2, or zener diode, Z, could be shorted.)
   b. If the output voltage is high, Q_1, Q_2, or Z could be open. There could also be an open in the feedback-resistance path, R_2 or R_3.
   c. If the input voltage is low, the input capacitor, C_1, could be open.
4. The individual transistors can be checked out by measuring their voltages while adjusting R_3 as in the preceding section.
5. The zener diode can be checked by applying dc voltage directly to its cathode and bringing it up to the zener voltage. The voltage should not increase after reaching its zener value.

Voltage regulators are usually packaged in IC chips. These will be taken up when linear integrated-circuit chips are discussed in Chapter 14.

10-2-3 Checking a Protective Circuit

Figure 10-3 shows a method of protecting the power supply and load circuits from excessive current caused by a shorted load or excessive line-voltage surges. A short in the load will increase the *BE* drop of Q_3 and turn it on. Q_3 will then short the *BE* junction of Q_1 and turn it off.

Less drastic changes in the load will be regulated in the normal way through the R_4–Q_2 circuit. Only excessive changes will cut off the current. R_2 is very small, about

TIP 10-2

A voltage regulator can be checked with a substitute dc supply connected to its input. Bring up the voltage slowly and keep the current well under the regulator transistor or substitute source rating. If the substitute voltage rises, check the voltages in the regulator and load circuits.

If the substitute voltage does not rise substantially as the current rises, there is a short. Disconnect the load circuits. If the voltage still does not rise, the regulator is shorted. Otherwise, check the voltages in the regulator circuits. Apply substitute dc voltages to various points in the circuit to locate the fault.

After repairing the regulator, reconnect the load and check it under reduced voltage.

2 Ω in this circuit, and it will not drop enough voltage at small current levels to cause Q_3 to counteract the action of Q_2.

The circuit is checked by measuring the input and output voltages of Q_1. Assuming that the input circuit is good, let's see what can be done if the output is low.

The first check would be to measure the voltage across R_2 or directly across *BE* of Q_3. If it is around 0.7 V, the load circuit has a short. Turn off the power and look for a short in the load circuit with an ohmmeter.

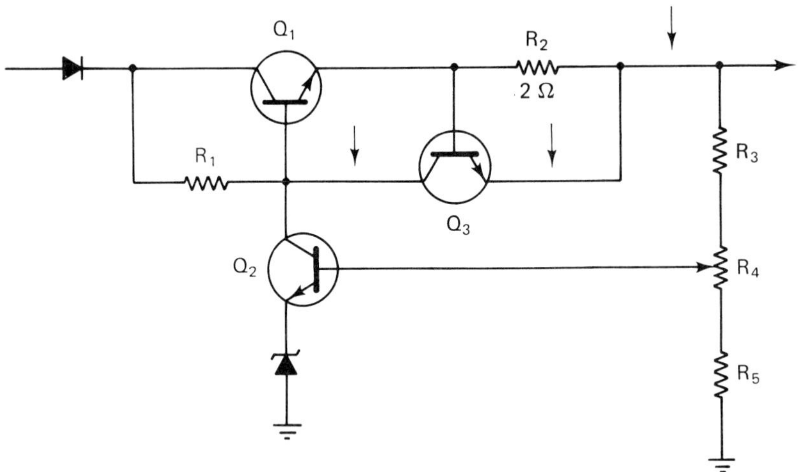

Figure 10-3 Q_3 in this circuit is controlled by the current and voltage on R_2. Excessive current and voltage on its *BE* terminals brings its collector down. This cuts off Q_1.

If you cannot find the short in the load circuit with an ohmmeter, use a variable dc power supply. Keep its output current at a safe value [$I_{safe} = (0.7/R_2)/3$]. You can now look for the short with a voltmeter.

You can use a series ac lamp to cut down the voltage and current automatically. You must jump a wire across the regulator. If voltage checks do not reveal the trouble, disconnect the load. If this does not bring up the voltage, the trouble is in the regulator circuit.

10-3 TROUBLESHOOTING A BASIC SCR CONTROL CIRCUIT

The silicon-controlled rectifier (SCR) shown in Fig. 10-4 is used to control ac voltages. It can pass current only when it is triggered by a voltage on its gate terminal, and once it conducts, it stays conducting even after the gate voltage goes down. It can only be turned off when its anode-to-cathode voltage ($V_{A\text{-}K}$) is turned off.

SCRs are normally used for controlling ac voltages, because they can switch the SCR off after each alternation when the voltage hits zero and trigger the gate on when the voltage goes positive. Figure 10-4 shows a simple control circuit in which the gate receives a small trigger voltage through the load. At this point, the ground side of the load has voltage because the SCR is switched off. When the trigger point is reached, the SCR conducts. The conduction will last until the source alternation reaches zero. The current waveform at the SCR shows what portion of the wave is cut off.

The duration of the cutoff and conduction times is determined by the variable resistor. Thus the amount of power delivered to the load is controlled simply by switching it off for short intervals of time. This is a very efficient way of reducing and delivering power. Recall that voltage-dropping resistors heat up and waste power. Although more efficient than resistors, transformers also use up power.

The SCR can be checked first with an ohmmeter as in Section 5-1-8 (Fig. 5-7), without disconnecting it from the circuit. Also check the gate resistors. If the ohmmeter reveals nothing positive, use a scope. Check the peak cathode-to-gate voltage. If it exceeds the specified gating voltage and you have no output waveform, the SCR is bad.

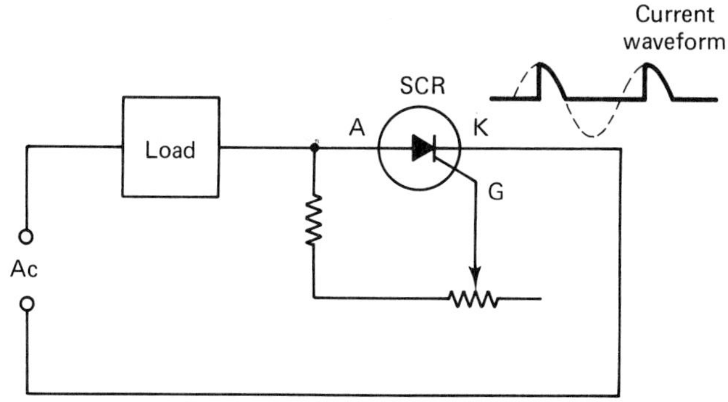

Figure 10-4 Basic SCR regulator circuit. The SCR cuts off varying portions of the ac half-cycles to reduce the power delivered to the load. The dashed parts of the wave are cut off.

The SCR can also be checked in-circuit (power off) with a metered dc power supply connected through a power resistor in series with the anode-to-cathode path (not in series with the load). The resistor should pass about the same amount of current as the load when it was working. That is, $R = V^2/P$, where V equals the ac line voltage and P equals the power rating of the load. With the dc power on, advancing the gate voltage should turn on the SCR, whereas turning down the gate voltage should not turn it off. Turn down the gate voltage, and turn off the dc supply. The SCR should now turn off. It should not turn on when the dc supply is turned on again.

Figure 10-5 shows a drastic method for protection against high voltages. It is called a *"crowbar" circuit*. Computer chips, in particular, cannot survive voltages much greater than 5.6 V. When the input voltage to this circuit rises too high, the SCR will short out the power supply. The power supply is protected by a fuse.

This is how the circuit works. When the voltage across the zener reaches a certain voltage, say 5.0 V, the zener will conduct and trigger the SCR. This will short out the supply line and save the expensive integrated-circuit chips.

To check this circuit for a short, proceed in the order shown below. Turn off the power before connecting or disconnecting a part.

1. First make preliminary checks with an ohmmeter because it is safest, fastest, and easiest.
   a. Make a hi–lo test of the zener diode.
   b. Check the resistor for an open.
   c. Check the SCR as in Section 5-1-8 (Fig. 5-7) but with nothing disconnected from the circuit.
2. If the quick resistance checks do not yield positive results, disconnect one end of the SCR and use a series test lamp to check the load circuit for shorts.
3. If the load does not have a short, or if you want to check the crowbar circuit first, reconnect the SCR and disconnect the load. Leave the series lamp connected and turn on the power. If the lamp lights, the circuit has a short.
4. To check the SCR for a short, disconnect the gate terminal. Turn on the power; if the lamp lights, the SCR is shorted. Otherwise, the gate circuit is bad.
5. To check the gate circuit, disconnect the SCR anode or cathode terminal. Short the gate resistor with a jumper. With the series lamp still in place, turn on the power and check the zener diode voltage; it should have the zener voltage across it.

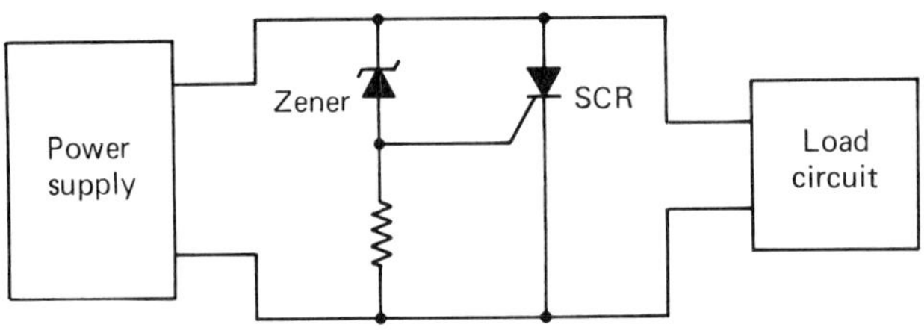

Figure 10-5 Crowbar circuit. Excessive input voltage will cause the SCR to conduct and bypass the load current. The power-supply fuse or protective circuit will then cut off the current.

10-4 A SWITCHING POWER SUPPLY

One advantage of the SCR regulator is that it conserves ac power by cutting off portions of each cycle. To cut down and regulate dc voltages, transistors have to be used. These transistors have dc current running through them continuously; consequently, they heat up and waste a lot of power. They have to be mounted on heat sinks. The switching power supply solves this problem. It has other advantages, as we shall see. Just as the SCR regulator reduces ac power by switching it off and on, the transistor-switching power supply turns dc current on and off. It switches the current rapidly and repetitively, to produce a pulsing wave. The pulses are then averaged and filtered to provide a lower dc output voltage.

Switching power supplies are now enclosed in integrated-circuit (IC) chips. Much of the information needed to troubleshoot these chips is given in this section. Pay special attention to the input, output, and control voltages in these circuits. IC chips are taken up in Chapter 14.

10-4-1 Checking a Basic Switching Power Supply

A basic switching supply is shown in Fig. 10-6. It shows a dc power-supply rectifier D_1 feeding a transistor switch Q_1. Transistor Q_1 is alternately turned on and off by a pulse generator or multivibrator that is connected to its base. (Multivibrators are taken up in Section 13-4.) The pulsing output of the switch is pumped into the inductor (coil) L_1, and charges its magnetic field repetitively. The coil repetitively discharges its voltage into the capacitor. The diode D_2 provides a return path for the coil discharge current. The capacitor is like a charged battery. It cannot discharge through the diode, so it discharges into the load circuit.

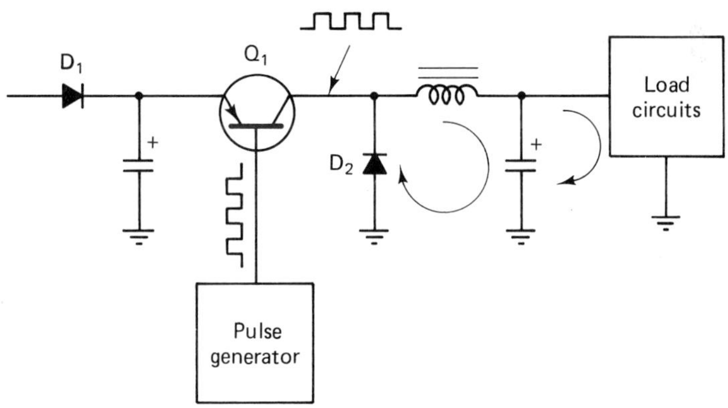

Figure 10-6 Switching power supply. The pulse generator turns Q_1 on and off repetitively to produce an average voltage at the emitter. The high frequency permits the coil, capacitor, and diode to work together to provide a dc output.

The Q_1 switch operates at a high frequency. This allows the use of smaller capacitors. The high-frequency coil, which has a ferrite core, can also be made smaller. The diode and transistor are high-speed types. Other types will not work.

Checking the Switching Power Supply with a Scope. The scope is very convenient for checking switching power supplies. Check for pulses at the base and collector of Q_1 in Fig. 10-6. No pulse at the base would indicate a fault in the pulse-generator circuit. No pulse at the collector would indicate a faulty transistor. An open diode would also cut down the pulse. There should be relatively clean dc at the output.

TIP 10-3

The output circuit of a switching supply can be checked with a scope. The input of the coil should have a rectangular pulse. The output should have very little ripple.

A shorted coil would show a large ripple at the input and output. An open capacitor would show a large ripple at the output.

Checking the Switching Power Supply with a Multimeter. Dc voltage and resistance checks will reveal drastic shorts and opens. Although the switching supply may not maintain its dc voltages in the same way, checking these voltages is not very different from checking dc voltages in the nonswitching supplies. You still have a basic dc voltage path running from the input rectifier to the load. You can apply the basic troubleshooting rules to this path. The individual parts are checked out in basic ways.

TIP 10-4

A switching-supply output inductor can be checked with a multimeter. It has very low resistance. The input dc voltage of the coil, as measured by a dc voltmeter, should not differ a great deal from its output voltage.

A dc power supply can be used to check out the load and pulse generator circuits. If these circuits work okay with a substitute supply, the switching transistor and coil circuits are bad. Otherwise, check these circuits out with the substitute voltages applied as needed.

10-4-2 Checking Switching Voltage Regulators

One common way of regulating the output of the switching supply is shown in Fig. 10-7. The output of the pulse generator shows switching pulses of varying durations. If the output voltage is too high, the on-time durations are shortened, and the average voltage is decreased to normal. If the voltage is too low, the opposite happens. The block diagram shows an error amplifier that controls the width of the pulses coming from a pulse generator. Incidentally, the triangular shape of the error-amplifier circuit block is standard for indicating an amplifier.

A key check that can be made on this circuit would be to vary the dc output voltage of the error amplifier while observing the output pulses of the generator with a scope. The width of the pulses should vary as you vary the dc voltage.

10-5 INVERTERS AND CONVERTERS

The power supplies we checked out so far converted ac to dc. Now let us look at two more types: the *inverter* power supply, which converts dc to ac, and the *converter,* which converts dc to dc—from one dc level to another. The inverter is used for operating ac equipment from batteries. The converter is used for boosting or reducing dc voltages with a minimum loss of power.

10-5-1 The Inverter

The basic theory for inverter circuits is simple. A dc voltage must first be converted into ac or pulses before a transformer can step it up or down. Figure 10-8 shows two transistors that switch each other off and on through the transformer windings to produce a pulsing output. The *LC* filter at the output converts the pulses into smooth sine waves.

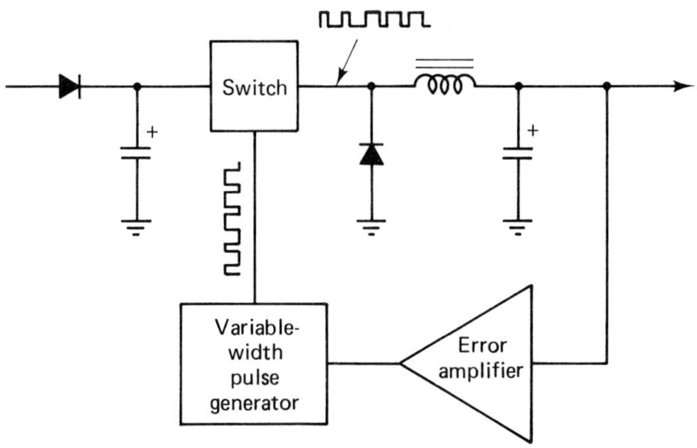

Figure 10-7 Switching voltage regulator. The feedback circuit varies the duration of the switching pulses to adjust the output voltage automatically.

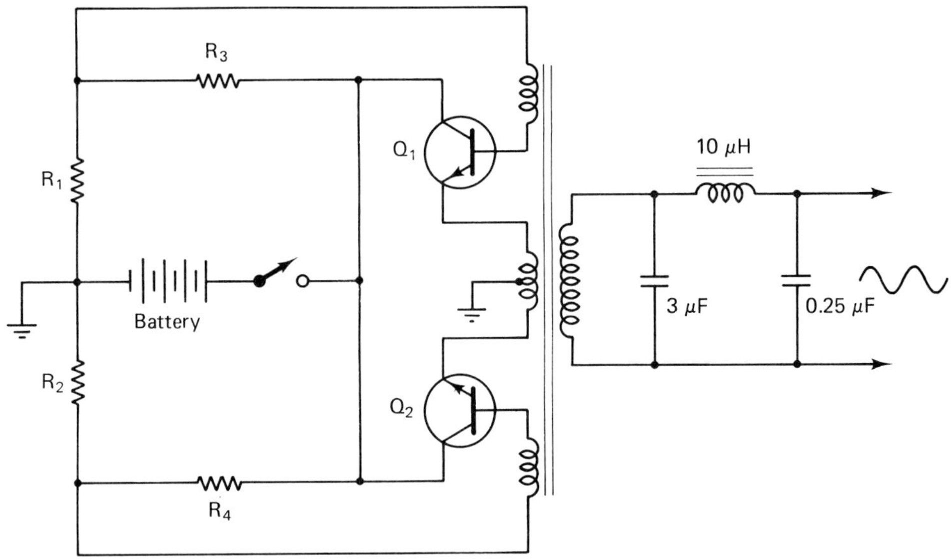

Figure 10-8 The inverter circuit converts the battery voltage to ac. The output *LC* circuit converts pulses at the transformer output into a sine waves.

Checking the Inverter. Troubleshooting the inverter is not much different from working on amplifiers. Check for battery voltage at the collectors. Check the dc base voltage; it should be less than 0.7 if the circuit is working (pulsing). If you get 0.7 V, the circuit is not pulsing, but the transistor is probably good. The other transistor is likely to be bad. The emitter should have 0 V dc. The corresponding transistor terminals should have equal voltage.

The transformer primary windings should have very low dc voltage drops and resistances. The secondary may have more resistance if the output is supposed to have a greater ac voltage than the input dc voltage.

As for the resistors, if any of them opens completely, the in-circuit resistance across it would increase substantially. In fact, if the other transistors are good, you will be measuring the sum of their resistances. All the other resistors should have resistances equal to their out-of-circuit values. (The battery is disconnected, and care is taken that the transistor does not interfere with your measurement, as described in Chapter 5.)

Of course, the battery voltage should be near maximum. Try out a new battery if the inverter is not working properly and you have less than 10% of normal voltage. A battery eliminator that outputs sufficient current can be used to check out the inverter.

Operation of the Inverter Circuit. The checks above should be enough to enable you to troubleshoot the inverter power-supply circuit. The following information may also be helpful. The operation of this inverter depends on two characteristics of transformers that we have not looked at yet. First, the direction of the secondary current of a transformer depends on the direction of the turns. In other words, the transformers can be wound so as to obtain the desired current directions or polarities. Second, the transformer has a *saturation* point beyond which there can be no increase of secondary current even though the primary voltage is still increasing. At saturation the primary circuit loses control over the secondary circuit. Also recall that transformers cannot work with steady dc voltages. Only changing voltages can be induced across them.

Note the two transistor amplifier circuits, Q_1 and Q_2. They are almost like the basic transformer circuit in Section 8-7. The only difference is that there is a transformer output at the emitter of each transistor. Another difference is that there is only one transformer. Actually, the emitter outputs are feeding the base inputs through the transformer.

Let us briefly run through the circuit operation. When we turn on the power, both transistors in Fig. 10-8 will have positive voltages on their bases. One of them, let's say Q_1, starts to conduct first. The current starts flowing through its emitter primary winding. A positive voltage is then coupled back to the base through the upper winding of the transformer. At the same time, a negative voltage is coupled to the base of Q_2 from Q_1 through the lower winding. As a consequence, the positive feedback voltage turns Q_1 on harder, and this also makes Q_2 turn off harder. In other words, one transistor is on now and the other is off.

Q_1 stays on until saturation sets in, and its emitter primary circuit loses control. When this happens, the base of Q_1 begins to lose some of its extra positive voltage. At the same time, the base of Q_2 begins to regain some of its original positive voltage, and it starts to turn on. Now since Q_2 is turning on, it is Q_2's turn to do the same thing to Q_1—to cut it off. This cycle is repeated over and over to produce a pulsing wave. This wave is coupled from the emitter to the output secondary.

10-5-2 The Converter

The inverter circuits can be used in converters to boost dc voltages to higher levels. They usually work at 60 Hz. This calls for heavy transformers and large filter capacitors. Converters, on the other hand, are for dc-to-dc operation; there is no need to limit their inverter sections to 60-Hz operation. They can operate at higher frequencies that permit the use of much lighter and smaller ferrite core transformers. The filter circuit components can also be smaller and lighter.

Converters can also utilize switching circuits. Basically, the converter can be designed by placing a transformer between the switching transistor and the output diode shown in Figures 10-6 and 10-7. The converters are checked out using the approaches described in this chapter as well as those in Chapter 6.

WHAT YOU LEARNED TO DO

In this chapter you learned to check the following power supply circuits:

1. A continuous-current voltage-regulator circuit.

2. A current-limiting circuit of a regulator.
3. A crowbar shutdown circuit.
4. A switching type of voltage regulator.
5. An inverter.
6. A converter.

QUESTIONS

1. In the transistor feedback regulator shown in Fig. 10-2, what would happen if the power transistor opens?
2. In the circuit in Question 1, what can cause the power transistor to burn out?
3. How can you check to see if an adjustable voltage regulator is operating?
4. What signal and dc voltages would you expect to see at the input and output of the inductor in a switching regulator?
5. Can you replace the transistor or the diode in a switching power supply with just any type of component as long as it can handle the voltage and current?

PRACTICAL TROUBLESHOOTING EXPERIENCE

Obtain a chassis that has a nonswitching voltage-regulator IC chip and a circuit breaker. If it has a fuse instead, you may choose to jump it with a circuit breaker of the same rating. If the fuse is good, disconnect it.

Use an isolation transformer. You will also need a series test lamp or variable transformer. Turn on the power and see if the circuit breaker trips. If the circuit breaker trips, start at step 1. If the breaker does not trip, start at step 4.

If you cannot find a chassis, obtain a schematic and make a "dry run." That is, trace the test points on the schematics as you read through the steps below.

1. If the circuit breaker trips, turn off the power and without resetting the breaker, check the input rectifier for a short with an ohmmeter. Then check the resistance between the input and output of the regulator for a short. Then check the output for a short. Replace the rectifier if necessary. If the resistance checks seem okay, proceed to the next step.
2. If the circuit breaker is tripped, reset it and apply reduced line voltage. If you are using a variac, turn up the voltage as high as you can without tripping the breaker. Check the input rectifier diode and replace it if necessary. Check the input and output voltages of the regulator. What could be wrong if the output voltage is higher than normal? What could be wrong with the regulator if the output voltage is lower than normal?
3. If you have lower-than-normal voltage at the output of the regulator, turn off the power and disconnect the line to the output pin; or else, disconnect the supply

line from the variable-resistor circuit, which should remain connected to the chip output. Turn the power back on and retake the output voltage. If the voltage returns to normal, what could be wrong?

4. If the circuit breaker does not trip, check the regulator input and output voltages. If there is no input voltage, what could be wrong? If there is no output voltage, what could be wrong?

11

Power Amplifiers and High-Voltage Circuits

EXPERIENCE REQUIRED

To make use of this chapter, you should know how to:

1. Check out multiple-transistor amplifier stages.
2. Trace dc signal paths from stage to stage (Chapter 9).
3. Check feedback paths (Section 9-5).
4. Work with positive and negative voltages (Section 2-5).
5. Check for shorted and open loads (Section 4-6).
6. Identify power and high-voltage components, such as power transistors, heat sinks, horizontal output transfomers, high-voltage rectifiers, and triplers.

11-1 POWER-AMPLIFIER CHECKPOINTS

Many power-amplifier troubles are associated with power-supply troubles. For instance, when a fuse blows, we learned to look for trouble in either the power-supply or power-output stage. We touched on this when we worked with basic power supplies in Chapter 6 and with advanced supplies Chapter 10.

Beginners should learn how to troubleshoot the power output stage early. It is widely used, very basic, easy to locate, and has a high failure rate. It drives devices that take up a lot of power. As a consequence, the output transistor runs hot and is prone to failure. It often has to be mounted on a metal platform called a heat sink, to draw off the heat.

The most common power-output device or load is the speaker. Other types are motors, relays, and electromagnetic units for video-display tubes. All these power-consuming devices have low resistances or impedances that pull down hard on the power source.

A simple type of power amplifier is one that uses transformer coupling. We saw one in Chapter 8 (Fig. 8-14). The transformer is used to match up the load characteristics with the transistor characteristics to deliver the maximum power. It has been found that maximum power is delivered to the load when its impedance equals the amplifier impedance.

To the troubleshooter, aside from the problems associated with the heavy load, this amplifier is similar to the basic transistor amplifier. You should be able to work on it with the information given in Chapters 4 and 8.

The tips and procedures for checking out the power amplifiers in this chapter should be studied very carefully. The knowledge gained here will help you troubleshoot many other types of circuits.

11-1-1 Complementary-Symmetry Amplifier Voltages

The complementary-symmetry amplifier and improvements on it have largely replaced the transformer-coupled amplifier. In Fig. 11-1, a pair of NPN and PNP output transistors are connected in what is called a complementary-symmetrical arrangement. The arrows show that when the signal voltage at the input of Q_1 goes down, the collector goes up. The bases of Q_2 and Q_3 also rise and their emitters must follow. When the input signal reverses, the opposite will happen. Thus Q_2 and Q_3 work together to produce an output signal.

> *Don't forget:* An arrow can represent either the normal voltage change of a signal at a particular instant, or it can represent a defective dc voltage that is stuck too high or too low. Note also that the arrows show how defective voltages can be transferred in the forward direction.

Notice the feedback from the output centerline. It sends back a voltage (dashed arrow) that opposes the input voltage. This will counteract distortion arising out of the circuit. For instance, if the output is a little too high at any instant, the feedback voltage at that instant will reduce it back to normal.

Note that the voltage on the centerline output splits the supply voltage in half. This allows maximum up-and-down voltage swing. Look for this key voltage when checking symmetrical amplifiers.

The speaker output terminals should have 0 V dc on them. A large electrolytic capacitor is used to block the centerline voltage. This dc voltage can damage or weaken the speaker. [For circuits that have a 0-V dc centerline, a blocking capacitor is not needed (see Fig. 11-4).] A speaker can withstand up to about 1.7 V dc without harmful effects.

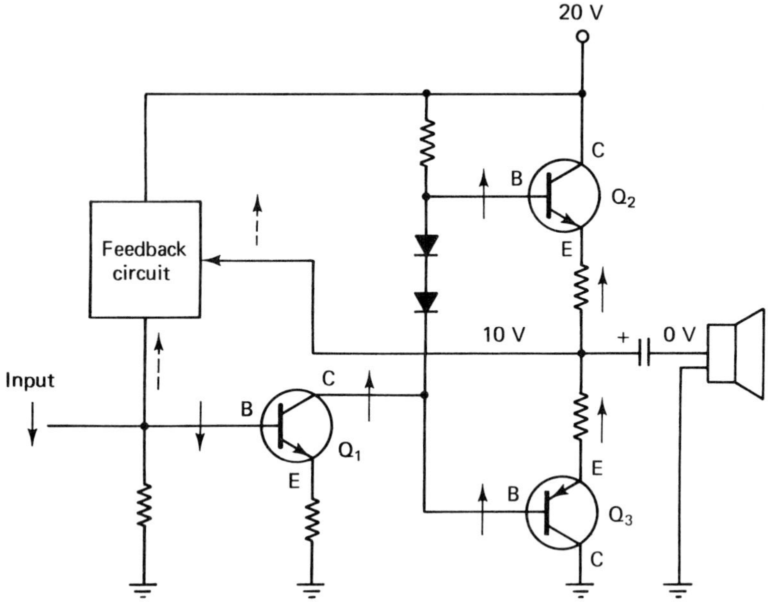

Figure 11-1 Complementary-symmetry amplifier. The centerline voltage cuts the supply voltage in half. The solid arrows show how a higher-than-normal dc voltage at the output can be traced back to the input. The dashed arrows trace the resulting feedback correction voltage.

TIP 11-1

The centerline dc voltage in a complementary-symmetry or similar amplifier should split the supply voltage in half. Its average dc value should not vary widely with changes in signal level.

TIP 11-2

The speaker terminals of an amplifier should have 0 V dc on them. Amplifiers that have no coupling capacitors may not have exactly 0 V dc.

11-1-2 Comparing Top- and Bottom-Side Voltages

The circuit in Fig. 11-2 is called a quasi-complementary-symmetry amplifier because the NPN output transistors are not complementary. It is widely used in audio ampli-

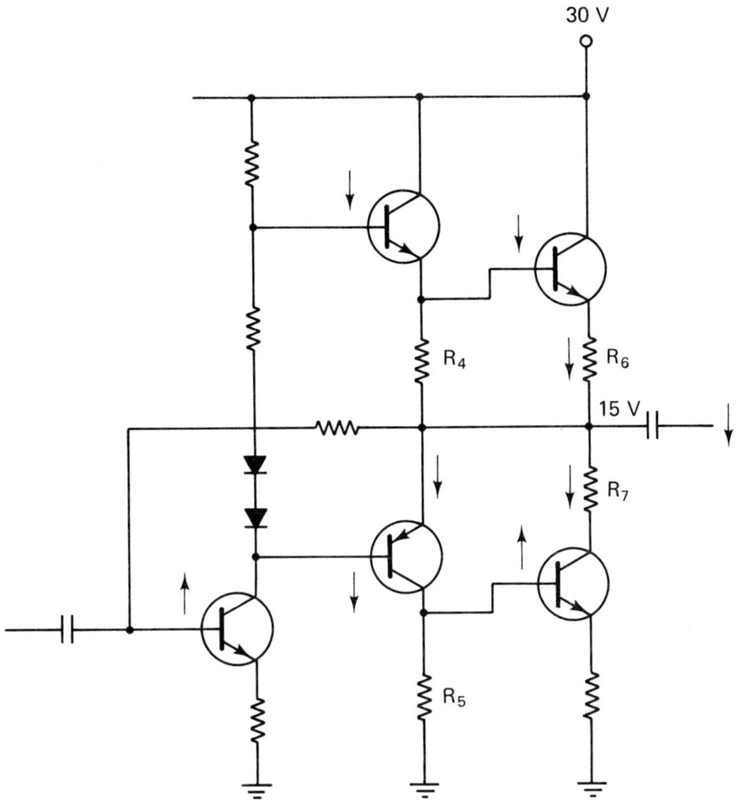

Figure 11-2 Quasi-complementary-symmetry amplifier. Note the direction of the arrows as you trace a fault from input to output. The voltages across the corresponding top and bottom parts should be equal.

fiers. The thing to remember about this or any other amplifier with a centerline is that the corresponding top and bottom parts have equal voltage drops. R_4 and R_5 should both have about 0.7 V drop with no signal applied. R_6 and R_7 should have much less voltage across them than do R_4 and R_5. The corresponding *BE* voltages should be fairly equal.

TIP 11-3

The voltages across the corresponding top and bottom parts of a push-pull or complementary-symmetry amplifier should be approximately equal.

11-1-3 A Positive–Negative Power Supply

Figure 11-3 shows a power supply that is used for the zero-centerline amplifier. It looks like the basic bridge rectifier. The difference is that the negative output has replaced the ground connection, and ground has been moved over to a center tap in the transformer. Troubleshooting it is not very different. You still have the basic sections to check—the filter, the rectifier, and the transformer or input section. Shorts and opens will have the same effect. A shorted capacitor will blow a fuse or a rectifier. An open capacitor will cause some loss of voltage and ripple or hum.

There is one problem you might come across when working on a circuit that has negative and positive voltages. Suppose that the ground line breaks at point X shown in Fig. 11-3. Notice that you would have a totally unbalanced condition existing between the positive and negative supply lines in some of the circuits. Needless to say, the voltages in these circuits would be upset. You may have negative voltages in normally positive circuits, and vice versa.

To explain further, if circuits 1 and 2 together have less resistance than circuit 4, they would pull the opened ground voltage up from 0 V to perhaps 5 V, as shown

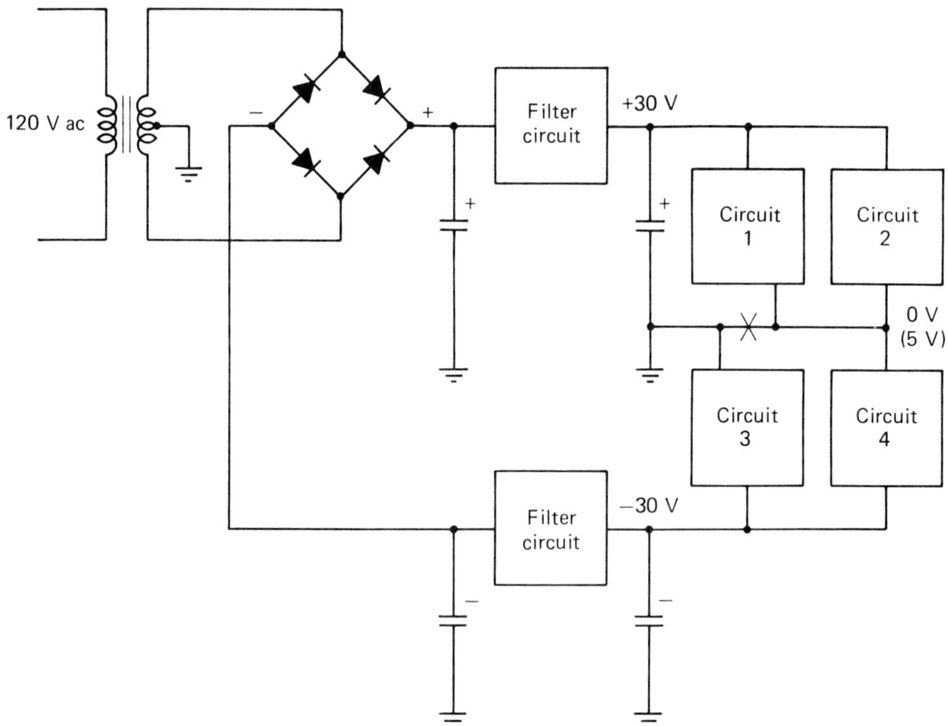

Figure 11-3 Bridge-rectifier power supply with positive and negative outputs that is feeding amplifier circuits 1 to 4. Note how a break (×) in the ground line upsets the dc voltages on three of the circuits.

in the schematic. Also, circuit 4, which is normally a circuit with all negative voltages, would have some positive voltages in it.

> TIP 11-4
>
> If you see wrong-polarity voltages in a system with positive and negative voltages, check the ground connections.

11-2 TROUBLESHOOTING A POWER AMPLIFIER

When troubleshooting power supplies, we divided the troubles into the shorted power supply and the open power supply. The same thing is true with shorted and open power amplifiers. In the case of a shorted power output amplifier, as soon as we turn on the power, the fuse will blow, and we cannot troubleshoot it. Obviously, we have to use a different approach when the fuse blows and when it does not.

11-2-1 The Preliminary Checks

Suppose that the main amplifier fuse is blown. This means that we have to find out whether the short is in the power supply or the power amplifier. First we look for the fuse going to the power amplifier. If it is blown, the power amplifier has a short.

Otherwise, check the power output transistors and power rectifiers with an ohmmeter. A third method is to use a series test lamp or a variable transformer and check the voltages.

Before you do anything else, however, note the following tip:

> TIP 11-5
>
> If one side of a stereo system is working, the common power supply is okay. Compare the voltages and resistances of the right and left amplifiers.

11-2-2 A Point-by-Point Failure Analysis

In Fig. 11-4 we are looking at the schematic of a stripped-down version of a quasi-compementary-symmetry amplifier; many of the necessary but nonbasic parts are

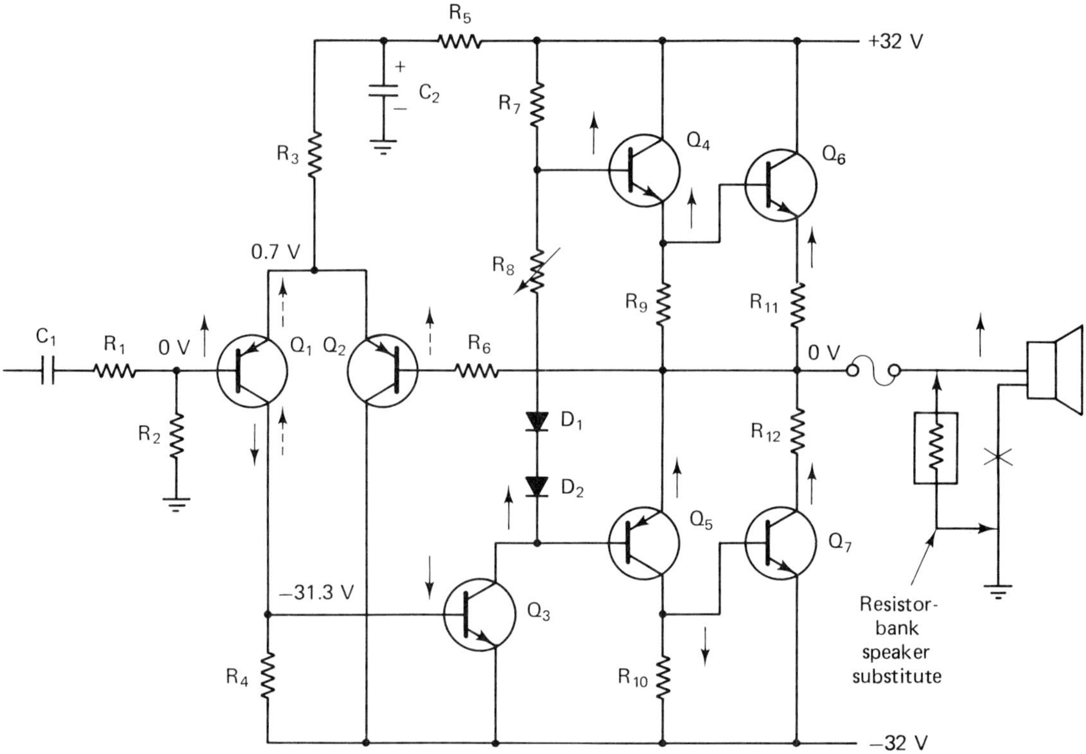

Figure 11-4 The differential amplifier in this popular circuit amplifies the difference between the feedback and the input voltage. It rejects interference signals picked up at the bases. The arrows show how to trace defective dc voltages.

missing. It has positive and negative supply voltages. Let's say that "down" is toward the negative supply line, and "up" is toward the positive supply line. This is easy to see in a typical schematic, because the negative voltages are drawn below the zero-volt centerline.

Let's start out with Q_6 and Q_4, the output and driver transistors. If Q_6 shorts from collector to emitter, the centerline voltage would be pulled up toward +32 V. This could put as much as 60 sustained volts on the bottom transistors, since they could have nearly +32 V on top and −32 V underneath. R_{11} has a very low value, so it may not drop enough voltage to save the bottom transistors.

If the driver transistor, Q_4, shorts, it would drive up the base voltage of Q_6 and cause it to effectively short out. R_9 may tend to drop some voltage but not enough to counteract the short. Without going into details, this would set off a chain of actions that could possibly damage the transistors.

If there is no fuse or thermal cutout to protect it, the speaker may be damaged. Some technicians replace all the transistors and check all the other parts in the circuit while the transistors are out.

A similar thing happens when Q_5 or Q_7 shorts. The centerline voltage could drop to nearly -32 V, and all the transistors could be subjected to nearly 60 sustained volts and excessive power drain or *BE* drop. The fuse or thermal cutout for the speaker should open.

> TIP 11-6
>
> In a complementary-symmetry power amplifier without protective circuits, if an output transistor shorts, it may be best to replace all the transistors at one time. While the transistors are out of the circuit, check out the other parts under full voltage. After replacing the transistors, test the amplifier under reduced voltage.

An open output or driver transistor will not cause as much trouble. Say that Q_6 opens. The centerline voltage will tend to fall, but feedback action will cause Q_5 and Q_7 to cut off and reduce the current. Q_4 can pull up on the centerline voltage without too much strain. Its emitter resistor is large enough to protect it, too. The speaker fuse or thermal cutout may not blow, although the voltage may go negative. The open transistor can be identified by the reduced voltage drop across its emitter resistor, R_9, R_{10}, R_{11}, or R_{12}.

Having checked out the more catastrophic faults at the output, let us now go to the input. The base of Q_1 should have nearly zero volts. If it is too high, C_1 or Q_1 could be shorted. Note the solid arrows: They show what would happen to the dc voltages throughout the amplifier if the base voltage is too high. Also, if Q_1 base is too high, its emitter voltage will rise and cut off Q_2. The feedback path will be disabled, and the centerline voltage cannot be restored. The circuit breaker for the speaker could open.

> TIP 11-7
>
> If the speaker circuit breaker of a quasi-complementary-symmetry amplifier opens, it could be caused by any defective transistor.

Getting back to the schematic, we see that the emitter of Q_1 should have about $+0.7$ V. If it is too high, any part or connection below it might be open. That is, Q_1, Q_2, and R_4 could be open. Also, C_1 could be leaky, causing the base voltage to rise and the emitter to follow. Finally, there could be something pulling up the centerline feedback voltage. Suspect an open R_6 connection among a lot of other things, the transistors being the main suspects.

If the emitter of Q_1 is too low, there could be a short below, an open above; or perhaps something is pulling down the feedback line. This something could be a shorted C_2, open R_3 or R_5 on one line, or the transistors on the centerline.

The collector voltage of Q_1 should be -31.3 V. This puts 0.7 V across the *BE* terminals of Q_3. Any fault ahead of it in the signal or feedback path will change this voltage. Q_3 itself could be defective.

If Q_3 shorts, it will pull down the centerline voltage through Q_5, and the fuse or thermal cutout to the speaker could open. The feedback path will be blocked at Q_3, and the centerline voltage will remain down. Recall that ''down'' is more negative. If Q_3 opens, the centerline voltage will rise and remain high.

Now let's see what can go wrong with Q_2. It handles the feedback signal from the centerline together with Q_1. The feedback correction voltage is passed through them to Q_3. Q_2 and Q_1 form a differential amplifier. Recall that a differential amplifier passes only the difference between the voltages at its inputs. In this case the difference voltage would be that between the signal and feedback voltages. Any noise or hum picked up by both inputs would be rejected.

11-2-3 Power Amplifier Checks

Before you do anything else, check the speaker. You want to avoid the situation where a repaired amplifier had been returned to the user, only to have it blown out by a shorted speaker or an incorrect connection. The opposite has also happened. A defective amplifier can blow out the technician's test speaker. Refer to Fig. 11-4 again for the following checks.

1. Check the speaker with an ohmmeter. Speaker resistances (not impedances) range from 4 to 15 Ω—about 7 Ω for most push-pull amplifiers.

> **TIP 11-8**
>
> Always ask for the user's speaker. The speaker could have been shorted. A shorted speaker can blow a new output transistor. A defective output transistor can blow the speaker.

2. If the speaker is bad, remove it and check the voltage on the speaker terminals for zero volts.
3. With the speaker or its substitute removed, check the resistance across the speaker output terminals. In this amplifier it should measure near infinity with a low-voltage ohmmeter. (In some amplifiers, there may be a dc path to ground from an extra feedback path.) If there is a short, the most likely culprits are the

TIP 11-9

Test the speaker with your ohmmeter on the R×1 range. Scratch the ohmmeter leads against the speaker terminals. You should hear scratches from the speaker.

TIP 11-10

To avoid damaging a speaker, you can substitute it with a resistor bank made up of ten 100-Ω 10-W resistors connected in parallel, or a single 8-Ω 100-W dummy transmitter antenna. If you are only looking for defective voltages, a 100-W incandescent lamp will do.

output transistors. However, you may have to check out other paths connected to the centerline. In actual circuits, you will find capacitors that are not shown in this schematic.

4. With the speaker and substitute removed, check the centerline voltage. If there are variable resistors in the amplifier circuit, adjust them to obtain the correct centerline and *BE* voltages if you can. If you have to turn the control to an extreme, or if you cannot get near zero-centerline voltage, the transistors are probably bad. Before checking them, however, check the supply-line voltages.

5. If the fuse blows, replace it and use a variable transformer to reduce the ac line voltage. Start out at 0 V. The positive and negative supply voltages should be equal. If they are not equal, check the power supply. Check the ground connections.

6. Check the output transistors, Q_6 and Q_7. They will probably check bad. Before replacing them, you have to check the circuit to make sure that they do not blow out again.

7. With the output transistors removed, check the amplifier voltages at full power. The centerline voltage should be close to 0 V dc. Check the base and emitter voltages as well as the *BE* voltages of the drivers, Q_4 and Q_5. If these voltages are good, you can put in the new output transistors.

8a. If the drivers check okay but the voltages are bad, check the Q_1–Q_2 and Q_3 circuits. Check the electrolytics first since they are the ones most likely to go bad. The schematic shows one of them. If it opens, the signal will deteriorate but the dc voltages will not be affected. (A scope or a substitution test will reveal an open capacitor.)

If a capacitor shorts, dc voltage checks will help you find it. If the input capacitor, C_1, shorts, the base voltage will rise. If there is a voltage drop across R_1, we can be fairly certain that C_1 is shorted. If C_2 shorts, the voltage above it will fall.

At reduced voltage if necessary, check all the resistor voltages. If R_1 opens, a jumper wire connected from the emitter of Q_1 to the input side of R_1 will not change the base voltage of Q_1. If R_1 increases resistance, the base voltage may not change. Check it with an ohmmeter if you have not done so already

If R_2 opens or increases resistance, there may be no significant change in dc voltage at the base of Q_1, so you will have to turn off the voltage and check it with an ohmmeter.

If R_3 opens or increases resistance, the voltage below it will fall. If it increases resistance, the voltage below it may not fall significantly, but the voltage above it will rise.

If R_4 opens or increases, the voltage above it will rise (become less negative) and there may be excessive *BE* voltage on Q_3.

If R_5 opens or increases resistance, the voltage below it will fall.

If R_6 opens or increases resistance, you may have to use an ohmmeter to discover it.

If R_7 opens or increases resistance, the voltage below it will fall. Any opposing tendency through the transistors will be switched off by the diodes.

If R_8 opens or increases resistance, excessive *BE* voltages on Q_4 and Q_5 could damage them, and Q_6 and Q_7 could be damaged, too.

If R_9, R_{10}, R_{11}, or R_{12} opens or increases resistance, the voltage drop across it will increase. Transistor actions may counteract certain voltage changes, but the resistor voltage drops will increase significantly in the end.

The diodes, D_1 and D_2, should each drop 0.07 V. If either opens, excessive *BE* voltages on the drivers, Q_4 and Q_5, could damage them. The output transistors, Q_6 and Q_7, could be damaged, too. These diodes usually come packaged in one unit. You may have to find out whether there are two or three in series—whether they should have approximately 1.4 or 2.1 V across them.

8b. If the drivers, Q_4 and Q_5, are bad, remove them and check the base terminal voltages. If these voltages are good, you can install the new drivers safely.

9. After installing the new drivers, check out the circuit voltages at reduced line voltage first, then at full line voltage. If the voltages are good, install the new output transistors.

10. After replacing the transistors, with the substitute resistance load for the speaker connected, run the amplifier at reduced voltage for 20 minutes. Monitor the centerline voltage. After a few minutes it should settle down to where its peak dc fluctuation ceases to creep up or down. If it does not settle down and fluctuate about a steady center voltage, something is wrong. We're talking about a few hundredths of a volt change.

Turn up the volume occasionally. Watch the centerline voltage when you do this. It should not change excessively.

11. Monitor the emitter and centerline voltages for several hours at full voltage. Watch for creeping changes.

TIP 11-11

If you do not have replacements for the two output transistors of a power amplifier, you can remove them and connect a 100-Ω 10-W resistor across each of the two collector-to-emitter terminals. The base terminal is left open. You can then check all the dc voltages in the circuit. You can do the same for the two symmetrical driver transistors with a 1000-Ω 2-W resistor. Look for symmetrical voltages on the output and driver transistor terminals.

11-3 DEAD POWER AMPLIFIER (FUSE NOT BLOWN)

You may have a dead power amplifier that has not blown a fuse. In most amplifiers you should get some kind of output, even if it is distorted. You have three likely causes for no output at all. One is an open power supply. Another is a bad speaker. A third could be an open output coupling capacitor. Refer back to Fig. 11-4 for the following tips and steps.

TIP 11-12

If you have no output from a stereo-amplifier channel, switch the right and left inputs. If you can get an output, the amplifier is okay.

Do not switch speakers. A shorted speaker can burn out an amplifier.

If you have no output from either channel, do not forget to switch input functions. For instance, if you have a signal when you switch from tape to AM, the amplifier is okay.

In addition to the tips above, the following checks should help you track down the faulty part.

1. Check for an open fuse or thermal cutout to the speaker with an ohmmeter. If either has opened, check the amplifier.

TIP 11-13

To check a dead amplifier, turn down its level control for a very low output, and touch the input terminals. If you hear hum at the speaker, the amplifier is okay. You may have to turn up the volume to hear the hum. Do not try this test if the amplifier does not have a level or volume control; you may damage the speaker.

2. Check to see if you have supply voltage at the power output transistors. They may be located externally on the back of the cabinet. One of them, Q_6, should have the full supply voltage on its collector casing. The other, Q_7, should have very little voltage. If you have no collector voltage on Q_6, a power-supply rectifier or an internal fuse is probably open.
3. Check for an input signal at the input capacitor. If you have a signal here, the preceding circuits are okay.

The following tip is for amplifiers that require output capacitors.

TIP 11-14

To check for an open output capacitor in an audio amplifier, turn off the power and rub the capacitor input lead with your analog ohmmeter prod. Ground the other probe. The range should be R $\times$ 1. If the capacitor is good, you should get a scratching noise from the speaker. The capacitor could be shorted; check it with a voltmeter, with the speaker disconnected.

11-4 DISTORTION, HUM, AND NOISE

Distortion and hum are frequent complaints. In these cases careful checks and adjustments are necessary. The symptoms for distortion are generally poor quality, overtones, or mushy and scratchy sounds. Distortion occurs most often in the power amplifier itself, whereas hum pickup and noise generation occur primarily in preamplifiers. Right now, we are looking at the power amplifier.

1. Distortion is usually caused by an unbalanced centerline voltage and bad *BE* voltages. The procedure is similar to that for shorted amplifiers, but closer measurements are required.

2. Hum can be caused by open bypass capacitors in the supply line. Hum can also be picked up by the feedback line when a bypass capacitor opens (see C_2 in Fig. 11-4). Jump the bypass capacitors one at a time with a good one.

   Jump the input of the amplifier to ground with a high-capacitance electrolytic. If the hum is reduced, the power amplifier is not the source of the hum. You cannot just short the input to ground because that would upset the dc operating voltages in the entire amplifier.
3. Noise is picked up mostly in the stages preceding the power amplifier. Noise is also picked up through the supply-line distribution lines. In this respect the procedures for troubleshooting noise are the same as for hum.
4. Noise has a higher frequency than hum, so it requires less capacitance and shows a faster response. Jump all the small bypass capacitors with an equal- or higher-value electrolytic capacitor. You might have to use short leads and tack solder the substitute capacitors in place. Long leads can pick up noise and cause oscillations.

The basic amplifier in Fig. 11-4 is designed to reduce noise and hum by means of its feedback path. If you look at any real schematic of an audio amplifier, you will recognize the basic similarity between it and our amplifier. However, you may see other small capacitors, resistors, and coils added on. Some are for suppressing noise, hum, and oscillations. These will have to be checked out, too. You learn more about these in Chapter 12 when we study high-frequency circuits.

11-5 READING A STEREO AMPLIFIER SCHEMATIC

Figure 11-5 shows the amplifier section of a Sony STR-11S stereo receiver. It has improved circuitry not found in our basic amplifier. We will look at the left channel only, since the right channel is identical. The line that is connected to R_{702}, R_{602}, R_{605}, and so on, is the ground line. Q_{703} has been added to stabilize the driver and output stages and to prevent distortion.

The fusible resistors, R_{712}, R_{713}, R_{714}, R_{717}, and R_{720}, protect the circuits against excessive currents. They, along with the circuit breaker CB_{701} and their counterparts in the right channel, must be replaced only by factory-authorized parts.

The zener diode, D_{701}, and the electrolytic capacitor, C_{702}, stabilize the operation of Q_{701}, Q_{702}, and Q_{703}. The Q_{701} emitter resistors, R_{704} and R_{705}, divide the signal to the amplifier pair Q_{706} and Q_{707}. Although the bottom transistor, Q_{707}, has a low collector-to-emitter voltage, it does amplify the current.

The tone-control circuits determine the amount of bass and treble fed back to Q_{702} and Q_{701}. It is a negative feedback circuit. This circuit can be checked out with an ohmmeter.

11-6 TROUBLESHOOTING A VERTICAL-OUTPUT STAGE

The vertical amplifier is used in video monitors to spread out the display vertically. When it weakens, the display flattens out. When it fails completely, the display

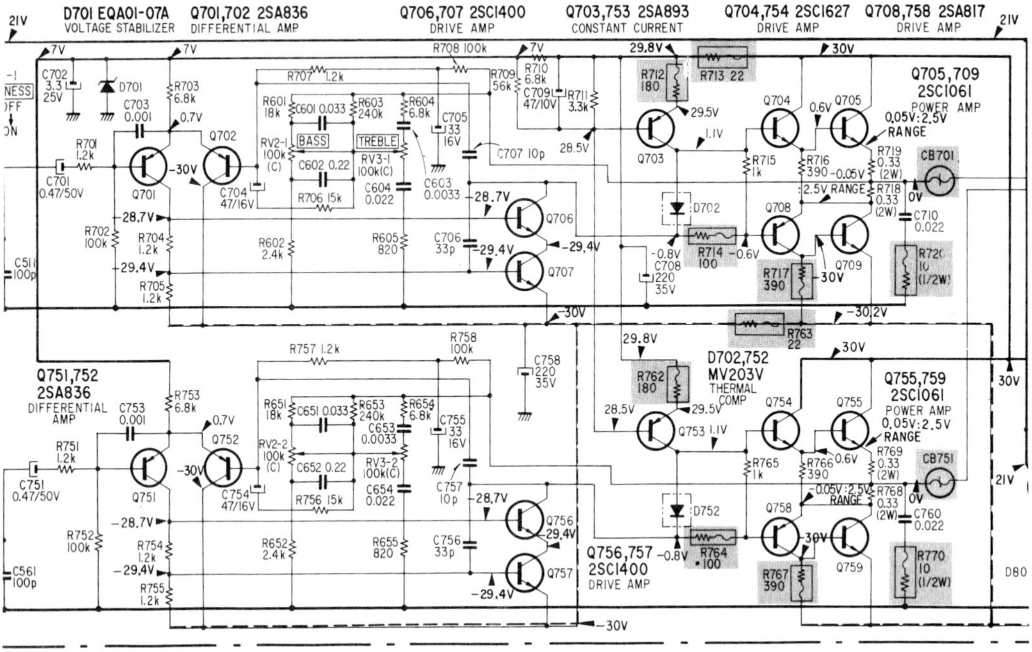

Figure 11-5 Schematic of the amplifier section of a Sony stereo unit. (Copyright Sony Corporation)

collapses into a bright horizontal line across the face of the display tube. These amplifiers are basically the same as audio amplifiers. Because of the high power requirements, they often fail.

11-6-1 A Basic Vertical Output Amplifier

The vertical amplifier in Fig. 11-6 inputs a pulsating vertical deflection signal. The output drives a powerful pulsating current through a large electromagnetic coil called the *yoke,* which is mounted on the neck of the CRT (see Fig. 13-4). This produces a magnetic force that drives a white horizontal line up and down the screen at the rate of 60 times a second. This rapid movement of the line covers the entire screen with a bright *raster* of approximately 500 horizontal lines.

You will notice some special connections, called *takeoff points,* that are typical of vertical output amplifiers in video equipment. A *blanking pulse* is fed to the video-display circuit to blank out undesirable lines in the picture. A *convergence pulse* is sent to a circuit that properly positions the various color elements in the display.

The vertical amplifier usually obtains its power from the horizontal amplifier section. There are several advantages in this, which we will learn about in the next section.

Troubleshooting the vertical output amplifier is not very different from troubleshooting audio amplifiers. Following are some hints that will help you to trace troubles to this amplifier and within it.

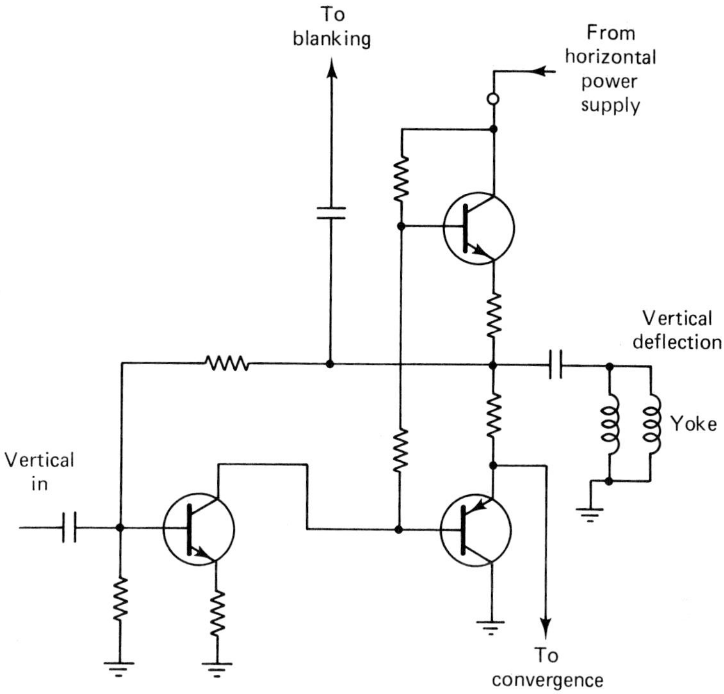

Figure 11-6 The vertical output amplifier, which is used in video display tube circuits, is very similar to the audio amplifier, and troubleshooting it is similar.

1. In somc systems a short in the vertical output circuit may disable the horizontal circuit (see the next section). This is because it derives its power from the horizontal circuit.
2. A horizontal white line across the display indicates a completely dead vertical section, not necessarily in the output circuit. The horizontal section that produces the horizontal white line is okay.
3. Lack of height in the display indicates a weak vertical amplifier, usually an increased-value collector resistor in an early stage.

The main thing to remember about this circuit is that it is basically the same as the audio amplifier in the previous sections. Here and there in various designs you will find a few extra parts, but you should not have much trouble adjusting. For instance, just as tone-control resistors and capacitors are added to basic audio amplifiers, you will find waveshaping resistors and capacitors added to vertical amplifiers.

One difference between the audio amplifier and the vertical output amplifier is the signal. The audio amplifier handles sine-wave signals. The vertical amplifier handles pulsing signals. Most of the pulse shaping is done before the signal reaches the output amplifier. In some vertical amplifiers, the top output transistor is designed to handle the high upward pulses, and the bottom output transistor handles a lower voltage. The centerline voltage no longer splits the supply voltage in half. If the supply voltage is 100 V, the centerline voltage may be 20 V intead of 50 V.

The following tip may be useful. It tells you whether the yoke or the amplifier is bad when there is no vertical deflection on the screen.

> **TIP 11-15**
>
> To check the vertical deflection yoke, apply about 24 V ac to it. This voltage can be derived from a standard power transformer through a 50-μF electrolytic capacitor. Assuming that you have a positive dc voltage on the yoke, the positive terminal of the electrolytic should be connected to the yoke input. The other transformer output lead is grounded. If there is no dc voltage on yoke, you can connect the transformer leads directly to the yoke leads. If the yoke is good, you will obtain a raster that will show a distorted pattern.
>
> If the yoke is good, the vertical amplifier is bad. Apply the same signal to the vertical amplifier input. You can keep injecting this voltage into the preceding stages until you find the bad stage. Ground the capacitor to discharge it after each check. Add a 1-kΩ resistor in series with the capacitor for extra protection.

You may find a suitable ac output at the power transformer. Just connect the electrolytic to it. In the old days the vacuum-tube heater voltage was used for this purpose. All you needed was an electrolytic capacitor with a jumper clipped to one of the vacuum-tube heater pins. This was also used as a signal injector for testing the video and audio stages. Do not use the CRT heater voltage for this purpose. The CRT is too expensive; do not fool with it.

We will return to the vertical amplifier in Chapter 13 when we take up TV and video-display circuits. We consider the symptoms of defective vertical amplifiers there.

11-6-2 A Vertical Output Amplifier Schematic

Figure 11-7 shows the vertical amplifier section of the RCA CTC-126A chassis schematic. This circuit provides excellent vertical linearity with a minimum of parts. At first sight you might think that it works like any basic vertical output amplifier, but note that the predriver, Q_{503}, cannot drive the output transistor, Q_{504}, because C_{508} is too small. The drive signal must be fed through the bottom output transitor, Q_{505}. This is different from the basic circuit, in which has the drive signal splits in two to feed the both output-transistor inputs.

Note also the extra 115-V supply which brings the collector voltage to 27.2 V, which is higher than the 23-V source above R_{520}. This, too, is different. The extra voltage is used to improve the timing of a portion of the vertical output signal.

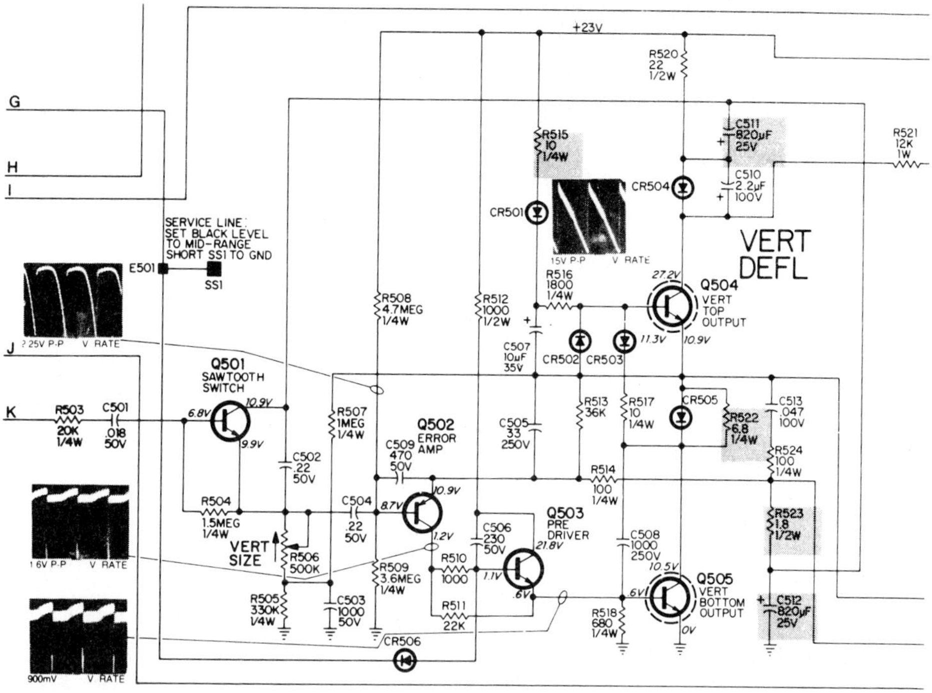

Figure 11-7 RCA vertical output circuit. (Reprinted courtesy of Thomson Consumer Electronics)

The switching actions of the diodes are the keys to the operation of this circuit. The various switching actions and the charge–discharge actions of the electrolytic capacitors, and their timing are quite complicated and difficult to follow. Fortunately, the schematic shows the resultant dc voltages. This makes things easier.

A key voltage would be the 10.9 V dc on the line at the emitter of the output transistor, Q_{504}. Note that this voltage is also present on the line at the emitter of Q_{502}. These two lines output to the yoke, which is connected across C_{513} and R_{524}.

A short above the 10.9-V line or an open below it will cause this voltage to rise. An open above it or short below it will cause it to fall. Defective base voltages on Q_{504} and Q_{505} will also change this voltage. This can be followed up by checking the voltages on Q_{503} and Q_{502}.

Dc defects in the Q_{501} emitter circuit would be blocked by C_{504} and substantially by the 1.5-MΩ resistor, R_{504}, but the collector voltage may cause problems. Introducing a substitute 10.9-V dc voltage through a 1-kΩ resistor at the collector may help.

Other useful measurements can be made on the diodes and electrolytics. Recall that a forward-biased diode should have 0.7 V across it. An electrolytic capacitor should have a higher positive voltage on its positive terminal than on its negative terminal.

11-7 THE HORIZONTAL OUTPUT AND HIGH-VOLTAGE CIRCUITS

The purpose of the basic horizontal amplifier in Fig. 11-8 is to drive a pulsing voltage into the electromagnetic yoke coil on the neck of the CRT. There are two sets of coils, one for the vertical and one for the horizontal pulse voltages. The magnetic force of the horizontal coil drives a white dot back and forth across the face of the CRT, 15,750 times a second, to produce a bright line. The vertical yoke drives the dot vertically to produce 525 horizontal lines for the display-tube raster.

As you can see, the horizontal amplifier is not very different from the basic transformer-coupled transistor amplifier. It usually has several extra jobs to do, one of which is to produce approximately 30,000 V to light up the CRT screen. Let's take a closer look at the amplifier.

The input transformer T_1 receives the horizontal pulse signal that originates from a special pulse generator circuit (see Chapter 13). The amplified signal at the collector of Q_1 is applied to the horizontal output transformer. Two basic signals are developed by the transformer. First is the horizontal deflection voltage, which is applied to the horizontal yoke Y_1. Second, it steps up the voltage to about 10,000 V. This is fed to the high-voltage rectification unit, D_2, which is a voltage tripler that develops 30,000 V for the picture tube. All parts in the high-voltage circuit are of special materials and construction to withstand the high voltage.

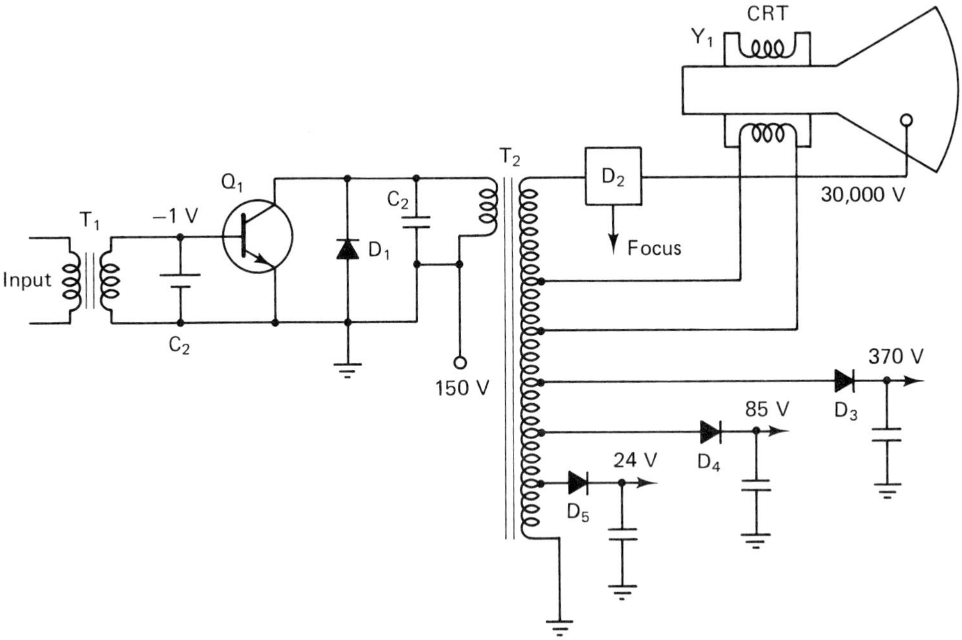

Figure 11-8 The horizontal output amplifier, which is used in TV and computer display tube circuits, is similar to the basic transformer-coupled amplifier. It also supplies dc power for several other circuits in the system (see D_3 to D_5).

The secondary of T_2 is also used to develop low voltages for various other circuits in the system. For example, it may produce 370 V for the vertical output amplifier or the CRT circuits.

A damper diode, D_1, is connected between the collector of Q_1 and ground to stabilize the response of the transformer. Coils and transformers have a tendency to generate extra signals when excited by pulse voltages. Diode D_1 dampens most of this unwanted signal, or *oscillation* as it is called.

Again, we can see that this is basically an amplifier circuit. The biggest difference lies in the ability of the output transformer to generate a high voltage. It causes a lot of trouble, as you can well imagine.

When we have no high voltage, we have to make some quick checks to see whether the trouble lies in the high-voltage circuit or in the horizontal output amplifier circuit.

11-7-1 Checking the High-Voltage Circuit

Noisy blue-white arcing results from high dc voltage. If you have it, that is good news. At least you know that the most expensive and hardest part to replace, the horizontal output transformer, is okay. You also know that the high-voltage tripler D_2 and horizontal circuits are okay. In most cases all you have are dirty surfaces that need to be cleaned. Dirty glass around the high-voltage connection of the picture tube (called the anode or belly button) is a common cause of noisy arcing.

> TIP 11-16
>
> **Caution:** Before working around the high-voltage anode button, you should discharge the CRT when the power is off. The CRT is actually a large capacitor that is capable of holding a high-voltage charge for a long time. Clip a jumper lead from ground to a long, thin screwdriver. Poke the screwdriver under the rubber "suction" cup insulator until you hit the metal button hole. Hold the screwdriver there for a few seconds before you withdraw it. You should repeat this after a minute or so, even if you did not turn the power back on. The CRT has reserve charges that work back toward the anode after it is first discharged.

If the glass around the anode is dirty, it is best to give it a thorough cleaning with a paper towel or tissue moistened with water or rubbing alcohol. Work from the inside around the belly button and out toward the graphite coating that surrounds it several inches away.

Excessive high voltage can also be caused by internal problems. The power-supply voltage regulator circuit can be misadjusted or bad. Turn it down, if necessary.

In some high-humidity locations, such as in Florida or Hawaii, it may be best to keep the supply voltage lower than the factory-recommended level.

Trouble in the power-distribution circuits can also give you excessive high-voltage problems. Look for hold-down resistors in the supply lines. Look for a short between two sections of a multiple-section electrolytic capacitor. The higher voltage from one section gets into the lower voltage section.

TIP 11-17

Excessive high voltage in CRT circuits, as indicated by sharp crackling noises and blue-white arcing, can be remedied by cleaning dirty surfaces around the high-voltage connections, cutting down excessive line voltage with a voltage regulator or series varistor, adjusting the voltage regulation circuit, replacing an open bleeder resistor, or replacing a filter capacitor that has a short between its high- and low- voltage sections.

TIP 11-18

There are several ways to check for dc high voltage: Use a high-voltage meter. Listen to it crackling at turn-on. Feel the hair on the back of your hand bristle when it is held near the face of the display tube. Hang tissue paper on the high-voltage line and see it move at turn-on. (**Caution**: Turn off power when removing tissue.)

In the old days, with vacuum-tube high-voltage rectifiers, you could draw a big zap with a screwdriver. What you did was to ground the shank against the high-voltage cage and bring the tip near an exposed high-voltage point. Zap! It's working!

> **Caution:** *Do not zap solid-state rectifier outputs. Generally speaking, solid-state devices cannot take sudden surge voltages, particularly when they are connected to big coils.*

Besides the noisy blue-white dc arcs, there may be quiet purplish high-frequency arcs, too. You will find them at the high-voltage transformer connection to the rectifier. If cleaning with solvent does not stop the arcing, the rectifier mounting should be replaced. Scraping the surfaces clean with a knife often helps remove the arcing.

TIP 11-19

To check the high-voltage rectifier, turn off the power, and with the proper precautions (Tip 11-16), remove the high-voltage lead clip from the CRT anode button and place it about an inch or two away. Use a 12-inch plastic stick. Turn on the power. If you get a silent purplish arc at the CRT anode button, the rectifier is shorted. If it crackles and flashes blue-white, it is good.

TIP 11-20

You can check for high-voltage pulses with a small neon lamp (type NE-2). Tape it to the end of a plastic stick and hold it near the output transformer; it should light up. Hold it about an inch away for every 10,000 V.

Touch it (the glass part) against the horizontal output transistor to check for ac there.

TIP 11-21

You can touch an exposed high-voltage point with the tip of a screwdriver and draw a silent purplish arc as you move the tip away. Do not ground the screwdriver! You should get about 0.3 inch of arc for every 10,000 V. Be careful not to create any arc paths on surrounding insulation. Do not hold it against an insulated part for support; you may char the insulation.

Incorrect voltages at the CRT socket can pull down the high voltage and cause a loss of raster. We'll take care of this problem in Chapter 13 when we check out video-display tubes.

So far, we have learned how to check the transformer T_2, the tripler D_2, and the CRT circuit for high-voltage problems. There are still other causes. We still have to look into the horizontal amplifier circuit.

TIP 11-22

To check for an internal CRT high-voltage short, turn off the power and disconnect the CRT socket. Turn on the power. Return of high voltage indicates a shorted CRT. Arcing from the socket to the CRT pins confirms the short. Crackling blue-white arcing indicates good high voltage. Silent purplish arcing indicates a shorted high-voltage rectifier.

11-7-2 Checking the Horizontal Amplifier

The horizontal output transistor Q_1 in Fig. 11-8 has a tremendous load. It must drive heavy currents through the yoke Y_1 and deliver enough power to obtain 30,000 V. Very often it is called upon by designers to provide voltage for the vertical amplifier and many other circuits (see D_3, D_4, and D_5). The high-horizontal frequency is well suited for a highly efficient and lightweight dc power supply. However, a shorted part in any of the circuits it feeds can blow the horizontal output transistor. When it goes out, it seems like everything has gone out.

In TV equipment, when you have a dead set and the fuse blows, it is either the power supply or the horizontal output amplifier, or both. Look for the horizontal transistor; it is easy to spot. It is the big metal one mounted on a heat sink.

1. If you have a dead system, check for voltage at the collector of the horizontal output transistor Q_1. To play it safe, use a neon lamp first. Place the lamp against the transistor; if it does not light, it is safe to use a meter. If you have dc voltage, the power supply is okay and the primary of horizontal output transformer is not open. The transistor is very likely to be bad (Fig. 11-8).
2. Check the horizontal-output-transformer resistances against the schematic. Your ohmmeter should give you clear-cut readings for open windings, because the coil resistances are small compared to other resistances in the circuit.

TIP 11-23

If you have a dead TV set, check the reaction time of the circuit breaker. If it cuts out immediately, the power supply or the horizontal output transistor, or both, are shorted. If it takes a little time, the circuits fed by the horizontal output transformer may be shorted. If you hear sound for an instant before the breaker trips, the sound circuits and everything supporting them, including the power supply, are probably good.

3. Check the diode rectifiers, D_3, D_4, and D_5, with an ohmmeter.
4. Check the horizontal amplifier base voltage. It should be negative. This is because the signal is rectified by the diode action of the *BE* junction of the transistor. If you have no voltage here, the signal source may be bad.

   If the base voltage is okay, the source is probably okay. Once in a long while, a far-off frequency might give a good dc reading, yet fail to develop high voltage. If you run into problems, check the frequency of the base signal with a scope. You should have about 15,750 Hz or 63 μs for TV circuits.
5. Remove the horizontal output transistor and check it with an ohmmeter. This is a very reliable test when the fuse or circuit breaker cuts out at turn on.
6. While the transistor is out, check the resistance from the collector terminal to ground. Measure both ways; you are measuring the power-supply rectifier resistance as well as damper D_1 resistance to ground. You should get infinite resistance on one of the readings.

   Check the base and emitter resistances to ground. If the base has infinite resistance to ground, the input transformer is open.
7. After replacing the transistor, check the equipment out at low line voltage with a series lamp or variable transformer. (If necessary, jump the power-supply regulator.) Hold a neon lamp or a scope probe on the outer rim of the horizontal output transformer. If it lights, you are doing fine. Run or "cook" the equipment for a while. Check the supply voltage; it should be about 70 V dc.

   If the neon does not light, use a higher-wattage lamp. The neon lamp should light up with around 80 to 90 V on the supply line. Hold the neon lamp close against the transformer.

   If you have a variable transformer, bring up its voltage gradually until the neon lights up. (Jump the power-supply regulator if necessary.) Cook the circuit with about 70 or 80 V dc on the supply line.

   If you have a scope, starting at 0 V ac with a variac, place the scope probe on the transformer rim, on the insulation. Do not ground the probe; leave the ground clip hanging loose. This makes the probe very sensitive. At a very low voltage, you will begin to pick up some uneven waveforms. You may have to pull the probe away if the waveforms get too big. At about 40 to 50 V ac the waves will suddenly change into clean 15,750-Hz pulses if the circuit is working. Cook the circuit at about 60 V ac.
8. If you still have no high voltage, turn off the power and feel around the transformer for hot spots. A hot spot indicates a shorted transformer. (It is safe; you have no high voltage and the power is off, right?)
9. Remove the damper diode, D_2, and recheck the amplifier at reduced line voltage. If the neon lamp now lights or the scope shows pulses, even if distorted, the diode is shorted. (**Don't forget:** We are not touching any connection directly with the scope.)

The checks above cover the basic parts in the horizontal output amplifier circuit. At this point the trouble would seem to be in the transformer, the yoke, or the circuits feeding off the transformer.

TIP 11-24

Horizontal pulses can be checked with an ungrounded scope probe held in the vicinity of the horizontal circuits. You should detect high-voltage pulses about 7 inches away from the horizontal output transformer. To check the horizontal output transistor, bring the probe up closer and touch the collector with the insulated part of the probe, if necessary.

TIP 11-25

If you obtain a good voltage at any secondary terminal of a horizontal output transformer, and bad voltages at another, the bad voltage is caused by an open terminal or winding. If it is an open winding, it must be above the good voltage point.

This also proves that the all horizontal-deflection circuits up to the output transformer are good.

10. Check the dc voltage at cathode of diode D_3. If there is no voltage on the cathode D_3, disconnect it. If the D_3 supply line is shorted, voltage should now return to D_4. A neon lamp may light if held near the flyback.

    Note: It is possible to have voltage at D_3 and not have high voltage at damper D_2 if the high-voltage winding just below D_2 opens.
11. Disconnect the yoke. If it is shorted, some voltage should return to D_4. The neon may light.
12. The high-voltage lead has already been disconnected. Disconnect the lead from the flyback to the tripler. If the neon lamp now lights up, the tripler is bad.
13. Disconnect the flyback lead to D_4. If the D_4 supply line is shorted, voltage may return to D_3. The neon may light.
14. Disconnect the flyback lead to D_5. If the D_5 supply line is shorted, the neon lamp may now light.

We have disconnected all the flyback secondary leads; nothing can load it down now. If it does not work, we have nothing else to blame. It is no good. To confirm this, substitute another flyback. Anything on hand will do for test purposes. Just connect the primary. The neon light should light.

15. Disconnect the ground lead of the transformer. All secondary leads have now been disconnected. Check the voltage at the ground lead or any lead. It should read 0 V. Otherwise, it is shorted to the primary.

11-7-3 HV Transformers with Built-in Triplers

The basic checks made here can be applied to all modern horizontal output and high-voltage circuits. Figure 11-9 shows a horizontal output transformer with a built-in tripler. The high-voltage diodes are built into the coil assembly and heavily insulated. The voltage developed by each coil is rectified and added to the next coil above it. Each coil and rectifier has only a third of the total voltage on it. The coils can be designed for minimum strain. All you have to do is to check for high voltage at the CRT anode. If you get no high voltage there, check the horizontal output transistor circuit and proceed as you did with the basic circuit. Also check the focus and screen circuits for shorts. If you have good voltages at any point in the secondary, the other points that are defective are probably open.

11-8 TROUBLESHOOTING A VCR MOTOR-DRIVE AMPLIFIER

Figure 11-10 shows a pair of cassette front-loading motor-drive amplifiers for a VCR. This circuit is a basic representation of motor-drive amplifiers in VCRs, the only difference being that it has no speed control. The motor is reversible. When it has 12 V on the forward terminal, it spins forward. When it has 12 V on the reverse terminal,

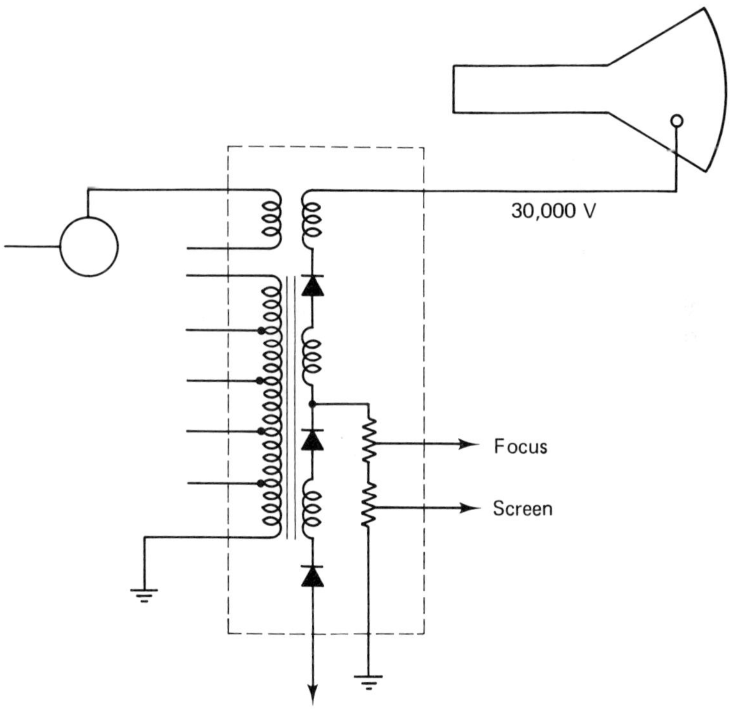

Figure 11-9 A horizontal output transformer with a built-in voltage tripler that generates 30,000 V.

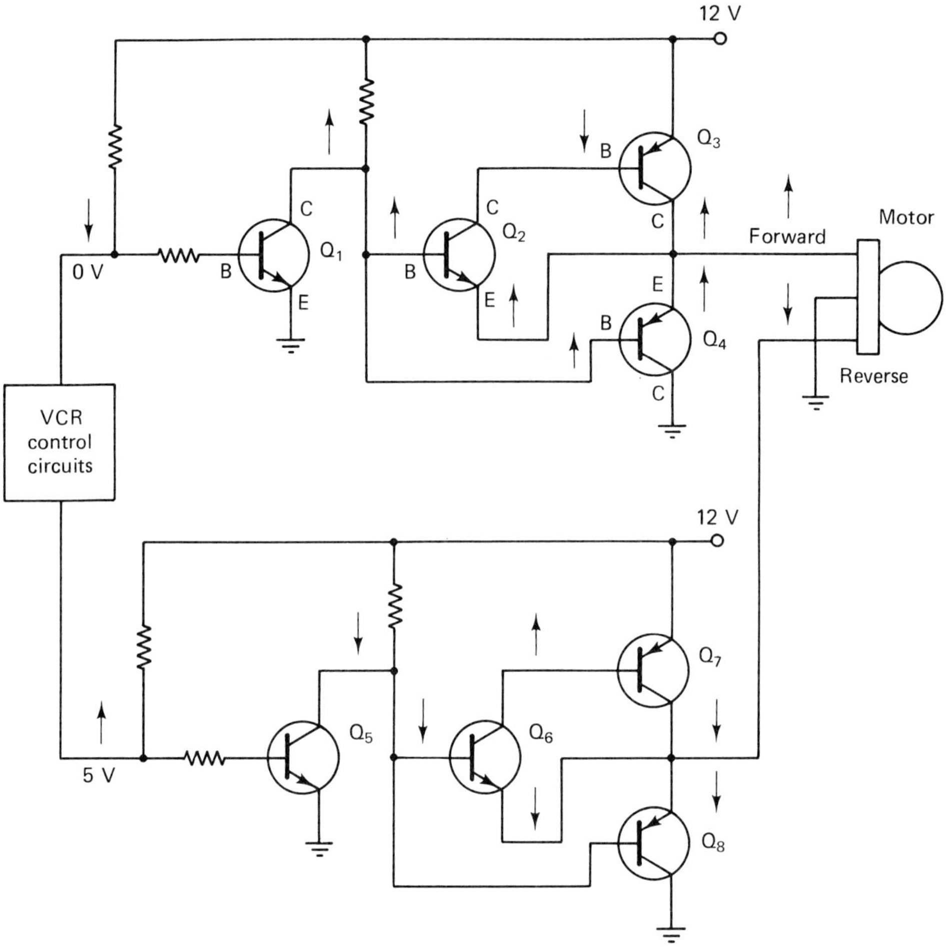

Figure 11-10 A pair of VCR motor-drive amplifiers that are actually more like power switches. The arrows indicate that the motor is running in the forward direction. The reverse-motor voltage is off.

it spins backward. When one amplifier is on, the other is off; they cannot be on at the same time. To stop the motor, both amplifiers must be off. The VCR control circuits turn the motors on and off at the correct times. The two amplifiers are exactly alike. Actually, what we have here is a couple of power switches. We'll look at one of them.

To run the motor, the input to Q_1 is held low (0 V). This keeps the Q_1 collector high and brings up bases of Q_2 and Q_4. Note the arrows. Transistor Q_2 has two outputs. Its collector is held low, and its emitter is held high. Transistor Q_3 has a low voltage on its base, and Q_4 has a high. The low voltage on the base of Q_3 keeps its collector down to 0 V. The low voltage on the base of Q_4 keeps its emitter high at 12 V. This voltage runs the motor forward.

However, do not forget that the reverse terminal of the motor must be disabled at this time. To do this, the VCR control circuits place a high (5 V) on the input resistor of Q_5. If you follow the arrows, you will see that a low (0 V) is placed on the reverse terminal of the motor.

To stop the motor, both terminals must have 0 V. This is done by putting a high on both amplifier inputs. The use of zero voltage to initiate an action is very common in control circuits. You need only to ground a terminal to start an action.

Troubleshooting this amplifier is a matter of knowing when the motor should be going forward, reversed, or stopped. You have to read the respective voltages for each mode, and of course you should know that the collector inverts and the emitter follows.

You can short an amplifier input to check its operation only if the other amplifier is off. Otherwise, you will be working at cross purposes. Imagine a motor running forward, backward.

You can check the motor by removing the drive belt and applying 12 V to it from an external source. You can apply the 12 V from the VCR power supply if you use a fuse or circuit breaker to protect it. In servicing VCR motor circuits, you must be aware that there are all sorts of protection circuits that stop the motor if anything happens to the mechanical system. This means that the trouble could be elsewhere, not in the motor or motor driver. This can be checked out by checking the base voltages of the driver amplifier.

The most common cause of front-loading failure is a worn drive belt. The motor itself is durable. You can check it by spinning it. If it spins freely, it's probably good.

TIP 11-26

A stuck or slow motor can be checked by spinning it. If it spins freely for a long time and slows gradually, it is good. If it spins like something is holding it back, and comes to a sudden stop after a brief slowdown, it is no good. It will not start under load. Drop some kerosene or some other kind of cleaning and lubricating fluid onto the exposed motor shaft. Then spin the shaft manually. Push it in and out; this helps get the cleaning fluid in to the bearing. Keep spinning and pushing and pulling until it spins freely.

Extreme care must be taken when lubricating VCR mechanisms. Don't get oil on your fingers or on anything but the sliding and spinning metal (not nylon) parts. A small amount of lubricant in the wrong place might be touched inadvertently and transferred to a belt. It does not take much to ruin a belt. Oil migrates; it spreads outwards from spinning parts. Clean everything; clean all the surrounding parts, movable or non-movable, with alcohol. Keep wiping your fingers. Follow the manufacturer's instructions.

We will study the control circuits for this motor later. Remember that it has to do three things: spin forward, stop, and spin backward. Furthermore, it has to do this in conjunction with the cassette loading and unloading sequence and a few other operations in other subsystems.

WHAT YOU LEARNED TO DO

In this chapter you learned to troubleshoot various kinds of output amplifiers. You learned that they were all basically alike. You will be able to work on all types of power amplifiers if you learned to troubleshoot:

1. A basic quasi-complementary-symmetry amplifier. It had positive and negative supplies, and it had a zero-volt centerline. You learned to check its feedback system.
2. A vertical output amplifier for a CRT display monitor. It is almost identical to the audio output amplifier.
3. The horizontal output and high voltage stage of a CRT display monitor. You learned to isolate the high-voltage circuit from the amplifier circuit.
4. A VCR motor drive amplifier (or power switch). You learned that this circuit differed from other amplifiers only in that it delivered either the maximum voltage or zero voltage.

QUESTIONS

1. What is the key voltage in complementary-symmetry amplifiers?
2. What should you look for when comparing the voltages in a complementary-symmetry amplifier?
3. In Fig. 11-4, figure out what would happen to the voltages on all the other transistors when Q_6 has a *CE* short. See what happens when each of the other transistors has a *CE* short.
4. What would happen to the centerline voltage if R_6 of Fig. 11-4 opens?
5. What part or parts of Fig. 11-4 could be bad if the centerline voltage is too high and the junction voltages R_3 and R_6 are below normal?
6. In Fig. 11-4, what would happen to the centerline voltage if you short the base–emitter terminals of Q_1?
7. In Fig. 11-4, what would happen to the base of Q_3 if the centerline voltage goes up?
8. In Fig. 11-4, what might happen if Q_6 shorts?
9. What would happen if you touch R_7 and it suddenly opens completely?
10. What key voltage is associated with distortion?

11. What parts and circuits would you check for excessive hum?
12. What differences can you find between the audio amplifier in Fig. 11-1 and the vertical amplifier in Fig. 11-6?
13. What similarities can you find between the basic transformer-coupled amplifier in Fig. 7-15 and the horizontal amplifier shown in Fig. 11-8?
14. If the transistor Q_1 of the motor driver in Fig. 11-10 shorts when there is 0 V on its base, what would happen to the motor if there were 5 V on Q_5?

PRACTICAL TROUBLESHOOTING EXPERIENCE

1. Obtain an audio amplifier with output transistors and a nonregulated supply. Obtain the schematics. If the amplifier is dead, check the power supply and the power-output amplifier with a series test lamp. Substitute the speaker with a 10 Ω 100-W resistor bank (or higher power rating, if required). Repair the power supply if necessary. Refer to Section 11-2-3 for help.
2. If the amplifier is shorted, pull out the output transistors. Check the centerline voltage and compare the voltages above and below it. If they are balanced and seem okay, recheck them at full ac line voltage.
3. If the voltages are good, install new output transistors and check the amplifier at low line voltage first. If all the voltages look good, check the amplifier at full voltage.

3′. If the voltages are bad, check the *BE* voltages of all the transistors. Record all voltages. Compare the base, collector, and emitter voltages of each transistor. Recall that if the base voltage is too low, the collector should be too high and the emitter too low.

    If the transistors seem good, check the electrolytics. Next, check the resistors and diodes. Replace the part or parts that you find are bad. A driver transistor is probably bad.

4. After replacing the parts, check the amplifier out at reduced voltage for a few minutes while checking the voltages. Then check the amplifier at full voltage for 20 minutes. Monitor the voltages.
5. While you are monitoring the amplifier, signal-trace it using a signal generator and scope. The waves may be distorted, but the tops and bottoms should look fairly alike. Reduce the signal generator voltage. If the signal looks good now and the dc voltages are okay, the amplifier is probably good.

    If they are not balanced, check the base, emitter, and *BE* voltages carefully. The top and bottom counterpart voltages should be the same.

6. Occasionally check an output-transistor emitter voltage drop. It should stay steady. If it does not level off as the amplifier warms up, something is still wrong. More careful checks have to be made. Consult the troubleshooting checks in this chapter and in Chapters 7 and 8.

    Bear in mind that if you use a tuner for a signal, the output transistor emitter voltage drop may vary with changes in the input signal level.

12

High-Frequency, Wideband, and Related Circuits

EXPERIENCE REQUIRED

To understand and use the information in this chapter, you should know how basic amplifier stages and power supplies are checked out. You should know that coil–capacitor (*LC*) filter circuits can pass the desired frequencies and reject unwanted frequencies. You should also be able to recognize the basic high-frequency components, such as small coils, transformers, variable capacitors, trimmers, and transmitter coils.

Knowing how to use the signal generator and scope would be a great help, especially as a learning tool. However, for most of the checks in this chapter the scope is not an absolute requirement. There is no circuit or device in this chapter that cannot be checked with a multimeter or a substitute jumper.

> *Note:* The circuits in this chapter provide a groundwork for understanding and troubleshooting the integrated circuit chips in Chapter 14, which have largely replaced many of the circuit functions in this and the preceding chapters.

12-1 RF TUNED AMPLIFIERS

Although the basic *LC* circuits we studied in Chapter 7 have largely been replaced by other types of filters (see the next section), there is plenty of older equipment around that needs to be worked on. *LC* circuits are commonly used in radio communications and TV receivers for tuning in the transmitters. Transmitters require them for maintaining their assigned frequencies. Without these tuned circuits, we would not be able to separate the stations. We would receive several stations simultaneously at the same point on the dial.

12-1-1 Troubleshooting Tuned Amplifiers

The tuned amplifier shown in Fig. 12-1 uses *LC* circuits to couple the signal between stages. Recall in Chapter 7 that a tuned circuit passes only one frequency or band of frequencies and rejects all others. C_1 and the primary of T_1 make up the input tuned circuit. The arrow between the windings of T_1 indicates that it is tunable. The secondary of T_1 picks up the desired signal and passes it on to the transistor. The output tuned circuit, C_3, and the primary of T_2 are tuned to the same frequency as the T_1 circuit. It further rejects the undesired frequencies. In some tuned circuits, the capacitors are adjustable. C_N is a neutralizing capacitor. It counteracts some undesirable effects to be discussed in the next section.

Troubleshooting dc voltages in this amplifier is not especially difficult. There should be no voltage drop across the transformer windings. This means that the transistor collectors have the full supply-line voltages on them. In receivers, the coil wires are very thin and tend to break at the lugs, where they are weakened by soldering. If the primary coil of an output transformer breaks, the collector voltage will drop to

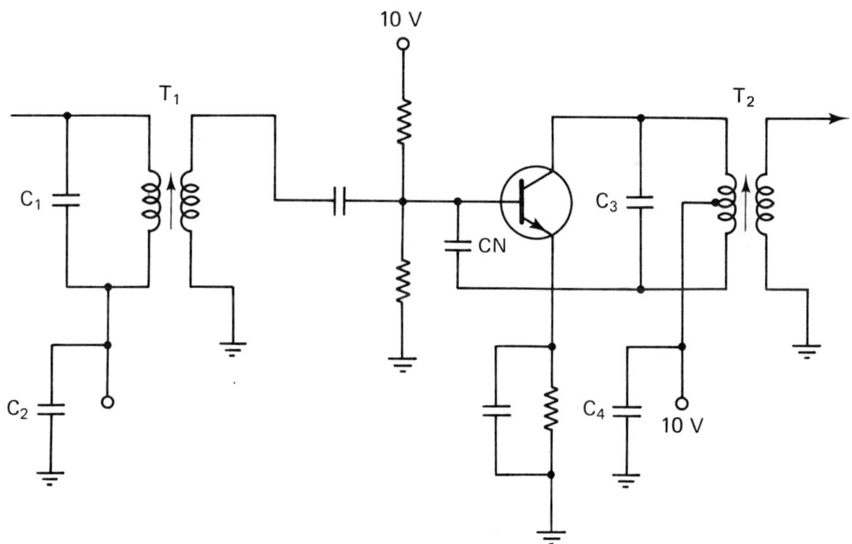

Figure 12-1 Tuned amplifier. The tuned circuits, T_1–C_1 and T_2–C_3 are tuned to pass a desired frequency. This circuit may be used to tune in a radio station.

zero. The other windings are easily checked with an ohmmeter. The voltage drops across the base resistors should be nearly in proportion to their resistances.

12-1-2 Undesired Oscillations

One problem with tuned-circuit amplifiers is oscillations. Under certain conditions voltage is unintentionally fed back from the output to the input and then reamplified and fed back again. This recycling process involves the stray capacitances that exist between the wires, conductors, and parts in the circuit, as well as the stray inductances in the wires and conductors. This feedback and recycling are repeated constantly, and fluctuating voltages are developed in the amplifier. You will learn a little more about these oscillations in Section 12-3.

These spurious oscillations have undesirable effects on amplifier performance. In radios you can hear whistles and howls, distorted sounds, or motorboating. In TV sets you can see streaks tearing across the screen, or repetitive blackening or visual sputtering at a motorboating rate.

Another source of undesirable feedback can be traced to the the electromagnetic field surrounding a coil. You get an idea of what it is like when you pick up waveforms in the vicinity of a circuit with an ungrounded scope probe. The probe does not have to touch anything. Just hold it in the air. The voltage is pretty high. To prevent this kind of feedback, the high-frequency coils are often shielded in metal cans.

There is still another way in which this insidious feedback can occur—through the transistor itself. It takes place through the *BC* junction, which serves as a capacitor between the output and input circuits. Under certain conditions the feedback voltage and the input voltage reinforce each other to produce oscillations. This is called positive feedback, as opposed to negative feedback, in which the output and input voltages tend to cancel each other. When tuned circuits are involved in the oscillations, the process is called *regeneration*.

The purpose of the neutralizing capacitor (C_N in Fig. 12-1) is to counteract the positive feedback through the *BC* junction. The voltages on the opposite ends of a transformer winding are always of opposite polarity. The neutralizing capacitor feeds a voltage that opposes the positive feedback voltage through the *BC* junction.

With respect to tuned circuits, there is still another source of unwanted oscillations. At the very high frequencies, any length of wire or conductor can serve as a coil. Any two wires, conductors or connections, can serve as a capacitor. In other words, we can have tuned circuits that can break into oscillation just about any place in the chassis or PC board. This type of unwanted disturbance is called *parasitic oscillation*.

A very common source of oscillations is an open decoupling capacitor. Capacitors C_2 and C_4 in Fig. 12-1 bypass the signals onto ground. If they open up, the supply line will pick up all types of voltages and feedback signals from one circuit to another and cause oscillations.

Some tuned amplifiers have dc feedback lines connected to their base input circuit. Dc feedback is taken up in Section 12-7. When the components or connections in these lines become defective, positive-feedback signals are transferred or picked up and cause oscillations.

There are still more causes of undesired oscillations. They can be caused by incorrect transistor operating voltages that produce distortion. Distorted operation creates new frequencies that can oscillate in the surrounding circuitry (Section 9-4-3). In relation to this, oscillatons can occur when connections or conductors open up to create new feedback and "tuned-circuit" paths.

Another problem you should look out for are the parts and leads that are moved from their original "factory-adjusted" positions. The circuit leads are often positioned or "dressed" at the factory to prevent oscillations, and sometimes, even to "tune" a circuit to obtain a desired frequency response.

The tips below are based on the observations above. Many of them can be applied to low-frequency circuits as well as to the high-frequency circuits in this chapter.

TIP 12-1

If oscillations occur in an amplifier:

1. Move your hand over the space around the suspected circuit, or touch the base of a transistor in the suspected stage with an ungrounded meter probe and see if you can affect the oscillation waveform on the scope or the sound at the speaker.
2. Move an ungrounded scope probe in the space around the suspected parts to search for the source. Cut down the scope gain as you get nearer the area of the oscillation. Once you get into the general area (it may be too big to tell you much), clip the ground lead and scope various signal points to find the source.
3. Note the dressing of the wires around the amplifier. Try moving them slightly with a plastic stick to see if you can affect the oscillation. If nothing happens, leave the wires in their original positions.

TIP 12-2

If oscillations occur in an amplifier, scope the supply line or jump the bypass capacitors. Also check the neutralizing capacitors if there are any.

Also check the bypass capacitors in the dc feedback line to the input circuit. (Dc feedback is discussed in Section 12-7.)

TIP 12-3

If oscillations occur in an amplifier, tighten the ground connections. Make sure that the coil shields are grounded. Scope the ground conductors and shields. Look for cracks in the PC board. Bend and twist the board gently to see if you can get rid of the oscillation.

In the following tip, you are instructed to detune an amplifier with a tuning wand having brass or ferrite slugs on its opposite tips. With this wand, you can detune the coil without disturbing its original adjustment. The inductance of a coil is determined not only by its windings; it is also determined by the material in its core. Ferrite increases the inductance substantially, and brass or aluminum decreases it slightly. You can make a "detuning wand" by gluing an old slug onto a worn-out alignment tool.

TIP 12-4

If oscillations occur in a tuned amplifier, detune the suspected coil by inserting a brass slug or a ferrite slug into the core. If the oscillation goes away, something is wrong with the coil itself or something associated with it. Make some other checks before adjusting the coil permanently.

TIP 12-5

To get rid of oscillations, check the transistor dc voltages. Replace incorrect-value resistors. Check the transistor itself. If bad, replace it. If nothing else is found wrong, replace the transistor even if it checks good.

12-2 VARACTORS AND SAW FILTERS

The basic tuned circuits that we covered used coils and capacitors that had to be tuned mechanically, either at the factory or by the user. In the latest equipment, varactors and SAW filters have practically eliminated, or greatly reduced, the need for mechanical tuning.

12-2-1 The Varactor Tuner

Electronic pushbutton tuning is made possible with varactor diodes. Any reverse-biased diode can act as a capacitor; furthermore, you can change its capacitance with a dc voltage. *Varicap* diodes or *varactors* are designed specifically for this purpose. Figure 12-2 shows one in a tuned circuit. When voltage is applied across its terminals, the diode capacitance changes, and with it, the frequency of the tuned circuit.

Except for the diode and its dc control voltage, the varactor-tuned amplifier is checked out basically like any other tuned-circuit amplifier.

> **TIP 12-6**
>
> The varactor, which is used for changing the frequency of tuned circuits, can be checked out like any other diode. It should give you a hi–lo reading.

In late-model tuners, the dc tuning voltage is provided by microprocessor-based (computer) circuits. The tuner unit has an input for the dc tuning voltage. The microprocessor circuits feed control voltages into the varactor circuits to select various stations. If necessary, consult the service manual or schematic for the correct voltages.

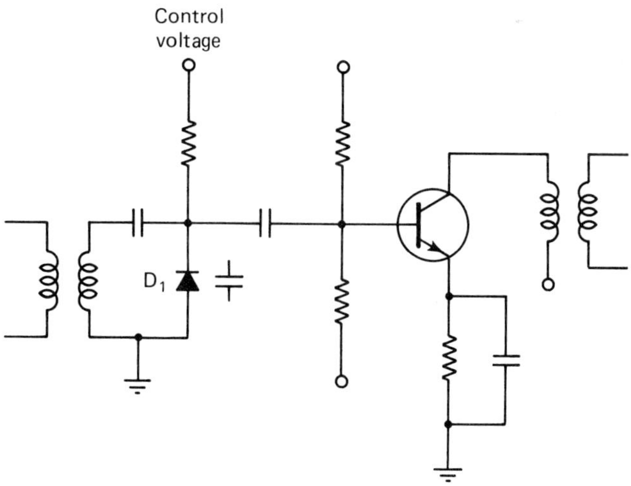

Figure 12-2 This amplifier tunes in a signal by adjusting the voltage on a varactor diode that acts as a variable capacitor.

TIP 12-7

You can check a varactor tuner by feeding it a substitute control voltage from a variable power supply. If you can pull in the stations as you change the voltage, the tuner is probably okay and the control circuits are probably bad.

12-2-2 The SAW Filter

The SAW filter has replaced many of the *LC* tuned circuits in amplifier stages (see Fig. 12-3). Certain amplifiers in radio and TV receivers have fixed tuned circuits. In TV recievers, they are required to pass a wide band of frequencies. SAW filters are manufactered to meet this requirement. They consist of an input section and an output section. When the input crystal receives an electrical signal, it generates surface acoustical waves (SAWs) on a crystal surface. When the output crystal receives these waves, it generates electrical signals that correspond to the desired frequencies. The crystals can be manufactured to pass only the desired frequency or band of frequencies.

You have probably encountered acoustical crystal devices already—at a much lower frequency, however. The inexpensive crystal earphone that you plug into your ear operates on the same principle as the input of a SAW filter. They both convert electrical energy into vibrational (acoustical) energy. You also encountered it in crystal phono pickups. They do the same thing as the SAW output section. They convert vibrational energy into electrical energy. This two-way electrical energy conversion

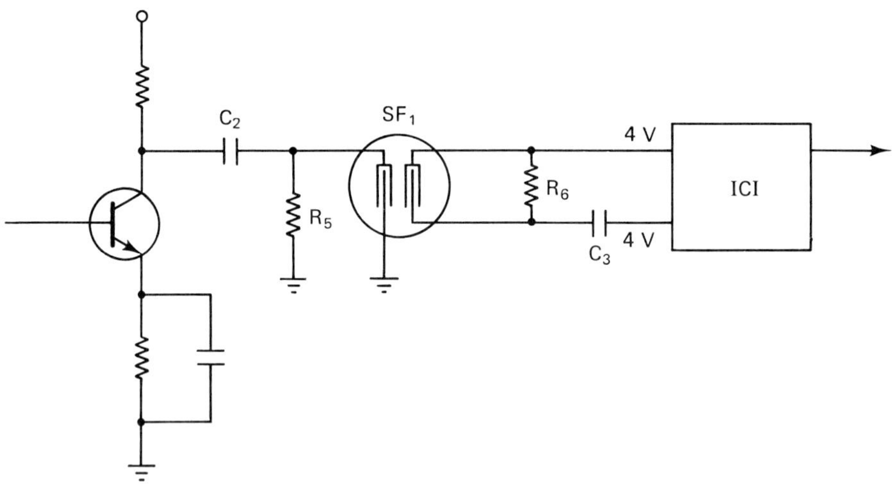

Figure 12-3 The SAW filter has largely replaced the *LC* tuned circuit in TV tuned amplifiers.

involves what is called *piezoelectricity,* the property of crystals to change form under electrical stress and to generate electricity under mechanical stress.

Figure 12-3 shows a SAW filter feeding an integrated-circuit chip. It inputs the signal from a simple voltage amplifier that passes all the desired frequencies as well as others that have to be rejected. The SAW filter SF_1 rejects the unwanted frequencies and passes only the desired band of frequencies. There are no tuned *LC* circuits associated with the IC chip.

The frequency response of a SAW filter that is supposed to pass what is known as the *IF frequency band* in TV receivers can be roughly checked with a signal generator and a 100-MHz scope. Inject the signal at the base circuit of Q_1, and vary it between 39 and 47 MHz. The output signal should be near zero from 39 to 41.25 MHz. As you increase the frequency, it should pass a band of frequencies roughly between 42 and 46 MHz. It should drop off to zero at 47 MHz.

You can make other tests with a dc voltmeter. The input of SF_1 should be 0 V. The output voltages should both measure 4 V dc. The resistances between the terminals of SF_1 out-of-circuit should be infinite. In-circuit, the input resistance should equal the resistance of R_5. The output resistance should equal that of R_6. The capacitors, C_2 and C_3, will enable you to get true readings with the ohmmeter.

12-2-3 Ceramic and Crystal Filters

The SAW filter is used to pass a wide band of frequencies. Other types of filters are used when less bandwidth is required. In Chapter 7 we learned that *LC* filters could do the job. Now we find that ceramic and crystal filters are used in place of *LC* filters.

Figure 12-4(a) shows a ceramic crystal filter that is used for FM IF amplifiers. Figure 12-4(b) shows a crystal filter circuit that provides extremely high selectivity. It is used in AM short-wave receivers that have to separate stations that are crowded closely together on the dial.

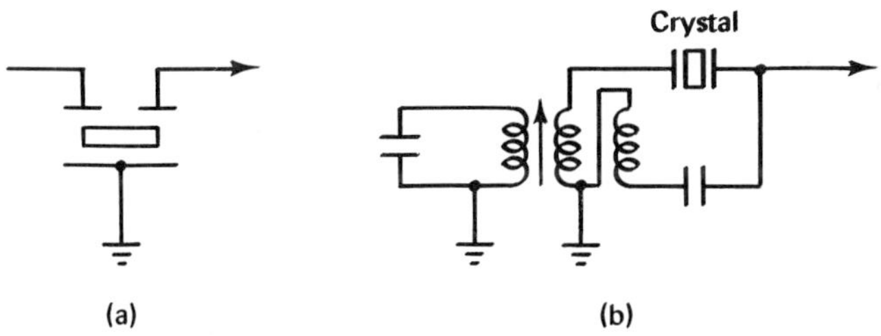

Figure 12-4 (a) Ceramic crystal filter. (b) Crystal filter for shortwave circuits.

12-3 OSCILLATORS

The oscillator is the source of all kinds of signals in the air at this very moment: AM, FM, TV, shortwave, UHF, radar, satellite, and so on. It has other important uses, too. It is an awesome little circuit. Once you get into the circuit, however, it is checked out just like an amplifier.

12-3-1 Checking Out a Simple Oscillator

In the preceding section we tried to get rid of oscillations in amplifier circuits. Now we are trying to get them. Oscillators are usually tuned *LC* circuits that are energized by an amplifier. Figure 12-5 shows an oscillator circuit.

The coil and capacitor network, or tuned circuit, is also called a tank circuit. The electrostatic field of the capacitor and the electromagnetic field of the coil keep bouncing the charges back and forth between them, to produce oscillations.

How Coils and Capacitors Oscillate. The electromagnetic and electrostatic fields are "elastic." When the capacitor discharges into the coil, the magnetic field stretches back and bounces the charge right back into the capacitor. Then the electrostatic field of the capacitor is stretched back and bounces the charge right back to the coil. Thus the charges keep bouncing back and forth. The frequency of the bouncing can be adjusted by changing the size of the coil or capacitor. It takes longer to stretch out the field of a big capacitor or coil than that of a small one. The shorter the stretch, the faster the bounce. The smaller the capacitor or the coil, the higher the frequency.

Tuned circuits can break into oscillation when energized by a single pulse. The problem is that the oscillations die out after a time. The only way to keep them going is to keep pumping pulses into them. This is the job of the amplifier and feedback circuit.

When the circuit is first turned on, the tuned circuit is energized and starts to oscillate. It outputs the oscillating signal to the amplifier. The amplifier then feeds back the signal to the tuned circuit and reenergizes it. The process keeps repeating itself.

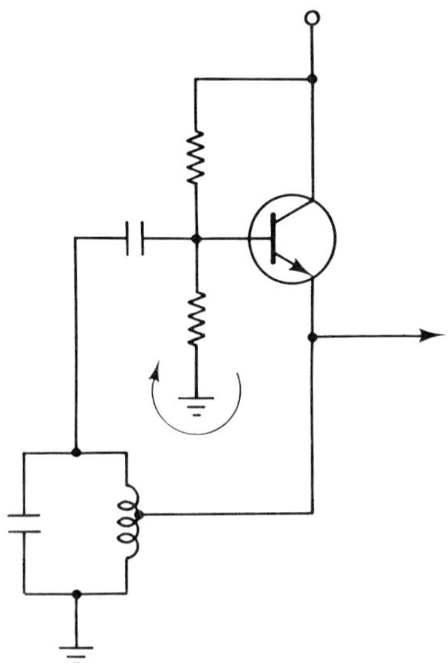

Figure 12-5 An oscillator is used to generate signal voltages. Basically, it is a tuned circuit that is fed energy back through an amplifier as shown by the arrow.

When tracking down the more obscure faults, we have to remember that the tuned circuit consists not only of the capacitor–coil network as shown on the schematic; it also consists of all the other conductors, objects, and stray fields around them.

Figure 12-6 shows various types of oscillators. The main differences among them are the ways in which the oscillating networks are hooked up and the path the feedback takes. The arrows in the illustrations show the feedback paths. Figure 12-6(a) shows capacitively coupled feedback. Figure 12-6(b) shows inductive coupling.

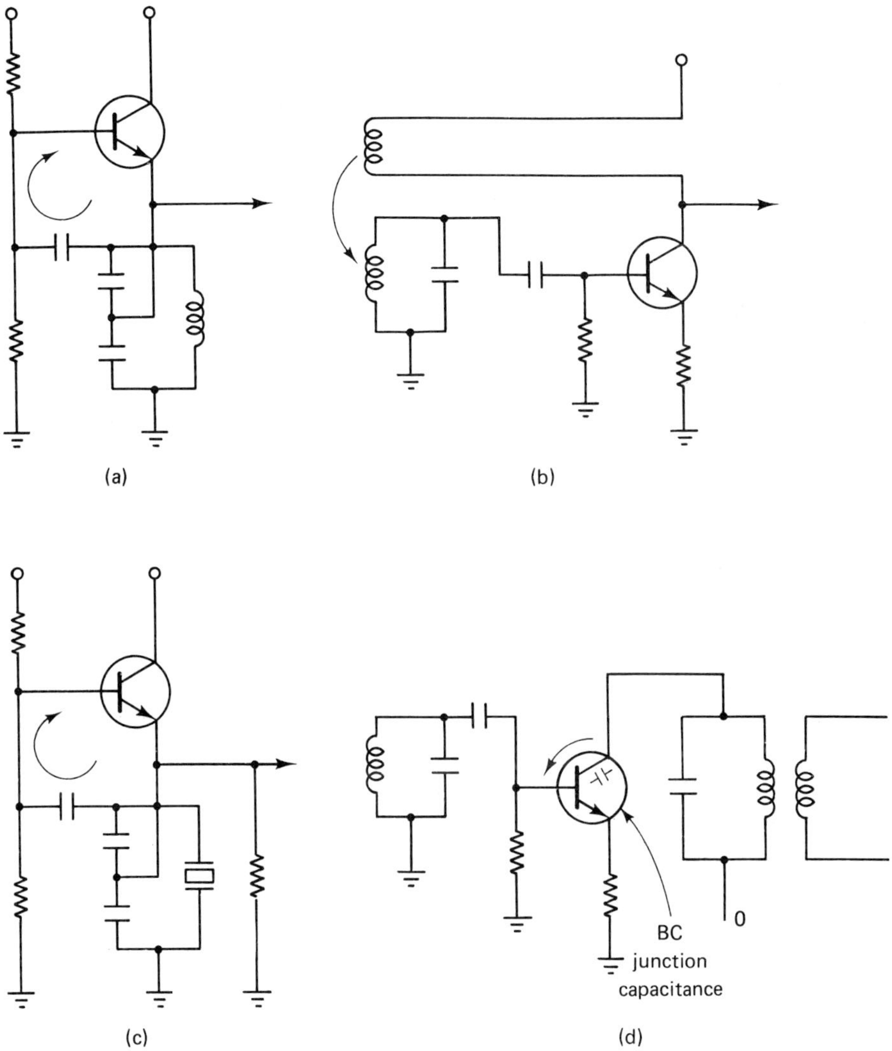

Figure 12-6 Various types of oscillators. They all have a tuned circuit feeding back an amplifier. The oscillator in (d) is fed back through the *BC* junction of the transistor. It looks like a tuned amplifier.

Figure 12-6(c) shows a crystal Y_1 in the tank circuit. When a crystal is energized, it oscillates at a fixed frequency. This keeps the frequency from drifting. Drifting is caused by slight changes in the capacitance and inductance values in the circuit as the circuit warms up.

In Fig. 12-6(d) we see what looks like an ordinary tuned amplifier. Recall that ordinary tuned amplifiers will oscillate under certain conditions. These conditions can be met by adjusting the tuned circuit off just slightly so that the peak of the output feedback voltage through the *BC* junction coincides with the input base voltage.

Generally speaking, the methods for isolating the trouble to the oscillator stage are the same as for isolating a defective amplifier stage. You trace the signal or lack of signal back to the defective stage. In addition, there are other approaches that depend on where and how the oscillator is used. We take them up as we get to the circuits in which they are used.

Within the oscillator stage itself, making dc voltage checks on the collector or the emitter is the same as for amplifiers. However, the base terminal can present a problem. Touching it with a probe might kill the oscillations.

Resistance checks in the oscillator circuit are the same as for amplifiers. The open coils are easy to detect. However, changed values are not as easy.

Oscillators can fail in several ways: They can go dead, have weak output, or change frequency. They can also have an unstable combination of these faults. The following tips should help.

TIP 12-8

The transistor is the most likely part in the oscillator to fail. Some transistors will not sustain oscillations, even when they check okay in a tester. If warming or cooling a transistor restores oscillation, it is probably bad.

TIP 12-9

You might be able to restore oscillation by moving the parts and leads around. You can also cool them or warm them up. If you can restore operation when you do this on a capacitor, it is probably bad. If it happens with a coil, try squeezing or spreading its windings or resoldering its terminal lugs.

TIP 12-10

A short in the stage following an oscillator can load down the oscillator and possibly kill the oscillations. Reduced supply voltage can also stop the oscillations.

TIP 12-11

In very high frequency oscillators, you can bend or squeeze the coil a little to get the right frequency. Moving the parts and leads can also change the frequency. If slight cooling or warming of a transistor or capacitor changes the frequency excessively, it is probably bad.

TIP 12-12

As a general rule, do not make an oscillator adjustment unless it is a user adjustment or a factory-recommended periodic maintenance adjustment.

Tips 12-11 and 12-12 seem to contradict each other. In Tip 12-11 you know that the coil inductance has decreased with age. This means that the oscillator frequency increases with age. As long as you know this and the desired output, there is no harm in ''squeezing a little.'' Furthermore, the change is very small and limited. If you fail to achieve a positive result, you know what you must do to get back to the original condition, if necessary.

It is a different story when you are adjusting a coil slug. You do not know in which direction to turn, and there is no limit to the amount of turn or number of turns you can make, so there is always a temptation and tendency to turn too far. In most cases you have to set up the right test equipment to find the direction in which the slug must be turned.

If the oscillator is the only circuit that needs adjustment, or if its adjustment is called for in a given procedure, the following tip can be followed.

TIP 12-13

You can use a signal generator and scope to check the oscillator. Turn off the power and connect the generator and a 10:1 scope probe across the tuned circuit. You should get a sharp peak at the specified frequency. Next, turn on the power and inject the signal into transistor base and check the output. You should get a full sine wave that peaks at the specified frequency.

12-4 THE SUPERHETERODYNE CIRCUIT

The superheterodyne circuit was one of the biggest advances in radio communications. Prior to its invention in the 1920s, radios used to squeal and howl. When you have more than two stages tuned to the same frequency, they energize each other and oscillations can break out very easily. Yet at least three tuned stages were required to really separate the stations and deliver enough signal power for satisfactory listening.

The problem was solved by converting the radio station frequency (RF) in the receiver to a lower one without destroying the information it carried. This lower frequency is called the intermediate frequency (IF) because it lies between the radio frequency and the audio frequency (AF). The IF tuned amplifier stages were placed between the RF tuned amplifiers at the input and the AF amplifier at the output.

In this setup, the low-frequency interference from the IF tuned stages could not be picked up and fed through the RF tuned stages. Thus unwanted oscillations could not take place. More important, the fixed tuned stages did not have to be tuned manually. They were preset at the factory. The old radios had ganged-tuned variable capacitors on a long shaft and had to be housed in long cabinets.

Figure 12-7 shows the basic superheterodyne circuit. Let us look at the inputs and outputs of each section. First we have the RF section (T_1–Q_1–T_4). It selects and amplifies the RF signal from the station. The selection is done by the tuning circuit (T_1–C_1). This is only the first stage of selection. More stages are needed to reject completely all the nearby stations.

The mixer circuit, T_3–Q_1–T_4, combines the RF station signal with the Q_2 oscillator signal. Transformer T_3 couples the oscillator signal to the emitter of the mixer Q_1. In AM broadcast radios, the oscillator frequency is always 455 kHz higher than the station frequency. Maintaining this difference is the key to superheterodyne reception.

This 455-kHz difference is maintained because C_1 in the RF tuned circuit is connected mechanically to C_2 of the oscillator tuned circuit. They are mounted on the same shaft and they maintain the same frequency difference as they are rotated. Through a process called heterodyning, the output frequency of T_4 is always the difference between the station frequency and the oscillator frequency, or 455 kHz.

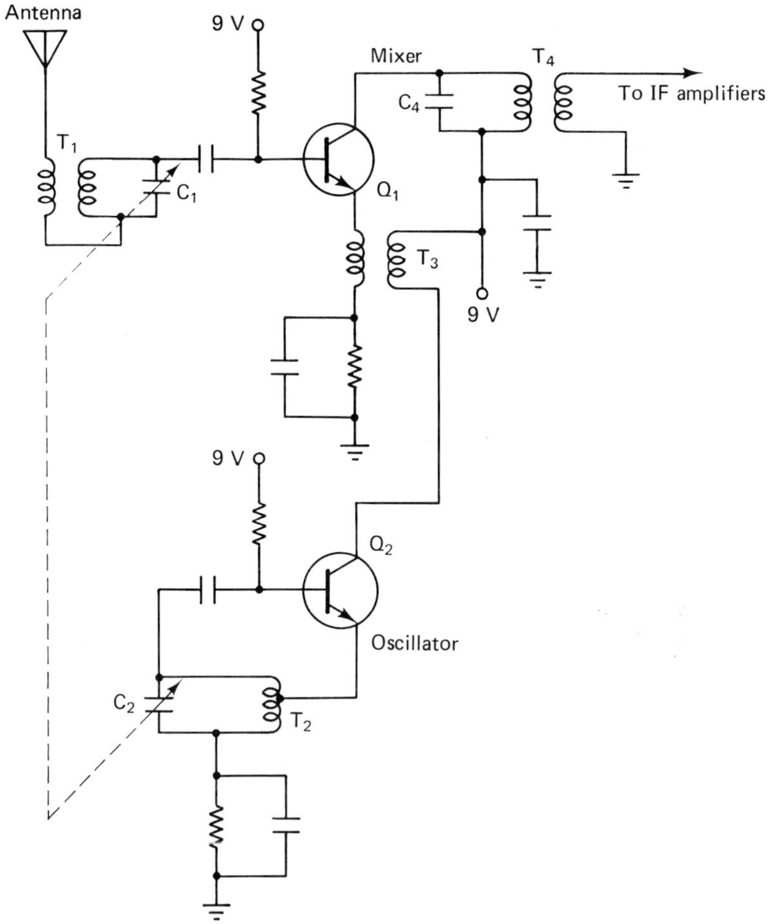

Figure 12-7 Superheterodyne circuit. The tuned amplifier Q_1 mixes the signals from the antenna and the oscillator to create a lower frequency, called the intermediate frequency, IF. The IF frequency remains the same for all frequencies coming in from the antenna.

Hence the tuned circuit (T_4–C_4) is fixed at 455 kHz. This difference frequency is called the intermediate frequency (IF).

A detailed knowledge of the heterodyning process that is responsible for the conversion from RF to IF is beyond our purposes. As troubleshooters, the main things we should know are the input and output frequencies of each stage and section and what can make them go wrong. Notice that both transistor circuits are like amplifiers, and we troubleshoot them as such.

For your information, the 455-kHz IF frequency is often used in broadcast AM radios. 10.7 MHz is for FM, and a band of 41.25 to 47.25 MHz with peaks at 43 MHz and 45 MHz is used for TV. In the older sets, special equipment was required to align these stages.

The combined oscillator and mixer stage is called the *converter.* To troubleshoot it, use the information gained in the sections dealing with tuned amplifiers and oscillators.

Modern receivers use varactor tuning in the RF stages and SAW filters for the IF stages. The SAW filters are fed by a transistor preamplifier. The output of the SAW filter is fed to an integrated multifunction chip. We discuss integrated-circuit chips in Chapter 14.

12-5 AM DETECTORS, MODULATORS, AND IF STAGES

In ordinary radio communications the RF frequencies range from 500 kHz on up. The audio frequencies (AF) range from only 20 to 20,000 Hz, or cycles per second; these very low frequencies cannot be transmitted economically. Radio transmitters transmit audio signals by impressing them on the carrier RF wave. The transmitter circuits vary the strength of the RF wave at the audio rate. This action is called *amplitude modulation* (AM). The receiver must extract the audio signal from the RF wave.

12-5-1 Checking an AM Detector

In an AM receiver, the detector (or demodulator, as it is sometimes called) has the job of extracting the audio wave from the AM wave. Figure 12-8 shows what happens to a station signal as it passes through the various stages of a radio. In television reception, these signal changes are basically the same. The amplitude-modulated RF wave is shown at the converter. The RF frequency is so high that the waves are too crowded to be seen individually. The wavy "envelope" that encloses the RF wave shows the audio modulation. The complete wave is unusable because the upper peaks will be canceled by the lower peaks. Either the top or bottom part of the envelope wave must be cut off to obtain the pure audio wave. This will be done in a later stage.

The output of the next stage shows the modulated IF wave. This wave is still unusable as is. One purpose of the IF section is to further reject the unwanted stations and select the desired station. It also amplifies the signal and passes it on to the detector stage.

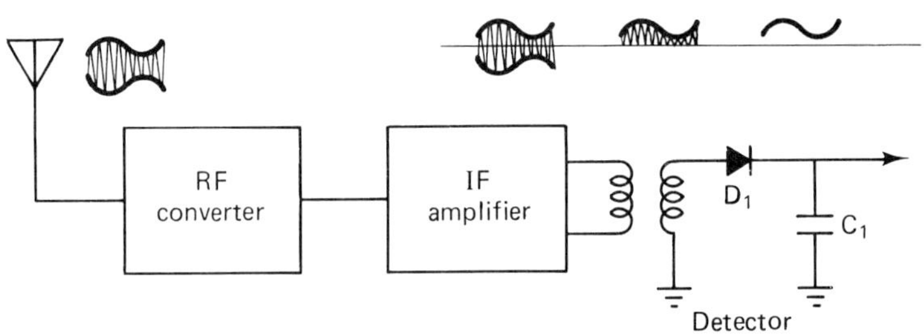

Figure 12-8 AM detector. It converts the IF signal into an audio signal that can be amplified to drive a speaker. Note the audio "envelope" that encloses the high-frequency wave. This envelope is extracted and becomes the audio wave.

The detector stage extracts the audio signal from the modulated IF signal. Notice that this circuit resembles the transformer power-supply rectifier and filter sections. One difference is that the components are specially designed for the higher frequency. There is one other difference. Let's look into this.

The diode D_1 rectifies the AM wave, so that only the top half of the wave gets through. This is just what a power-supply rectifier would do to an ac voltage. Next, the capacitor C_1 filters the wave. What happens here is similar to the power supply, but not exactly. Filtering is only partial in this case. C_1 is large enough to filter out the 455-kHz IF portion of the pulses, but not large enough to filter out the lower-frequency audio pulses of typically 1 kHz. The result is almost a pure audio wave.

This wave is fed to an audio amplifier. The natural frequency response of the audio amplifier and speaker filters out the remaining RF component. In fact, an earphone connected to the detector will filter out the RF and pass the audio only. The same is true with your ears.

TIP 12-14

If the detector is working, you should have a dc voltage at its output. If you do have this dc voltage, all the preceding tuned circuit stages are probably good.

The diode and capacitor can be checked out with an ohmmeter. The capacitor can be checked more quickly with a scope. The presence of RF at the output would indicate an open capacitor.

12-5-2 The IF Amplifier

Figure 12-1 is actually an IF amplifier. The two tuned circuits can be tuned to the IF frequency. The tuned circuits are actually enclosed in rectangular metal cans to keep them from picking up or causing interference and feedback. They are called IF transformers. The capacitors are usually built in to the same shielded unit with the coil. At the detector, the IF transformer unit may also enclose the diode.

In regular AM broadcast-band radios the stages are tuned to a very narrow band of frequencies centered around 455 kHz. In FM receivers, the amplifiers are tuned to a wider band of frequencies centered at 10.7 MHz. This is because they must pass the signal equally as its frequency varies above and below 10.7 MHz. In TV receivers, the bandwidth must be 6 MHz. For this reason the IF stages in older TV receivers were stagger tuned. Each stage was tuned to a slightly different frequency.

Aligning these stages used to be a big job that required special equipment. Now we have SAW filters that do not require alignment (see Fig. 12-3). For the most part we troubleshoot the IF stages as we would the tuned amplifer stages in Section 12-1. To troubleshoot a SAW filter stage, see Section 12-2-2.

12-5-3 The AM Modulator

We have seen how an AM wave was demodulated in the receiver. Now we will see how it is modulated in the transmitter. At the transmitter, the RF carrier wave originates in an oscillator stage. It is amplified and modulated before it is sent to the transmitting antenna. Figure 12-9 shows a basic modulated RF amplifier. The collector voltage of the Q_1 amplifier is varied at an audio rate. As can be seen, the modulated amplifier and the modulator are both basically amplifiers. The first is an audio amplifier. The second is an RF amplifier.

> *Note:* Of course, this is only a stripped-down version of a real transmitter circuit. However, the biggest difference is not in the complexity of the circuits. Rather, it is in the physical construction of the parts. They must conduct high currents and meet strict frequency and interference standards. Other than that, you can see that there are two familiar-looking amplifiers. Troubleshooting the modulator is basically the same as troubleshooting an audio power amplifier. The RF amplifier is basically the same

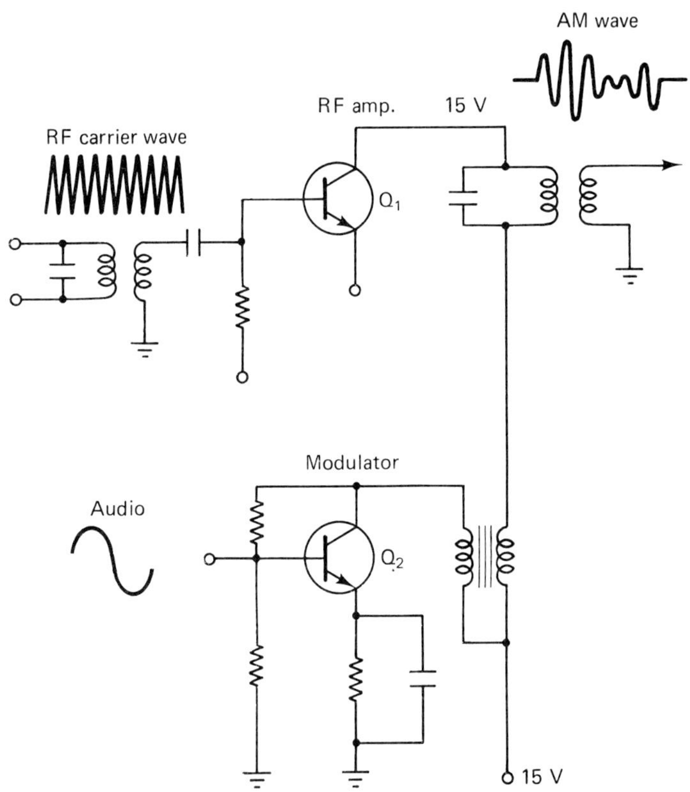

Figure 12-9 AM modulation circuit at the transmitter. The audio amplifier or modulator varies the strength of radio-frequency amplifier to produce the amplitude-modulated (AM) wave.

as the tuned amplifier we saw in Fig. 12-1. However, the circuitry and adjustments required to meet the strict frequency, power output, and modulation standards are beyond our purposes.

12-6 FM DISCRIMINATORS AND MODULATORS

A frequency-modulated (FM) wave is less affected by interference than is an AM wave. The FM wave maintains a steady amplitude, but its frequency is varied at an audio rate. The FM receiver responds to the varying frequency, which carries the audio information. It does not respond to interference and noise, which are mostly variations of amplitude.

12-6-1 Checking the FM Discriminator

The FM discriminator, or detector as it is sometimes called, extracts an audio voltage from the FM signal. To give you an idea of how this can be done, an AM detector will respond to an FM signal if the last IF stage is slightly detuned. The signal will vary in strength as its frequency varies around the resonant frequency of the IF tuned circuit. These variations take place at an audio rate. Of course, this is very inefficient and impractical. Actual FM discriminators are much more efficient.

The FM discriminator in Fig. 12-10 converts the FM signal from the IF amplifier into an AM signal for the audio amplifier. When this circuit receives a signal, the diodes in the circuit will develop dc voltages that can be measured with a voltmeter.

The filter circuit consisting of R_3 and C_2 is called a *deemphasis circuit*. It cuts down the high frequencies that were emphasized at the transmitter. This action is explained in the next section when we look at the transmitter.

> **TIP 12-15**
>
> In the Foster–Seeley FM discriminator (Fig. 12-10) and the similar ratio-detector discriminator, there should be equal dc voltages on the two diodes and their resistors, if any. The output dc voltage should be zero.
>
> If all the preceding stages are okay, you should have dc voltages across the individual diodes or resistors.
>
> If one diode has no voltage but the other does, use an ohmmeter to check the diode and its half of the secondary winding.
>
> If you have excessive high frequencies at the output, check the deemphasis capacitor at the discriminator.

Interesting Sidelight. Figure 12-10 shows a schematic of an FM discriminator. The action is very complex and has to do with the timing

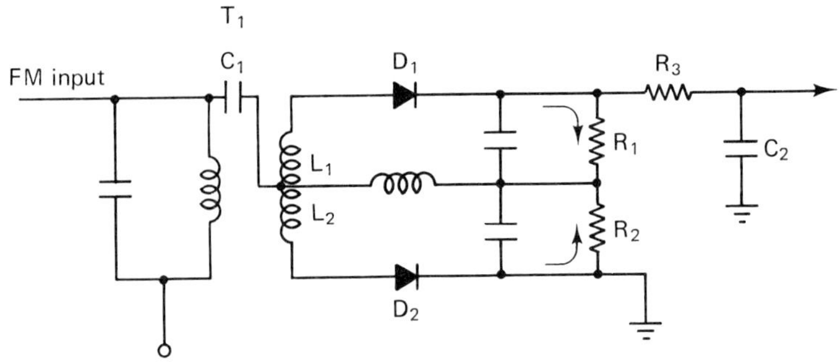

Figure 12-10 The FM detector or discriminator converts the FM station signal into an audio signal that can drive the speaker after it is amplified.

(phase relations) of the signal waves that are output from the capacitor C_1 and the secondary windings, L_1 and L_2. For instance, if the peaks of the L_1 and C_1 waves come in at the same time (in phase), they will combine to produce a high output at D_1. If the peak of the L_1 wave is delayed (out of phase), the output will be low. The tuned transformer primary circuit has the capability of advancing and delaying the peaks (shifting the phase) of the L_1 and L_2 signal waves. This advancing and delaying takes place at an audio rate, so the voltage at D_1 changes at an audio rate. A similar thing happens with C_1, L_2, and D_2.

The diodes rectify their respective signals. Recall that the voltages at the ends of a transformer have opposite polarities. When the L_1 side is positive, D_1 conducts, and when the L_2 side is positive, D_2 conducts. The rectified outputs across R_1 and R_2 add up to produce a strong output signal.

The foregoing may not be a complete explanation, but as troubleshooters, we need to focus mainly on the input and output signals and the dc voltages. We should also know that the circuit is called the FM discriminator, or more specifically, the Foster–Seeley discriminator.

12-6-2 Checking the FM Modulator

FM modulation at the transmitter can be accomplished with a varactor. Figure 12-11 shows a simple FM modulator circuit. Not surprisingly, we see another amplifier. In this case the tuned circuit with feedback makes it an oscillator. Varactors have been placed in the tuned circuit as added capacitors that determine the frequency of the oscillator. The frequency of the tuned circuit changes in accordance with the audio voltage impressed on the varactors.

The *RC* coupling circuit at the output serves as a low-frequency filter to emphasize the higher frequencies. This will overcome the effects of atmospheric interference, which are mainly at the higher frequencies. The higher frequencies, including the noise, are deemphasized at the receiver.

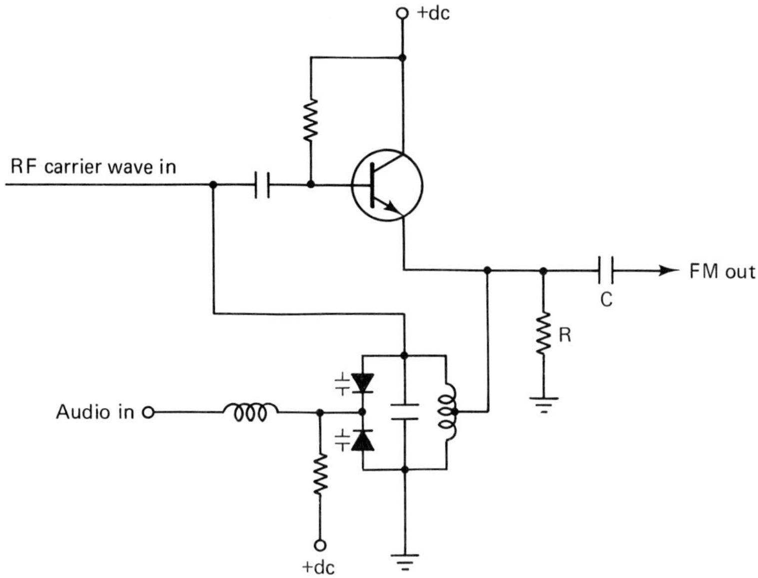

Figure 12-11 The frequency of the RF oscillator in this FM modulation circuit is varied at an audio rate. The output is an FM wave.

Note: To troubleshoot this circuit, check the transistor voltages. The tuned circuit (the tank) can be checked with a scope. A varying voltage from a dc power supply can be substituted at the dc bias input to see if the frequency can be changed. The diode and varactor diodes and the coil can be checked with an ohmmeter. They do not have to be removed from the circuit for this test. The final transmitter check involves advanced equipment and procedures.

12-7 FEEDBACK CONTROL CIRCUITS

Modern electronic circuits require automatic, self-regulating controls. The feedback circuits we saw in Chapters 1 and 11 are similar to this kind of control (Figs. 1-6 and 11-1). In general, self-regulation is accomplished by monitoring the output of a system. A sampling of the output voltage is fed back to the the feedback control circuit, which develops a correction voltage. This correction voltage is fed back to the input of the system. The fundamentals learned from the following examples can be applied to a wide range of equipment.

12-7-1 The Automatic Volume Control (AVC) Circuit

You can control the gain of the amplifier by varying its *BE* voltage. The amplifier has good gain when its *BE* voltage hovers close to 0.7 V. When the *BE* voltage gets down to about 0.4 V, the output waveform begins to flatten out and there is less gain.

Similarly, when the *BE* voltage increases to about 0.8 V, the response flattens out and there is less gain. In Chapter 8 we called this flattening out "distortion," and we tried to get rid of it. Here we feed in a very small signal, so there is less distortion. In most cases the distortion is of no consequence.

The point is: You can control the gain of the amplifier by operating it at various *BE* levels. You have the most gain when the *BE* voltage is closer to the midvoltage level, the least gain when the *BE* voltage moves away from it. Thus we have what is called a *variable-gain* amplifier. If the system can feed back a control voltage to the *BE* terminals of this amplifier, we have a self-regulating gain control.

The simplest type of feedback control is the automatic volume control (AVC) circuit of a radio (see Fig. 12-12). The detector diode D_1 does double duty. It produces rectified audio and AVC voltages. For audio, the diode output is only partially filtered by C_1. For AVC, the signal is more fully filtered by the bigger capacitor C_2. This dc voltage will vary with the strength of the signal. It is fed back to control the *BE* voltage of the variable-gain stage.

A stronger-than-normal signal will produce a more positive AVC voltage that will increase the *BE* voltage of the controlled stage and thus reduce its gain. A weak signal will do the opposite. Thus we have automatic gain control.

Note that the average dc voltage at the collector will change when the signal strength changes. If the feedback voltage is positive, as in this case, the collector dc voltage when measured with a voltmeter will fall when the signal strength increases.

The simple AVC circuit is not difficult to troubleshoot, because it only adds a resistor–capacitance filter to what already exists. It is primarily a matter of comparing voltages with the schematic plus a few special tips. The symptoms for defective AVC are weak signal, excessive differences in volume between stations, and squealing or excessive noise.

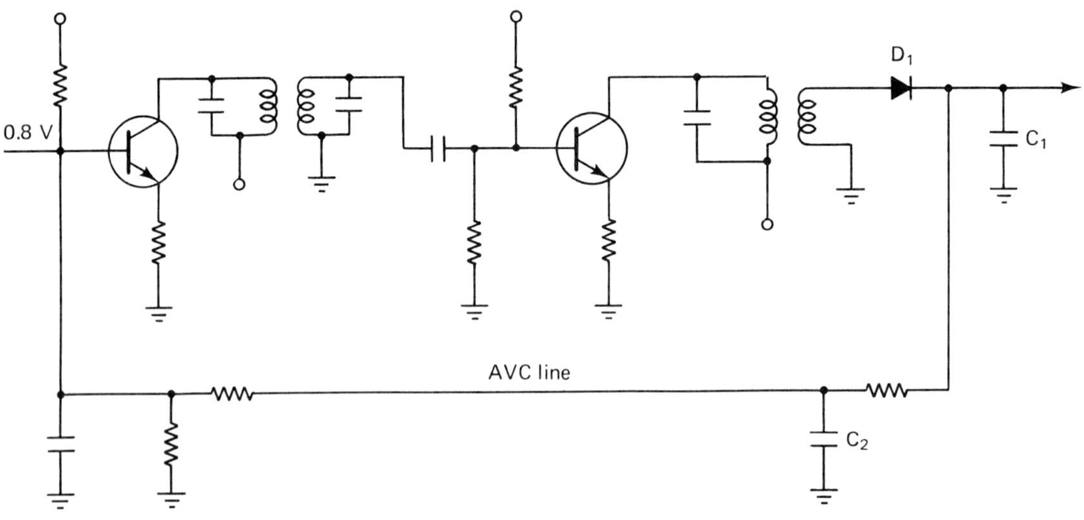

Figure 12-12 Automatic volume control (AVC) circuit in a radio. The correction voltage is derived from the detector stage.

TIP 12-16

The main check for AVC is to monitor the AVC voltage with a voltmeter as you change stations. The dc voltage should decrease between stations and increase when a station is tuned in.

Also check the collector voltage at the variable-gain stage. It should vary as you change stations.

TIP 12-17

An open AVC bypass capacitor can cause hum pickup, squeals, or weak signal. This capacitor has a low-voltage rating, so it may be of lower quality. Always suspect these types. They can be checked by jumping them with a good one. The scope may reveal an open capacitor when the AVC line picks up interference.

TIP 12-18

If the output of the detector has increased dc voltage when a station is tuned in but the AVC output does not, the AVC capacitor or resistor is probably defective.

Early television receivers also used this type of automatic gain control (AGC), but these have been supplanted by more sophisticated feedback circuits involving other parts of the receiver. We check out television AGC circuits in Chapter 13. Other types of automatic control circuits are also investigated there.

12-7-2 Automatic Fine Tuning (AFT)

The automatic fine tuning (AFT) circuit shown in Fig. 12-13 is for keeping the tuned-circuit frequency from drifting. A tuned circuit will drift as its inductance and capacitance values change with temperature and voltage variations. Special temperature-compensating capacitors are used, but additional control may be needed.

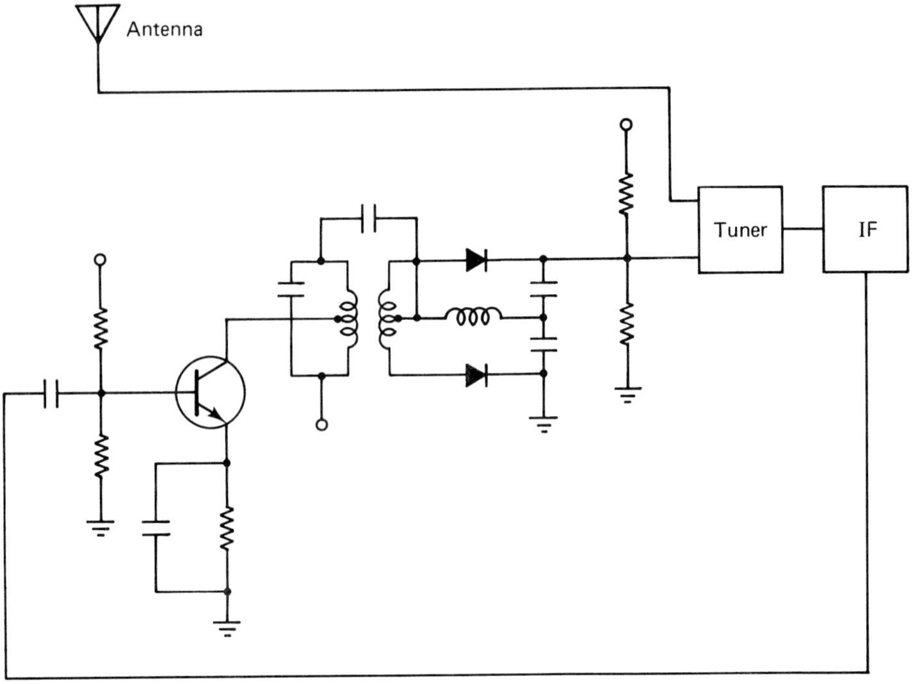

Figure 12-13 Automatic fine-tuning (AFT) circuit. Similar to the FM detector, it detects a change of frequency and develops a correction voltage.

This AFT circuit uses a discriminator that changes output voltage with frequency changes in exactly the same way that the FM discriminator did in the preceding section. The main difference is that the frequency changes are much slower now. This AFT discriminator is tuned to the IF frequency. At this frequency, a certain voltage is output to the tuner, which contains a varactor circuit (see Fig. 12-2). When the input signal drifts away from the frequency, the discriminator will output a correction voltage. This voltage will shift the tuner back to its original frequency.

This circuit can be checked for open windings with an ohmmeter. The diodes, too, can be checked with an ohmmeter. If the stations can be tuned in with a substitute voltage, the AFT circuit is probably bad. A voltmeter can be used to check the resistor voltages. Except for the emitter resistor, they should all drop voltages in proportion to their series resistances.

> *Note:* To find out whether the trouble is in the AFT circuit or the tuner, see Tip 12-16 in Section 12-2-1, which shows how to use a substitute voltage in place of the AFT voltage. To check the AFT circuit itself, see Tip 12-15 in Section 12-6-1, which deals with the FM discriminator.

12-8 THE BLOCK DIAGRAM OF A RADIO RECEIVER (AM OR FM)

We have checked out the basic circuits found in a radio receiver. Let's put them all together so that we can get a better picture of what is going on (see Fig. 12-14). Now we can trace the signal from antenna to speaker.

The antenna receives all the signals in the area. The converter selects the desired signal and converts it to an intermediate frequency. The IF amplifier passes only the IF frequency and further rejects all other frequencies. The detector converts the IF signal to audio. The audio section amplifies the signal power and feeds it to the speaker. The speaker converts the electrical signal into a sound signal.

The AVC line runs back from the detector to the converter or IF stage. This is true for both AM and FM receivers. The AVC keeps the volume from fading or coming on too strong.

The block diagram for the video and sound sections of a TV receiver is similar to that for the radio. The other sections bear some similarities, too. If you can trace the signals and voltages in a radio, you should not have too much trouble with TV receivers.

The block diagram will help you isolate a bad section by noting the output symptoms. For instance, if you have a rushing noise and no station can be tuned in, the problem is probably in the tuning or converter section. The rushing sound is atmospheric noise picked up by the antenna and amplified by all the receiver sections.

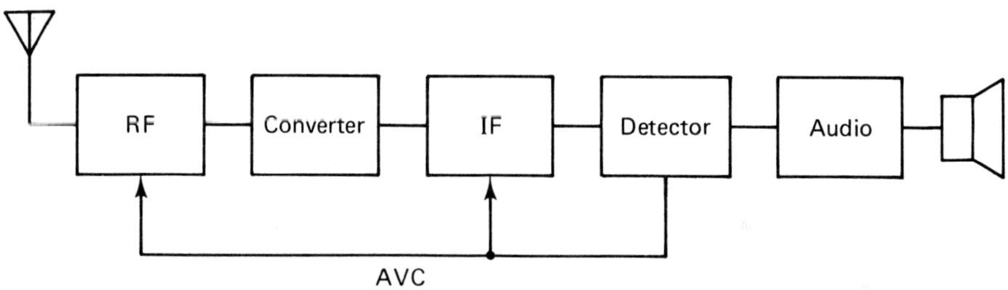

Figure 12-14 Block diagram of a complete radio receiver.

12-9 VIDEO AMPLIFIERS

Video amplifiers for display monitors and TV receivers must be capable of passing and amplifying a very wide range of frequencies, from 30 Hz to 4 MHz (or more for computer monitors). The basic amplifier cannot possibly do the job. At the high frequencies, the stray capacitances in the circuit will short out the signal to ground. Figure 12-15 shows a basic video amplifier, with peaking coils, L_1 and L_2. These coils boost the response at the high frequencies.

L_1 is grounded through C_1 for signal voltages. When the frequency of the signal increases, some of it will tend to be bypassed by the stray capacitances in the circuit,

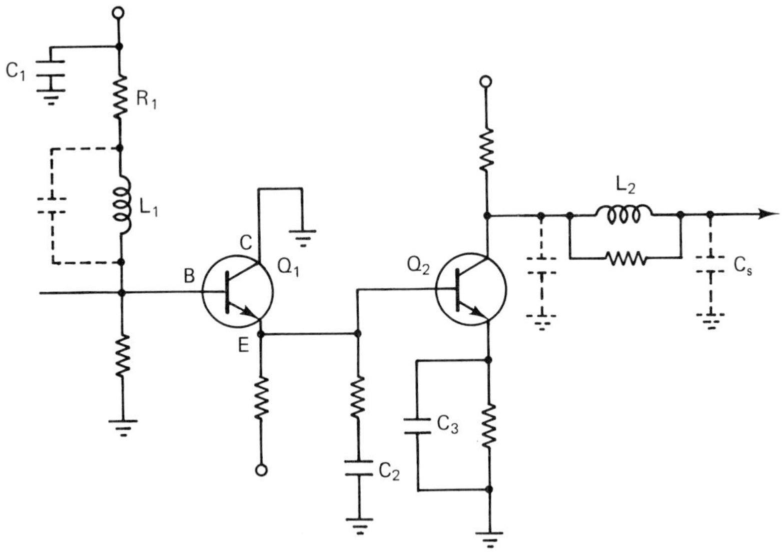

Figure 12-15 Video amplifier with coils to extend the response of the amplifier at high frequencies.

represented by the dashed capacitor. This is counteracted by the increased reactance and consequent voltage increase across it. Also, L_1 is designed to resonate with the stray capacitances at the higher frequencies.

Similarly, L_2 forms a tuned circuit with the stray capacitances to ground. At the high frequencies, the voltage across the capacitances builds up to a maximum at resonance. This boosts the output of the amplifier.

Amplifiers also tend to lose gain at the low frequencies. This is prevented primarily by direct coupling. Capacitive coupling would attenuate the low frequencies. Another method of comparatively boosting the low frequency is provided by C_2. It cuts down the high frequencies to the level of the low frequencies.

C_3 can boost amplifier gain at the higher frequencies. Its capacitance is made small so that at the low frequencies it has no effect. As we learned in Section 8-5-2, an amplifier needs an emitter bypass capacitor for maximum gain. In this case the emitter capacitor has a maximum effect at the higher frequencies.

Checking the Video Amplifier. The video amplifier is checked out like any other amplifier. The coils are easily checked when they open; their voltage drops and resistances increase drastically. If L_1 opens, the base of Q_1 will fall. If L_2 opens, the output dc as well as signal voltage will be lost. (In most cases the CRT gets its dc voltage from the collector of Q_2.) And, of course, an ohmmeter works fine for finding open coils. Shorting an open coil will also restore the picture. It may not be a good picture, but at least you know that the coil is open.

This brings us to the symptoms shown by defective peaking coils. If they open, the picture may be lost completely, as in the case of L_1 and L_2. A shorted peaking coil will result in a loss of detail in the picture.

An open C_2 will attenuate the low frequencies. It will result in a loss of contrast across wide areas of the picture. Dark rooms will look lighter. An open C_3 will lower the high-frequency response and may noticeably blur the edges of objects.

WHAT YOU LEARNED TO DO

In this chapter you learned to troubleshoot:

1. Tuned circuits, with voltmeters, ohmmeters, and scopes.
2. Tuned amplifiers, oscillators, mixers, and modulators in much the same way as you would the basic amplifier.
3. Varactor-tuned circuits by checking their tuning voltages or by substituting the tuned voltages. You learned to check out the varactor diode with an ohmmeter.
4. SAW filter circuits by checking them with a voltmeter, ohmmeter, and signal generator and scope.
5. AM and FM detectors, with an ohmmeter, voltmeter, and scope.
6. The AVC and AFT as diode circuits.
7. Peaking coils and compensation capacitors in video-amplifier circuits.

QUESTIONS

1. What can happen to a radio if its input tuned circuit goes bad? What can happen to a transmitter if its tuned circuits go bad?
2. What determines the resonant frequency of a tuned circuit?
3. What are the causes of undesired oscillations in an amplifier?
4. Draw a block diagram of a radio. Where is each of the following located in a radio: the converter stage; the oscillator; the detector? What is the function of each of these circuits?
5. How does an AM detector circuit differ from a power-supply rectifier circuit?
6. What key voltages and resistances should you check in an FM discriminator?
7. Briefly describe the operation of an AVC circuit.
8. How does the AFT circuit action differ from an FM discriminator circuit action?
9. What added part in a video amplifier extends its high-frequency response? What part extends its low-frequency response?
10. What can cause a loss of fine detail in a TV display?

PRACTICAL TROUBLESHOOTING EXPERIENCE

Repairing small radios is usually not economically feasible. However, it is a good learning experience. Many of the checks here can be applied to more elaborate equipment. You need an inexpensive pocket radio with transistors and IF transformers for this job. If you have a schematic, fine; if not, that's okay. Whether the receiver is working or not, let's start at step 1.

1. Check the battery voltage while it is in the circuit with the switch turned on. For clean sound at full volume, you should have nearly 10% of the rated voltage. For our purposes, if you have fairly good sound at low volume, it is okay.
2. If replacing the battery does not restore operation, shut off the power. Remove the battery and check the resistance across the receiver terminals for the battery. It should measure over 10.0 kΩ after the bypass capacitors have charged up.

   If the resistance seems okay, reconnect the battery and measure its voltage. If the voltage is okay, you may have an open somewhere. If it is low, something is shorting out. In either case, proceed with the following tests. If the voltage is low, turn on the power only when taking voltage tests.
3. If you have no sound, turn up the volume and hold the speaker close to your ears. If you hear no noise at all, the problem is probably in the audio stages.
4. If you have no noise at all, disassemble the radio. Do not pry things apart. Press and slide the plastic cover off. If the solder side of the board is exposed, turn on the power, touch various points at random with your finger, and see if you pick up anything from the speaker. If you do, the speaker and audio output section are probably okay.
5. If you cannot pick up anything by touching various points, perhaps the speaker is bad. Check it with an ohmmeter. One side of the speaker is grounded; the other side has a connection to the earphone plug. (Check to see if a defective earphone plug is disabling the speaker.)
6. If the speaker is okay, turn on the power and randomly scratch various points near the speaker with an ohmmeter probe tip, with the other lead grounded. The ohmmeter battery should generate scratchy noises from the speaker. If it does, the output capacitor and possibly the output transistor are okay.
7. If the ohmmeter scratch test produces no sound, randomly check for voltages at various points on the board. If you cannot find any voltage at all, follow the supply paths from the battery terminals and check out the voltages at all points until you find out where it is lost.
8. If you find a voltage point, see if you can find any voltages on the most-available transistor, preferably the one nearest the speaker.
9. Now comes the hardest part. If you cannot get at the output transistor, you may have to pull out the board. Let's hope you have an easy job of it. Sometimes the board can be opened like a door with the wires acting as hinges. Find the output transistor; it is near the big output capacitor, which has a lead connected to the speaker.

   Now check the voltages on all the audio transistors as needed. They are the ones that are not between the square metal cans; they are connected to the volume control. If the voltages are bad, check the transistors, resistors, and capacitors with an ohmmeter.

10. Perhaps the trouble lies in the stages before the audio section. Find the detector stage. It is usually near the IF can closest to the volume control and farthest from the tuning circuit. Monitor the dc voltage on the detector diode as you turn the tuning dial. If the voltage varies, the preceding stages are probably okay.
11. If you cannot find anything with the multimeter, it is now time to use a signal tracer, scope, or signal generator. Check for signal at the volume control and each audio transisor. Also check each capacitor to see if it is passing or bypassing a signal as it should. A bypass capacitor has one side grounded.
12. Check around the antenna and variable capacitor circuit for a signal, with scope, generator, or tracer. You can use your finger as a generator, too. If you can pick up a signal when your finger is touched to a point, at least the oscillator transistors are oscillating. Otherwise, there is no IF signal. Only this signal is likely to get through the IF amplifiers. This does not mean that the oscillator circuit is good; it means only that the transistors are capable of oscillating.

    If you can pick up a rushing sound by touching the tuning circuit but cannot pick up any stations, the oscillator could be bad. This means that noise signals having an intermediate frequency are getting through.

    To find the RF input circuit, the antenna can be identified as a big flat or cylindrical ferrite piece with thin wire wound around it. It is actually a part of the input tuned circuit. The variable capacitor is the unit that is mechanically connected to the tuning dial.

    On the scope you will be able to pick up the modulated signal with the time-per-division set to the 1.0-ms range. It may not look very much like the waveform shown in Fig. 12-8, because the audio envelope will vary. During steady portions of the sound, you will catch the resemblance. If you view the same waveform at 2 μs per division, you will see a carrier, but the amplitude variations will be blurred.
13. Look for bad voltages around the converter circuit. The oscillator can be checked out by injecting an unmodulated RF signal into the oscillator top side of the oscillator coil. Vary the signal about 455 kHz above the spot on the dial. If you get the station, the oscillator is bad.
14. The IF transformers can be checked out with an ohmmeter. The collector coil should pass voltages to the collector. The base coil may have an internal series capacitor, so an open across the can terminals may not indicate an open coil.
15. Check the dc voltages on all the transistors. Check for signal on all the transistors until you find where it is lost.

13

Monochrome and Color Display Tube Circuits

EXPERIENCE REQUIRED

To use the information in this chapter, you should have some prior knowledge of:

1. Video, horizontal, and vertical amplifier troubleshooting.
2. Feedback circuits.
3. Transistor input and output signals and dc operating voltages.
4. Transistor circuit defects.
5. Checking basic electronic parts.
6. How symptoms point to defective circuits.
7. Pulse signals.

It would be helpful if you can visually identify the various components in a TV set or video monitor, such as the picture tube, yoke, and flyback transformer.

> *Note:* To shorten our discussion in some of the later sections, we concentrate on the circuit inputs and output. If you need to work on the individual parts in these circuits, refer to earlier sections or chapters.

13-1 MONOCHROME CRT DC OPERATING VOLTAGES

The cathode-ray tube (CRT) rivals the telephone and the radio speaker as the most widely used electronic output device for conveying information. It is found everywhere—in television sets, video monitors, computer terminals, video-game parlors, and supermarket checkout counters.

The troubleshooter also finds it an important source of information. Except for audio troubles, it displays practically all the symptoms necessary for troubleshooting TV and video and for some computer-terminal symptoms, too. These symptoms often lead you directly to the bad part.

As always, we start with the output device and work our way in. The monochrome CRT is a very complex device that displays five types of trouble symptoms: brightness, focus, deflection, video, and sync.

> **Caution:** The glass tube is strong enough to withstand tremendous pressure since all the air has been pumped out of it. There is ''nothing but vacuum'' in the tube. However, great care must be taken when handling it. Breakage will ''implode'' the glass and shoot it around in all directions.

13-1-1 Brightness and Focus Voltages

First, we look at the output device symptoms. Most often, we will see a brightness problem. Next, we check the voltages on the device. The display tube requires six voltages for brightness. Some of these are adjustable. Figure 13-1(a) shows the schematic symbol for a monochrome CRT. Starting from the left, the CRT elements are called the heater, H; the cathode, K; the grid, G_1; the screen grid, G_2; the focus, F; and the second anode or high-voltage anode. All these terminals except the high-voltage anode come out as pins on the back of the neck, as shown in Fig. 13-1(b).

The high-voltage anode should have up to 20 kV for black-and-white television tubes. (**Caution:** Keep your hands away from the high-voltage lead.) In smaller tubes, it may be as ''low'' as 9 kV, but that is still very high. The anode terminal is on the bell part of the CRT. It is connected to a metallic coating on the inner surface of the glass. This forward part of the CRT receives a very high voltage that speeds the electrons out of the gun assembly as shown in Fig. 13-2.

Electrons are extremely tiny particles, so small that they are invisible even under a microscope. Since they are negatively charged, they are attracted by positive voltages. They speed toward the face of the CRT in a narrow beam. The inner surface of the face is coated with a fluorescent substance that lights up when struck by the electron beam. This forms a bright dot on the face of the CRT. Electromagnetic devices around the neck of CRT will move the dot around to create a display (see Fig. 13-4).

The cathode, K, has a low voltage, somewhere between 20 and 50 V above ground. Its job is to give off electrons that are shot toward the face of the CRT. It is coated with a special material that gives off electrons when heated.

> **Electron Flow versus Conventional Flow.** You may be confused about the flow of current both in and outside the CRT. All matter has positive and negative charges. An electrically neutral material has an equal number of positive and negative charges in it. If you remove a negative

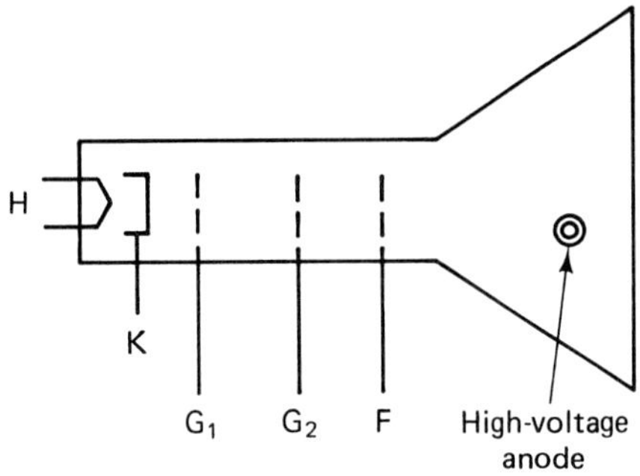

(a)

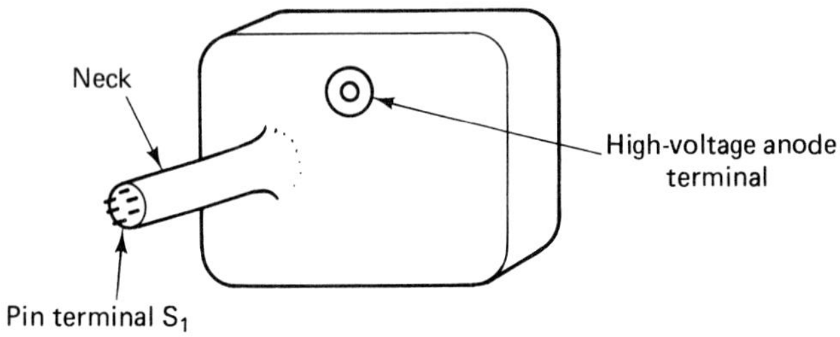

(b)

Figure 13-1 (a) Symbol for CRT and its terminals. (b) Rear view of a CRT showing the neck and pin terminals as well as the high-voltage terminal.

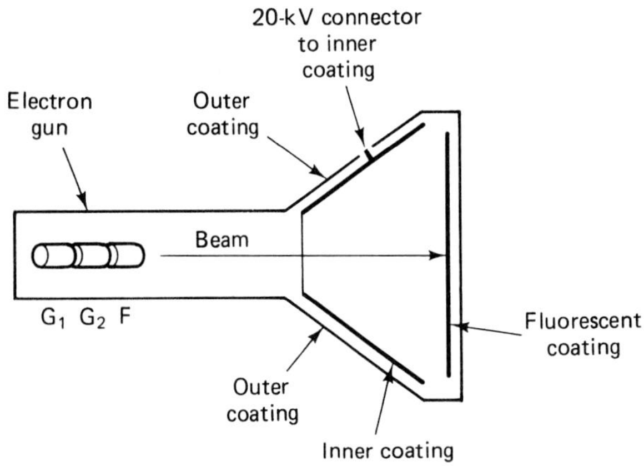

Figure 13-2 Internal structure of CRT. The gun shoots out an electron beam that lights the face of the tube with a bright dot.

charge (a given quantity of electrons, −1 V, for instance) from a material, it acquires a positive charge (+1 V, for instance). Conversely, if you add or put back a negative charge to this positively charged material, it loses the positive charge. The anode of a CRT has many times more positive charges than the cathode. With this in mind, let us look at the flow of current in a CRT.

It is true; there is a flow of negative electrons (electron current) from the cathode of the CRT to the anode. However, there is also an apparent flow of positive charges (conventional current) in the opposite direction. This happens because every electron that reaches the anode removes a positive charge. This positive charge can be thought of as moving to the cathode, because a positive charge appears there for every electron that leaves it. The power supply can be looked upon as forcing this positive charge back to the anode.

The heater, H, voltage is very low, about 6 to 11.5 V dc or ac across its terminals. This voltage is taken from batteries, power supplies, or from the horizontal output transformer.

The grid, G_1, voltage is usually about 40 V lower than the cathode. It is used to control the intensity of the electron beam. When its voltage goes down, it cuts down the amount of electrons that can get through. When its voltage rises, it allows more electrons to get through. Thus you can vary the intensity of the beam, and consequently the brightness of the dot, by varying the voltage on the G_1 grid.

The term *grid* is inherited from the corresponding structural element in the vacuum tube, the predecessor of the transistor. In very early tubes it was actually a thin wire grid through which the electrons had to pass. The CRT elements have functions that correspond to their vacuum-tube namesakes, but they are constructed like cylinders. When they are lined up, they have the appearance of the barrel of a gun. This gives rise to the term *electron gun*.

The screen-grid, G_2, voltage is relatively high, about 200 to 400 V. The screen grid helps draw out the electrons from the cathode and speed them up. It is also called the accelerating anode. Most often, it is simply called the *screen*.

The focus-grid voltage is usually higher than the screen. However, its voltage ranges from as low as 0 V to as high as 790 V, depending on the construction and placement of the focus cylinder. Its job is to narrow the beam and push stray electrons back into it. The stray electrons would create a fuzzy area around the moving dot at the face of the CRT.

TIP 13-1

Loss of brightness may be caused by the following: insufficient high voltage at the anode button, low grid voltage, high cathode voltage, low screen voltage, low focus voltage, and insufficient heater voltage.

Occasionally, an excessive grid voltage or loss of cathode voltage will cause excessive current flow and overload the CRT and black out the display.

TIP 13-2

Poor focus or fuzzy picture is caused by insufficient or excessive CRT focus voltage. Sometimes, it is caused by insufficient high voltage. This can be isolated by adjusting brightness control. If the width of the picture changes as you turn the brightness control, you have insufficient high voltage. This symptom is called *blooming.*

13-1-2 Summary of Typical Monochrome CRT Voltages

The following CRT voltages will help you determine whether a CRT is okay. Consult the schematic or service manual if you need more exact voltages. All voltages except the heater are measured with respect to ground.

Monochrome CRT

Heater	11.5 V dc or rms, or 44 V p-p horizontal
Cathode	30 to 80 V dc
Control grid	−15 to 40 V dc
Screen grid	100 to 280 V dc
Focus grid	0 to 100 V dc
Anode	7.5 to 20 kV

The condition of the heater is almost always determined by its brightness or color. It should be heated up to an orange color. Dark orange indicates insufficient heat (red is too dark to see). Bright orange or yellow indicates too much heat.

13-2 THE CRT VIDEO SIGNAL VOLTAGES

The video signal voltage is usually applied to the CRT cathode (see Fig. 13-3). Note the shape of the input wave; we discuss it later. It should be about 35 V peak to peak at the cathode. The CRT dc cathode voltage also comes from the video amplifier. This means that you have direct coupling. If something goes wrong with the video amplifier, the dc cathode voltage will be affected.

The video signal causes the cathode voltage to fluctuate. As the dot flies across the screen from left to right it changes brightness in accordance with the cathode voltage. In other words, it varies in accordance with the picture information. When it flies back, the signal blanks it out. Let's find out what makes the dot fly back and forth.

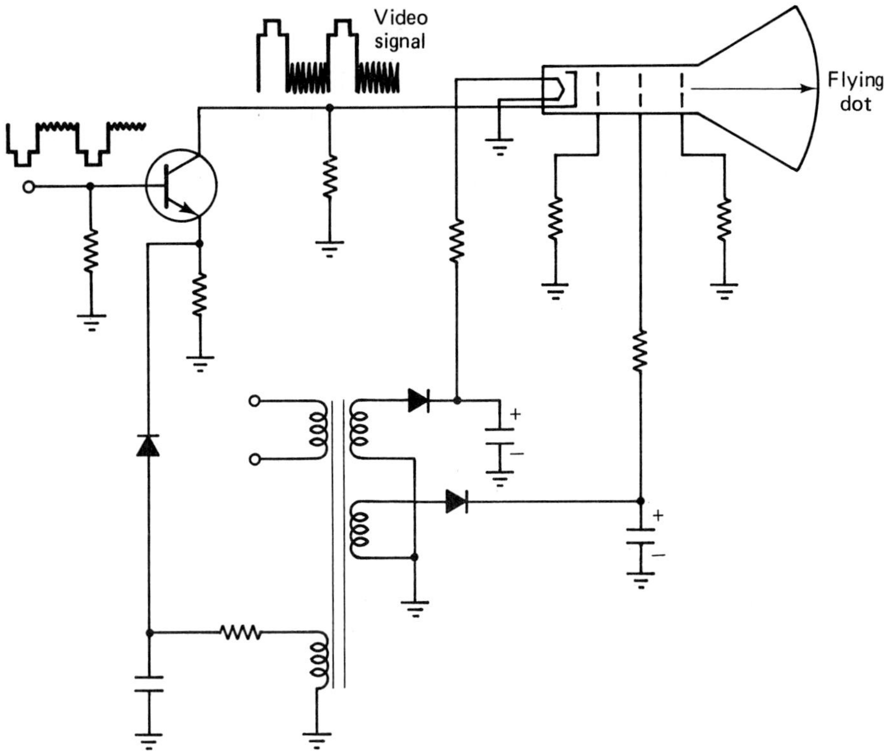

Figure 13-3 The video signal is applied to the cathode of the CRT. It varies the intensity of the electron beam. The heater and screen grid voltages are derived from the horizontal output transformer.

13-3 THE DEFLECTION YOKE VOLTAGES

The yoke is a set of electromagnetic coils mounted on the neck of the picture tube (see Fig. 13-4). Its changing magnetic force moves the electron beam sideways and up and down. There are two sets of windings. The horizontal winding moves the beam

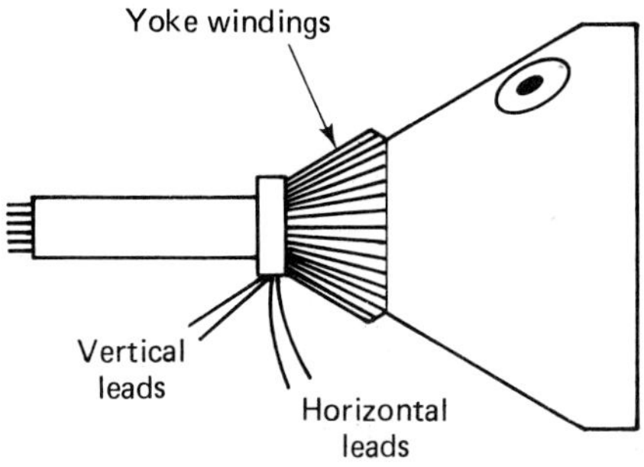

Figure 13-4 Placement of the yoke on the CRT. The yoke develops a magnetic field that deflects the beam across the face of the tube.

sideways, and the vertical winding moves the beam up and down. Together, they form a *raster* that covers the whole screen.

The windings are powered by pulsating deflection voltages from the vertical and horizontal amplifiers. Insufficient vertical voltage causes insufficient raster height [see Fig. 13-5(a)]. Insufficient horizontal voltage causes insufficient width [see Fig. 13-5(b)]. No vertical voltage at all results in a single horizontal line across the CRT face [see Fig. 13-5(c)].

Zero horizontal voltage, however, does not necessarily result in a vertical line because there is usually a loss of high voltage and a total loss of raster. If you do see a vertical line, the yoke is probably bad [see Fig. 13-5(d)].

Loss of both horizontal and vertical deflection voltages results in a dot in the middle of the CRT face. This can happen in systems that do not derive their high voltage from the horizontal output transformer (Section 11-7). They use a separate high-voltage transformer and driving circuit.

Figure 13-5(e) shows a display that is squeezed at the top of the display and stretched out at the bottom. This symptom indicates vertical amplifier distortion. Look for incorrect bias voltage on the base and emitter terminals of the vertical amplifiers.

TIP 13-3

A thin white line across the middle of the CRT screen is caused by a loss of vertical deflection voltage or by a defective yoke. Insufficient height is caused by weak deflection voltage, usually from a weak vertical amplifier.

A display that is stretched out or squeezed at the top or bottom indicates vertical amplifier distortion. Check the base and emitter circuits.

TIP 13-4

A thin white vertical line in systems that derive their high voltage from the horizontal circuit is caused by a defective yoke. Insufficient width is caused by weak horizontal deflection voltage.

As noted above, the vertical and horizontal yokes are each made up of two sets of windings. The vertical yoke has one winding that is responsible for one side of the picture and another winding for the other side. Similarly, the horizontal yoke has one

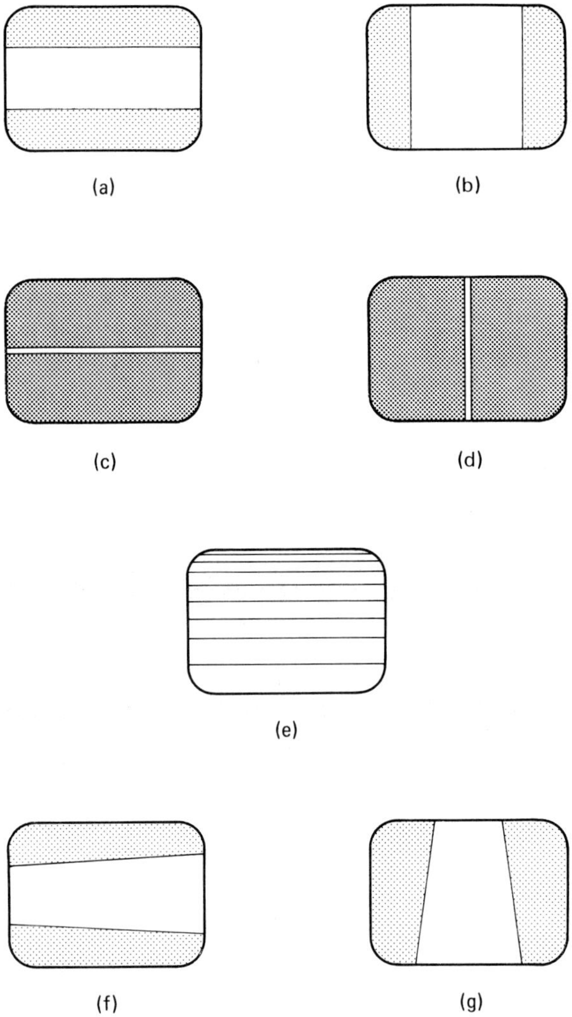

Figure 13-5 TV deflection circuit symptoms.

for the top and one for the bottom. If one of the windings shorts partially, the deflection on its side would decrease. The result would be a trapezoidal raster. Figure 13-5(f) shows the results for a shorted vertical yoke, and Fig. 13-5(g) shows the results for a shorted horizontal yoke.

> TIP 13-5
>
> A trapezoidal raster is caused by a defective yoke.

If you look for the yoke in a schematic, you will not find it drawn on the CRT. It is drawn elsewhere, usually near the deflection amplifiers, but not always. It is depicted as a set of ordinary coils.

Troubleshooting deflection circuits involves work on power amplifiers, high-voltage amplifiers, and multivibrators. The amplifiers were covered in Chapter 11, which dealt with power and high-voltage output stages. Next, we look at the sources of the pulses that drive these amplifiers. They are the multivibrators, popularly called oscillators.

13-4 MULTIVIBRATORS

Thus far we have focused mostly on signals that were basically ac sine-wave voltages. That is, they changed smoothly from one voltage level to another. RF, IF, and AF voltages are basically sine-wave voltages. The voltages required to drive the vertical and horizontal yokes are actually pulses, not sine waves. Let's look at the circuits that generate these pulses. They are basically transistor switches.

13-4-1 Transistor Switches

Switches produce pulse signals. They repetitively turn voltages on and off. We learned about manually operated switches in Chapter 5. One of the important things we learned was that when a grounded switch is turned off (opens), its high-side voltage goes up; and when it is turned on, its high-side voltage goes down (see Section 5-3, Fig. 5-14).

In Chapter 8 we studied the basic transistor switch (Fig. 8-17). When its base voltage goes up, it is switched on and the collector voltage drops to zero. When the base voltage goes down, it is switched off and the collector voltage rises. The circuit is checked out just like an ordinary amplifier. It has many uses. One use is in the multivibrator for generating pulses. Pulse generators were touched upon in Section 10-4 when we worked on switching power supplies.

13-4-2 Multivibrator Voltages

Figure 13-6 shows a multivibrator circuit. Sometimes it is called an oscillator. It is composed of two transistor switches turning each other on and off. The repetition rate, or frequency, of the pulses thus generated depends on the size of the capacitors and resistors in the base circuit.

Let's go over some of the highlights of the multivibrator circuit action. Basically, it depends on the switching action of the transistors and the charging of the two capacitors. During a part of the action, one of the capacitors, say C_1, is charged up through the *BE* path of Q_2 to ground. During another part of the action, the positive side of C_1 is grounded when Q_1 collector shorts (saturates). This puts the negative terminal of C_1 above ground. This negative voltage switches the other transistor, Q_2,

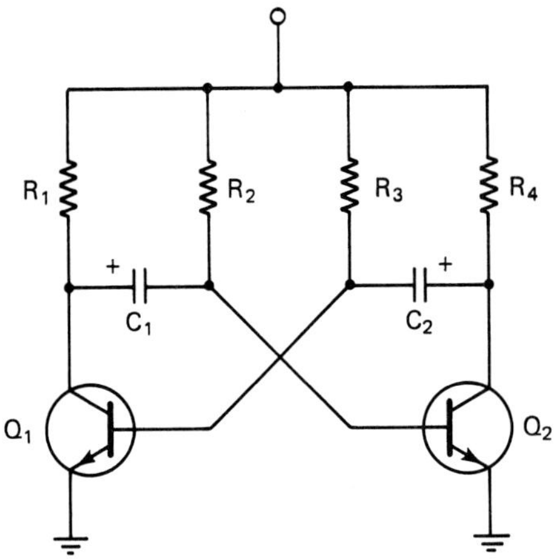

Figure 13-6 The multivibrator generates deflection pulses.

off. So Q_2 is off when Q_1 is on. Further investigation will show that this process is repetitive. The duration of the cutoff time depends on the value of the capacitor and the base resistor.

From the troubleshooter's standpoint, it is important to remember that the base voltage is negative part of the time. This holds the collector voltage high part of the time. If something goes wrong, the negative voltage is lost and the collector voltage will not be held high. This causes the dc collector voltage to be lower than normal.

TIP 13-6

If the multivibrator dc collector voltages are lower than normal, the multivibrator is defective. If the dc base voltages are more positive than normal, the multivibrator is defective.

TIP 13-7

A leaky multivibrator capacitor will increase the frequency. For vertical multivibrators, this means that the display will roll up. For horizontal multivibrators, you will see lines slanting down from left to right. You may also see horizontal movement. A defective transistor could also increase the frequency.

TIP 13-8

An increased-value multivibrator base resistor will decrease the frequency. A defective transistor could also decrease the frequency.

Sometimes a variable resistor is placed in series with a base resistor to control the frequency. Also, a voltage from an external circuit may be fed to one of the bases of the multivibrator to control its frequency. A more positive voltage will tend to increase the frequency (see Section 13-5-5).

TIP 13-9

If a multivibrator has a variable control resistor connected to a base resistor, note the adjustment position. If it is adjusted for minimum resistance, one of the base resistors could have increased value, or a more negative voltage from an external source may be influencing the base voltage.

If it is at the other extreme, you could have a leaky capacitor. Or a more positive voltage may be coming into the base from an external circuit. Also look for a defective transistor when the variable control is at either extreme.

TIP 13-10

If a multivibrator base resistor opens completely, the base voltage will stick low and the collector voltage will stick high.

Aside from the special tips given above, the multibrator is checked out like any other pair of transistor amplifiers. Check the collector voltage first, then check the base and emitter voltages. Take the resistance readings. Check the transistor in the circuit.

13-4-3 Blocking Oscillators

The blocking oscillator may also be used to generate vertical and horizontal pulses. It is a cross between a regular tuned-circuit oscillator and a multivibrator. It is called an oscillator because it looks like one (see Fig. 13-7). It is like a multivibrator because it has a charging capacitor C_1 and grid resistor R_1 that determines its frequency.

TIP 13-11

If a blocking oscillator fails, it will lose its negative base voltage and its collector voltage will fall below normal.

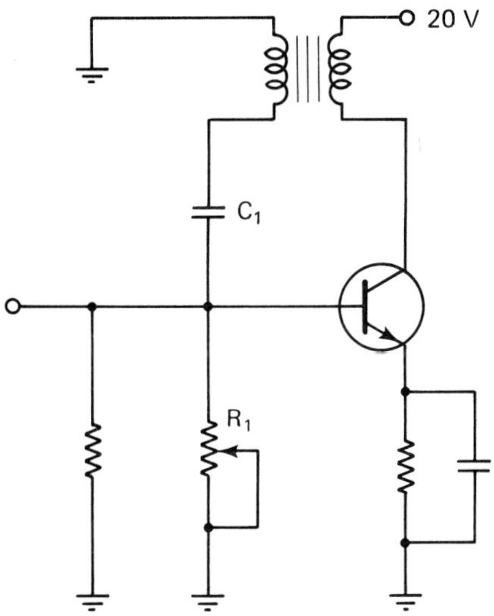

Figure 13-7 Blocking oscillator.

13-5 SYNCHRONIZATION CHECKS

When two circuits or systems are made to work together with perfect timing, we say that they are synchronized. Synchronization is required for many types of electronic applications. In the video system, the vertical, horizontal, and video circuits must operate together in perfect timing. We deal with it in this section.

13-5-1 The Basic Synchronizing Pulse Action

The various circuits in the video-display system must be synchronized; otherwise, the display will be scrambled up. The video signal contains synchronization pulses for this purpose. As was seen in Section 13-4, during each cycle a transistor is switched on when its base voltage goes positive. Basically, the incoming ''sync'' pulse works by kicking the transistor on just before its base voltage has enough positive voltage to switch it on by itself. Each incoming pulse starts a new cycle. The horizontal pulse comes in 15,750 times a second, and the vertical pulse comes in 60 times a second. Since the video signal carries the sync pulses, the vertical and horizontal multivibrators are forced to work in time with it.

13-5-2 Sync Separator or Clipper Voltages

We learned that the sync pulses come in on the same signal with the video. This combined signal is called the composite signal, or simply the video signal. In the receiver the sync pulses are separated out from the video by the *sync separator* or *clipper.*

Figure 13-8 shows the clipper circuit. It shows two of a long train of sync pulses arriving between two intervals of display signals. Every interval between the sync pulses has a display signal. Notice that the tips of the pulses extend above the picture level. The sync separator clips off the picture part of the signal. Only the sync tips show up at the output.

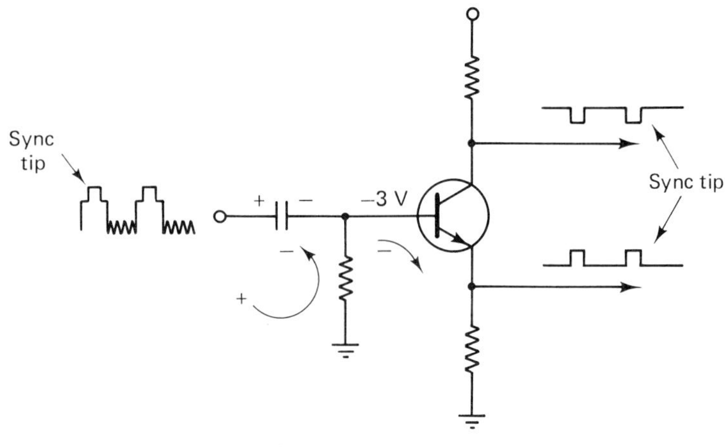

Figure 13-8 The sync separator or clipper extracts the sync tips from the composite video signal. Check for a negative dc voltage on its base.

Only those parts of the signal that are high enough to overcome the negative base voltage can get through. Only the sync tips get through. The negative voltage is developed by the rectifying action of the *BE* diode junction. The arrows show the charge and discharge current of the capacitor.

The output pulses from the separator are fed to both the vertical and horizontal deflection circuits. Each of these deflection circuits has its own circuit for processing the sync pulse.

TIP 13-12

If you have negative voltage at the sync separator base, the transistor is probably okay, and so too, the video and sync signal up to that point. All signal stages up to that point are also probably okay.

TIP 13-13

If the display shows both vertical and horizontal sync problems, the problem lies at or before the sync separator. If it shows only vertical sync problems, check the vertical sync circuits (see the integrator). If it shows only horizontal problems, check the horizontal sync circuits.

13-5-3 Vertical Integrator

So far we have separated the sync pulses from the video signal. Now we have to separate the vertical pulses from the horizontal pulses. A group of vertical pulses, shown in Fig. 13-9(a), comes in at regular intervals on the same train as the horizontal pulses and video signals shown in Fig. 13-8. The vertical pulses are separated by the vertical integrator circuit shown in Fig. 13-9(b). It looks like a low-pass filter circuit. It does work something like a filter. Most of the pulses on the wavetrain are horizontal pulses of short duration. These will be filtered out by the integrator. The vertical pulses come in 60 times a second. They come in as a group of longer-duration pulses, 1 to 6, as shown in Fig. 13-9(a); these will not be filtered out. They will be ''integrated'' by a special charging action of the capacitors into a single pulse as shown at the output of the integrator in Fig. 13-9(c). The shorter-duration pulses, 7 to 12, bring the pulse down from its peak. The duration of the whole integrated pulse is 400 μs. You will not be able to see a vertical pulse of this shape on the scope because it will be triggered or locked in by the short-duration horizontal pulses on the same wavetrain. Figure 13-9(d) shows the same long-duration pulse when viewed with the scope set to view the vertical signal. It is a clean-looking pulse.

Interesting Sidelight. The following is a more detailed description of the integrator action. Every $\frac{1}{60}$ of a second a group of pulses in the sync train that are wider at the top than the rest of the pulses reach the integrator. This group of 12 pulses, shown in Fig. 13-9(a), is called the vertical interval. The vertical integrator circuit, shown in Fig. 13-9(b), converts these pulses into a single pulse. When the first pulse comes in, the capacitors charge up partially. The next pulse comes in before the

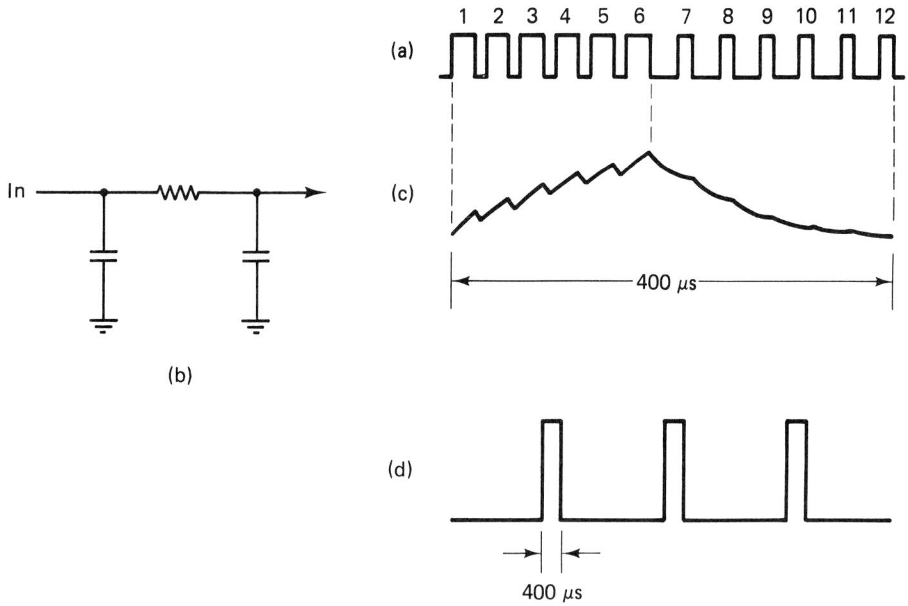

Figure 13-9 (a) Vertical-sync interval from which a vertical sync pulse is developed. (b) Vertical integrator circuit. (c) Vertical-sync pulse. (d) The same vertical-sync pulse viewed with a scope set to a slower rate.

capacitor has a chance to discharge, so it builds up the charge a little more. The following four pulses do the same thing and build up the pulses to a peak. The next six pulses are narrow and far apart. This gives the capacitor enough time to discharge a little between pulses. All this took only 400 μs.

If you look at the same pulse with the scope adjusted for 60 Hz, it will look very nice [see Fig. 13-9(c)]. The pulse is about 400 μs "wide" in both waveforms. These nice-looking sync pulses are fed to the base of one of the vertical oscillator transistors. Each incoming pulse kicks the transistor on. Thus the vertical deflection is forced to keep time with the video signal.

TIP 13-14

Vertical sync can be timed by adjusting the vertical hold control. The display should not roll up unless you turn the hold at least a quarter turn. When it does roll up, it must roll up fast, not slowly. If the display can roll up slowly, the sync signal is probably weak.

TIP 13-15

You can check the vertical sync signal by rolling down the dark horizontal bar. Adjust the contrast control until you can see a darker pattern inside the dark bar. The darker portion is the vertical sync signal. If you have good contrast between these parts of the signal, the sync signal is good. The trouble lies after the sync separator.

TIP 13-16

The source of poor vertical sync can also be found by adjusting the horizontal hold control. If the horizontal holds well, the source of the vertical sync trouble lies after the sync separator. See Tip 13-21 for checking the horizontal hold control.

Before ending this section we should mention something about the vertical blanking. All the sync pulses or tips lie on top of the blanking pulse. It does not do a perfect job under all conditions. When it is not working right, you can see the zigzag line that is traced out by the flying dot as it flies back to the top of the screen. To prevent this, a pulse is taken from the vertical output circuit and fed to the video output amplifier to cut it off during the retrace period.

TIP 13-17

If you see a faint white zigzag line going up the raster, check the blanking pulse circuits.

13-5-4 The Horizontal Sync Pulse

The horizontal sync pulse is taken off at the same point as the vertical sync pulse. It branches off to a differentiator circuit, which looks like the high-pass *RC* filter we saw in Section 7-7-1. Figure 7-11 shows what happens to a square wave that is fed into a differentiator. A similar thing happens when you feed a shorter-duration sync pulse through the same circuit with a smaller capacitor. This sharpened pulse is then fed to the horizontal circuits, where it serves to keep the horizontal oscillator synchronized.

13-5-5 Automatic Frequency Control

Early television sets had plenty of problems. The horizontal sync signal from the station just could not do the job when conditions where less than perfect. The pictures used to bend and break up all the time. Most often, you would see it fluttering at the top. The automatic frequency control system (AFC) helped solve this problem. This circuit controls the base voltage of one of the multivibrator transistors.

We already saw how a multivibrator hold control changes the base voltage. We also learned that an increase in the *BE* voltage speeds it up, while a decrease slows it down. This control voltage can be derived from an automatic feedback system.

The AFC circuit taps a portion of the output and compares it with the sync signal to produce a correction voltage. If the multivibrator is running too slow, the control voltage goes up. If the multivibrator runs too fast, the control voltage goes down.

Figure 13-10 shows the circuit. Two sync pulses are taken from Q_1 and coupled to the the AFC diodes, D_1 and D_2. The two sync pulses are rectified and meet at the junction of the two diodes to produce 0 V. A sawtooth pulse from the horizontal out-

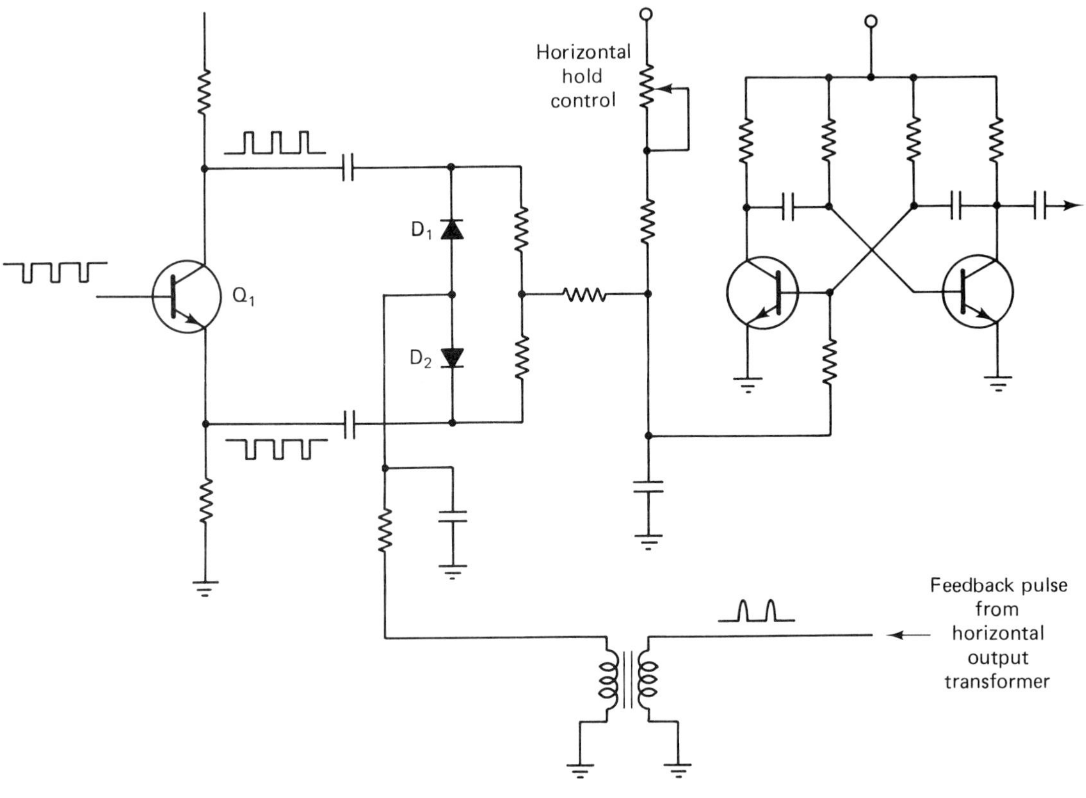

Figure 13-10 The automatic frequency control (AFC) circuit compares the feedback pulse with the sync pulse and develops a correction voltage for the horizontal multivibrator.

put transformer is also rectified and fed to the same point. This establishes the normal voltage at that point when the oscillator is running on time. When oscillator timing shifts a little, the output and feedback pulse is going to shift and the normal voltage is changed. This changed voltage is the correction voltage. It is filtered and fed to the base of a multivibrator transistor.

TIP 13-18

To check for sync and AFC action, read the input dc voltage on one of the diodes. If the voltage changes as you go off channel, you are getting signal. Reverse your meter polarity if necessary. Try both diodes.

If you do not get any change of voltage, either the sync is not getting through or a diode is bad. Check the diodes with an ohmmeter. They should have have hi–lo and equal resistances.

TIP 13-19

To check the AFC action in horizontal circuits, read the dc input to the base of the multivibrator as you adjust the horizontal hold control. Adjust and try to obtain a full picture. When the picture shifts to the right (your right), the voltage should go more positive. When it shifts to the left, the voltage goes less positive.

TIP 13-20

In AFC circuits, you can short across the two diodes to check the multivibrator. Adjust the horizontal frequency control to see if you can make the lines slant both ways. If you can, the fault is in the AFC circuit.

13-5-6 Some Horizontal Sync Checks

You can tell whether the cause of poor horizontal sync lies in the oscillator or the AFC by making some simple checks on the horizontal oscillator.

TIP 13-21

To check the horizontal oscillator, adjust the horizontal hold controls. If you can make the lines slant both ways, first one way and then the other way, the oscillator is okay.

If you see a lot of lines, the frequency is very far off. Make your adjustments so as to reduce the number of lines.

If you can reduce the number of lines to only a few diagonal lines, the sync is weak. It should lock in with more than, say, four lines.

The horizontal sync pulse along with the blanking pulse is sent all the way to the CRT cathode. The blanking pulse is utilized to blank out the flying dot when it is supposed to fly back to the left. Ordinarily, you will not be able to see this pulse. However, when the sync is weak, you will be able to adjust the horizontal hold control to shift a vertical bar into the CRT display area. This dark bar is produced by a series of horizontal blanking pulses running from the top to the bottom of the screen. If you see a darker bar running down the middle of the blanking bar, you are looking at the horizontal sync tips riding on the blanking pulse. This indicates that you have a good sync signal all the way to the CRT. The problem lies in the path between the horizontal oscillator and the sync takeoff point at the video amplifier. The video amplifier itself is okay.

TIP 13-22

If you have poor horizontal sync, carefully shift the picture over to one side a bit with the horizontal hold control. You will see a dark vertical bar. Adjust the contrast control until you see a narrower darker bar running up the center of the bar. This indicates good sync up to the CRT. You are losing it somewhere between the sync takeoff point at the video amplifier and the horizontal oscillator.

If you have several slanting bars, they will be narrower, but you should still able to see the darker part of the pattern.

As we said, you will not be able to see this bar when the sync is working. A good AFC system will pull the picture in when the frequency is quite a bit off. When the frequency is far off, you will see many slanted lines. When the frequency is not as far off, you will see fewer lines. A good system should pull the system in when there are about six lines. A few lines more or less is okay.

TIP 13-23

If you can pull in a picture with horizontal hold control while there are still about six slanting bars, the horizontal AFC is working properly.

Since both the horizontal and vertical depend on the sync signal, if one of these circuits is good, the sync signal must be good. If one circuit is good and the other is bad, the trouble lies after the point where the good sync signal splits off into the bad circuit.

TIP 13-24

To locate the source of weak horizontal sync, check the vertical hold. Turn the vertical hold control both ways to see how solid it holds. If the vertical holds solid, the horizontal sync trouble lies after the point where the vertical and horizontal sync signals split off.

13-6 DC AND SIGNAL VOLTAGES ON THE COLOR CRT

When checking and following up on color symptoms, we divide the signals into luminance and color. We further divide the color into dc and ac signal voltages. Each will give us an idea of where to look for the trouble.

13-6-1 The DC Voltages on the Color Display Tube

The color CRT support circuits are much more complicated than the black-and-white circuits. To begin with, the anode voltage for color CRTs is much higher. It ranges up to 39 kV. This requires new circuits for x-ray and fire prevention, not to mention insulation and spacing problems. The focus element sometimes requires several thousand volts. This places special requirements on the socket design. The focus terminal can be identified by the thick wire leading up to it and the heavy insulation around it. Also, it is the pin that is separated from the other pins. Sometimes there is a ball-type spark gap in the socket at the focus terminal.

In addition, the socket has to accommodate more pins with higher voltages than their monochrome counterparts. The color CRT has three "guns": a red gun, a green gun, and a blue gun (see Fig. 13-11). Each gun is composed of a cathode, a grid, and a screen grid. Various combinations of these primary colors can produce all the colors of the rainbow as well as black and white. Red and green together, for instance,

TIP 13-25

Poor focus in large-screen color CRTs is often caused by arcing around the focus connection in the CRT socket.

produce yellow. This may seem strange to someone used to mixing paints. You'll get used to this after you have adjusted a few service controls for color. A combination of all three colors produces white. Black is produced when all three guns are cut off.

The cathodes usually have about 250 dc volts, the grids about 150 dc volts, and the screens about 790 dc volts. In the early versions, these voltages were all independently adjustable. In some late versions, the grid guns are tied together internally and come out on one pin. The same is true for the screen and focus guns. The three cathodes, however, come out separately and are independently adjustable.

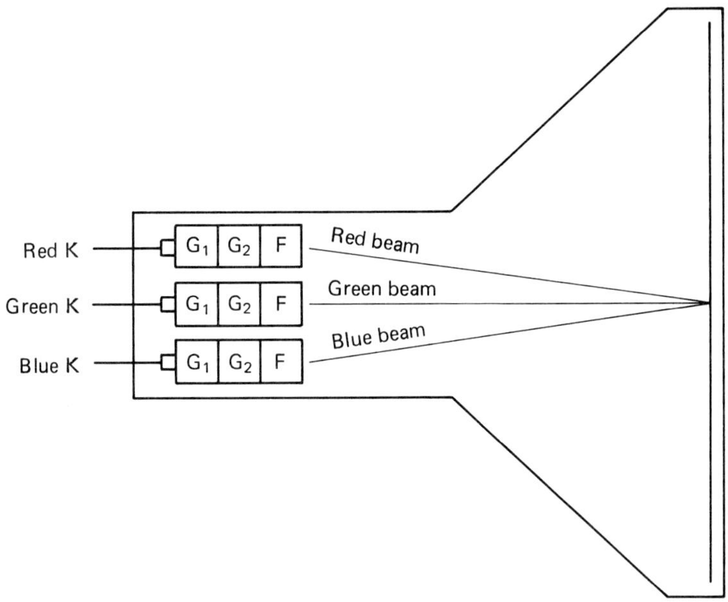

Figure 13-11 The color display tube has three guns, one for each primary color.

13-6-2 The Color Signal Paths

Modern color circuits are nearly all enclosed in integrated-circuit (IC) chips. To troubleshoot them, you should have some knowledge of the signal paths inside and outside these chips. This will help you remember the names and functions of the inputs and outputs of these chips.

In the block diagram in Fig. 13-12, the three color signals to the CRT—red, green, and blue—come from three corresponding color output amplifiers in the output block. They are fed into the three cathodes. These signals are about 130 V peak to peak.

The composite color signal from the transmitter contains the information necessary to produce all three colors. From the troubleshooting standpoint we can divide the signal paths into color and luminance (black-and-white) paths. They separate at the first video amplifier. The luminance signal goes to the luminance amplifier section, usually called the video section. The color signal goes to the color IF amplifiers (or chrominance amplifiers). They both end up at the color demodulator section. A delay line is placed in the luminance path because the color signal has to go through more stages in the color IF section.

TIP 13-26

If you have a black-and-white picture but no color, check the color signal path from its takeoff at the video amplifier to the color output circuits. The delay line is a good place to start.

If you have color but no black and white, check the luminance signal path from the video takeoff to the video output circuit. The symptom for a color display that lacks black and white is a faded color picture with blurred but definable outlines.

The color demodulator circuit also receives a signal from an oscillator. The timing of this signal is controlled by a crystal and a sync signal from the color sync amplifier. The color sync signal is contained in the composite color signal at the video amplifer. It is a series of eight small pulses that ride on the blanking pulse just behind the deflection sync pulses. Since its position and voltage level are very close to that of the deflection sync pulse, a loss of sync could also mean a loss of color.

TIP 13-27

If the vertical and horizontal sync are good, the color sync signal is very likely to be good up to the color takeoff point at the video amplifier.

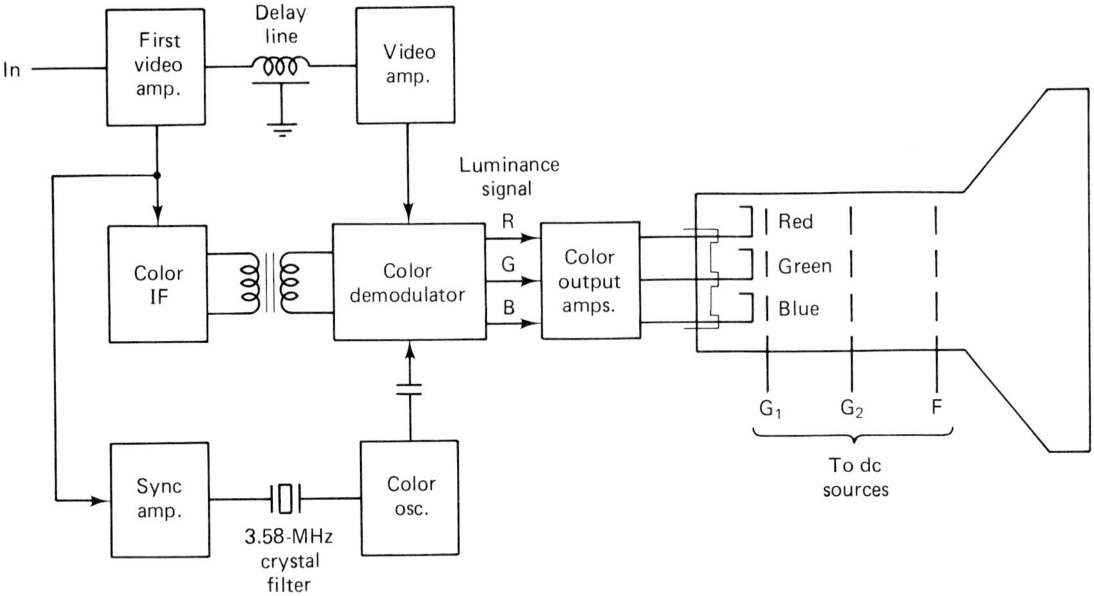

Figure 13-12 Block diagram of the luminance and color circuits for the color display tube.

The color, luminance, and color sync signals combine to produce the three separate red, green, and blue signals at the demodulator output circuits. A detailed explanation of the color-demodulation section is beyond our scope. Roughly speaking, the basic red, green, and blue color voltages in the signal come in on different parts of each cycle. With the right timing provided by the oscillator, the demodulator is able to separate and combine each color signal voltage with the luminance signal to output the three color voltages.

13-7 TROUBLESHOOTING COLOR PROBLEMS

As noted above, if you have black and white, look for trouble in the color sections. At this point, troubleshooters further divide their problems. They divide color problems into dc problems and signal problems. This is a good way to divide and conquer, because the dc symptoms point to certain circuits and the strictly signal symptoms point to certain other circuits. The dc paths from the CRT cathode go back only to the demodulator (see Fig. 13-13). (This is a simplified version of the RCA schematic in Fig. 8-20.) The dc path is blocked by coupling capacitors or transformers in the demodulator or color IF circuits. Notice that the color output amplifiers are dc coupled. Some CRTs have separate grid and screen controls for each color. They have dc control paths feeding each grid and screen terminal.

Unlike dc paths, color *signal* paths go all the way back to the first video amplifier. They take two routes, one through the color IF stages and the other through the color sync and oscillator stages. Let us take on dc color problems first.

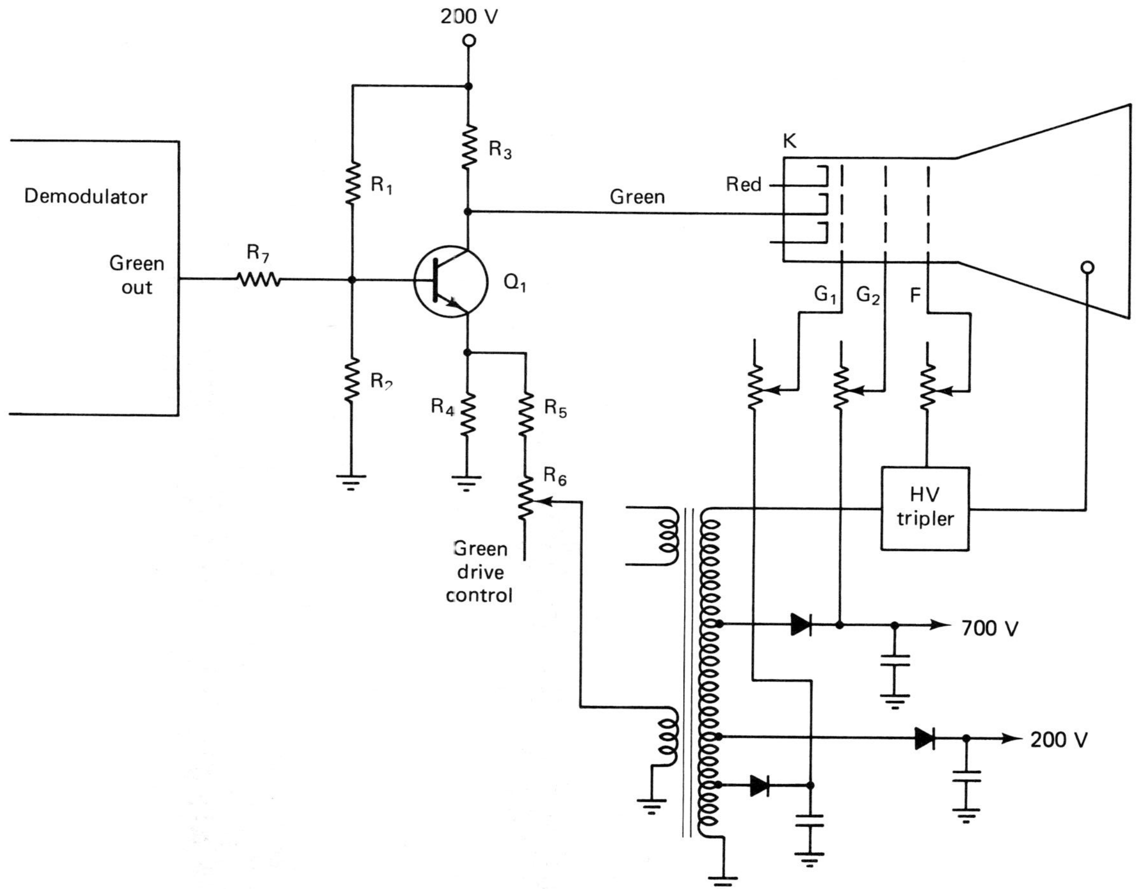

Figure 13-13 Output amplifier circuit for one of the color signals. Note also the dc voltage sources for the various pins.

13-7-1 DC Color Symptoms

To separate signal color symptoms from dc color symptoms, you must know how various colors are formed on the CRT screen. You can learn this by playing around with the screen or cathode controls. Turn down one color to simulate what happens when one dc color circuit goes bad. Each color gun element has its own dc circuits.

Note the following symptoms or tips; they will come in handy when troubleshooting color circuits and when adjusting service controls.

> TIP 13-28
>
> Loss of red results in a blue-green (cyan) display. This is caused by excessive red-cathode voltage, or insufficient red-grid or red-screen voltages.
>
> Blue plus green makes cyan (a blue-green hue). This is similar to mixing paints.

> TIP 13-29
>
> Loss of green results in a magenta display. This is caused by excessive green cathode voltage, or insufficient green grid or green screen voltages.
>
> Red plus blue makes magenta. This is similar to mixing paints.

> TIP 13-30
>
> Loss of blue results in a yellow display. This is caused by excessive blue cathode voltage, or insufficient blue grid or blue screen voltage.
>
> Red plus green makes yellow. This is different from mixing paints. You'll get used to it.

To find out whether you have a signal problem or a dc problem, note the following: If all the guns are getting the correct dc voltages, you will have a black-and-white display when the color control is turned down. If you have a color problem but the black and white is okay, the trouble is in the color signal path. The trouble is located ahead of the dc paths.

TIP 13-31

Dc color problems can be distinguished from signal color problems by turning the user color control down to obtain a black-and-white raster. If you have a dc color problem, you will not be able to get a black-and-white display.

If you can get a black-and-white display and you have a color signal problem, the trouble lies ahead of the direct-coupled output amplifier stage. It lies before the last blocking capacitor or transformer in the signal path to the CRT.

13-7-2 Troubleshooting a DC Color Problem

Always start your troubleshooting from the outside and work your way in. This is usually easiest way to go. Let's take one of the colors; the others require the same procedure. Let's say that the green dc path is giving us trouble. Refer to Fig. 13-13 for the following checks.

Check 1. Turn down the green dc-cathode control R_6. If you cannot get a magenta screen, something is wrong with the CRT or the dc paths to the green cathode, *K*.

Check 2. Measure the green cathode, *K*, voltage. Excessive cathode voltage, will give you a magenta raster.

A loss of green cathode voltage will cause a bright green raster. The green will override the other two colors. The picture may be lost or very faint or blurry.

TIP 13-32

If you have a dc color probem, the CRT or its socket could be bad. The socket assembly may have built-in spark gaps to protect the CRT gun elements from excessive voltage spikes. These occasionally short out, either from foreign matter or heat expansion. You might be able to take the socket apart and clean it out. You can also tighten up all the pin connectors.

If the socket is good, the CRT is bad. Get a new socket to make doubly sure. Check the suspected CRT with a tube checker before you install a new CRT.

Check 3. Check the green output transistor (Q_1) dc collector voltage. If it is bad, check the rest of the voltages.

Check the 200-V source. It is usually at the horizontal output transformer. Check for an open diode.

An open collector resistor R_3 will lower the collector and cathode voltages. This will cause excessive green.

Check 4. The green output transistor can have a bad dc input voltage. A higher-than-normal base voltage will turn the raster green. A loss of base voltage will turn the raster magenta. An open R_1 will lower the base and raise the collector. This will cause a loss of green, which would produce a magenta raster. An open R_2 will cause excessive green.

An open R_5 or R_6 in the emitter circuit will cause an excess of green cathode voltage. This will cause an excess green raster. An open R_4 will cause excessive green cathode voltage and a loss of green raster.

Check 5. Check the green-demodulator output circuit dc voltages. An open base resistor R_7 will cause a loss of green signal. If the black-and-white raster is okay when the color intensity control is turned down, the loss of green signal will show loss of yellows and greens in various objects in the picture. Normally green objects would have a magenta color. Reds and blues will be okay.

However, if turning down the color intensity control produces greenish or magenta raster due to a change in the dc voltage at the base of Q_2, the signal fault will be harder to see. Turn R_6 to obtain a black-and-white raster, then turn up the color control and see if you can distinguish a loss of green signal as above.

We have gone over the green dc-voltage checks. The other colors are checked in the same way. If you look at the schematics, you will see that all three color circuits are pretty much alike.

13-7-3 Troubleshooting Color-Signal Problems

As noted previously (see Fig. 13-12), as far as the troubleshooter is concerned, there are two paths for the color signal: the color IF path and the color oscillator path. While the paths may have some symptoms in common, they have other symptoms that differ.

Troubles in the color IF path cause weak color or loss of color. The colors do not change hue. They maintain good balance; that is, no color predominates. You can turn down the color control and have a good black-and-white display.

Troubles in the color sync and oscillator path can also result in loss of color, but there are also symptoms that are strictly associated with it. For instance, the color hues can be off, the colors may keep changing, or you can have rainbow patterns.

Moving rainbows in the display show up when the color oscillator goes off frequency. The further off the frequency, the faster the rainbows move. If the frequency is far enough off, the color will move so fast that you lose it. It is like when you spin a rainbow-colored top; it turns white.

TIP 13-33

If the color is weak or missing and the black and white does not change its tint when you turn the color intensity control down, check the color IF amplifiers. There are various names for these amplifiers; they are the ones that feed the demodulators.

TIP 13-34

If your picture elements are missing greens or have too much of magentas, check the green demodulator output.

If your color picture elements are missing reds or have too much of cyans (blue-green), check the red demodulator output.

If your color picture elements are missing blues or have too much of yellows, check the blue demodulator output.

TIP 13-35

If the color changes, or is off, or has moving rainbow stripes or areas, check the color oscillator and control circuits.

TIP 13-36

If the color is missing and the raster is slightly tinted, check the color-oscillator control circuits. Turning down the color control to obtain a black-and-white raster may help you detect the tinting. If you can notice the difference, the trouble is very likely to be in the color sync or oscillator circuits.

13-8 CHECKING LUMINANCE SIGNALS IN COLOR CRT CIRCUITS

The luminance signal is the black-and-white video signal in a color system (see Fig. 13-13 again). This signal goes through the video amplifiers after the color signal is separated from it. It goes through the delay line. In Tip 13-26 we learned that if we had a color signal but no luminance, we looked at the video signal path from the delay-line input to the video output. If you have no color but black and white, look into the color circuits.

If you have no picture at all, just a blank raster, the problem could be before the color and video takeoff point.

> Tip 13-37
>
> In a color TV set, if you have a raster but no picture, the problem is in the video amplifers, the IF amplifiers, or possibly the tuner or AGC.

13-9 COLOR CRT: NO-RASTER AND FOCUS PROBLEMS

The no-raster problem is the most common problem in color display equipment. It is most often caused by a loss of high voltage which is associated with the horizontal output circuit (Chapter 10). It can also be caused by a bad CRT. In either case it is a big job.

13-9-1 CRT Pin Voltage Sources

By "no-raster" we mean a screen that is completely dark. The primary cause of no-raster is no high voltage. Since we took that up in Chapter 11, let us look at other causes.

No-raster can also be caused by incorrect voltages on the CRT pins. As far as the cathode pins are concerned, this can happen only if all three voltages are bad. We must look for something that knocks all three of them out simultaneously. The chances are practically zero for all three to go out independently of each other at the same time. In Fig. 13-13 the three grid and screen elements are tied together internally, so you see only one grid pin and one screen pin. However, some CRTs have three grid and three screen pins, one for each color. In the latter case, you will have to check the common voltage source for the three grid or the three screen pins when you have brightness trouble.

1. The screen pin, G_2, can get its voltage from either a low-voltage winding on the horizontal transformer or off the high-voltage tripler, which may have subsidiary outputs for the screen and focus pins.

2. The grid is sometimes connected to ground. At other times its voltage is taken from the horizontal output transformer. Sometimes the grid is tied to a negative supply. A loss of negative voltage causes excessive brightness and some loss of high voltage. A blooming raster may be the result. We have more on blooming in the next section.

The CRT cathodes usually have separate dc color signal paths going back to the demodulator. Occasionally, there is a common path to the video amplifier. There may be brightness and contrast controls in the video circuit that simultaneously control the three cathode voltages.

3. Check the collector source voltage of the three color-output transistors. A loss of this voltage will bring down the voltage on all three cathodes. This may cause excessive current on all three guns and load down the high voltage so much that the raster may go out. Sometimes this condition gives rise to excessive brightness.
4. Check the focus voltage. Some types require a few thousand volts to obtain sharp focus. It is possible for a bad focus voltage to cut off the raster.

13-9-2 Poor Focus

Poor focus is usually caused by a defect in the focus circuit. The focus voltage is taken from the high-voltage block or tripler in Fig. 13-13. Check the resistors in the circuit. If you have to take voltages, use the high-voltage probe.

The CRT socket can develop leakage paths that can pull down the focus voltage. The focus voltage is very high and often arcs across insulated surfaces to ground.

Sometimes poor focus is caused by an internal defect in the CRT. Before you replace the tube, make sure that all the CRT pin voltages are good. Sometimes insufficient high voltage at the anode can cause poor focus. Sometimes excessive grid or insufficient cathode voltages can cause a loss of focus.

Insufficient high voltage can be distinguished from poor focus by turning the brightness up and down. Increased brightness loads down a defective high voltage so much that the display might not get brighter. Also, the picture width increases because the high voltage is pulled down. The electrons are slowed down and more easily deflected by the yoke. This symptom is called blooming.

> TIP 13-38
>
> A loss of focus may be caused by insufficient high voltage. Insufficient high voltage can be detected by blooming when the brightness control is turned up.

13-10 PRELIMINARY CRT CHECKS AND ADJUSTMENTS

The high cost of a display tube requires that you make an estimate before proceeding with the job. The problem is divided into two stages. First, you look for the most obvious defects that leave no doubt in your mind that the CRT is defective. Next, you have to make some voltage tests for the more doubtful symptoms.

13-10-1 The Obvious CRT Defects

The symptoms that tell you immediately that the CRT is bad are very obvious. A tube with internal arcing, either the dc light blue noisy arc or the quiet purplish ac type, is definitely bad. Tubes with a purplish glow in them are also bad. An inner coating of white substance on the glass indicates a hairline crack in the glass.

It is normal to see burn marks in the neck. During the manufacturing process the impurities are burned out. However, marks on the unlit face of the CRT are definitely bad and irreparable.

Another easily observable symptom is the failure of the CRT heater element to light up. Check the continuity between the heater terminals. If you have not memorized their pin numbers, look for the pins that are next to and on both sides of the key. In old sets, the heater leads are twisted and of the same color, usually brown or black.

If the heater is not open, we go into the second step of the the diagnostic procedure.

13-10-2 Further CRT Checks

We must now decide whether the trouble is in the CRT or in the circuits. There is a strong possibility that the CRT socket or the heater ground lead connection is bad. These are easy to check, so check them out early.

The heater voltage may be derived from a tap in a horizontal output transformer winding. This circuit could have an open. There is still the remote possibility of a short in the CRT heater.

One symptom that nearly always indicates a bad tube is a silvery picture. The very lightest part of the picture is lighted up with a silvery appearance, whereas the rest of the picture remains darker than normal when the brightness and contrast are turned up.

There are other symptoms that indicate the possibility of a bad picture tube, but they also point to faults in the circuits. These include dark picture, blurry picture, and one color missing. Sometimes one color predominates so much that it overrides the other colors, so only one bright pure red, green, or blue is seen. These symptoms require further checking.

Check 1. Adjust all the controls and try to obtain good sound and good black-and-white display raster.

Good sound and dark raster indicate the possibility of a bad CRT. However, check the CRT voltages to make sure. On the other hand, bad sound indicates other defects besides CRT and high voltage.

The ability to obtain a color raster but not a white one can indicate a CRT fault. However, check the dc color circuits first.

An ability to obtain various hues between two basic colors strongly indicates that the fault is not in the CRT as far as these two colors are concerned.

Check 2. If you have no raster and good sound, check for high voltage.

Check 3. Measure the pin voltages that can cause the observed symptoms.

For a no-raster condition all three cathodes must have bad voltages for the raster to go out. For this reason, look for a common source point for all three voltages if you have no raster. Also, a loss of grid, screen, or anode voltage will cause the raster to go out. In some older tubes the grids and screens have independent controls for each color. In these cases look for the common supply point for these voltages when the raster goes out.

For a color raster problem only the cathode of one gun need be bad. In older tubes only a pin voltage for one gun element need be bad.

If all voltages are okay and if you have a bad CRT symptom, either the CRT or the socket is bad.

Check 4. Remove the socket. Insert your probe into the pin holes to take the voltages. A special needle probe will be needed. The other pins should have normal or higher-than-normal voltages. If you get no voltage at all, the socket is bad. If you get equal voltages on the cathodes and normal or higher-than-normal voltages on the other pins, the socket is good and the CRT is very likely to be bad.

Use a socket adapter for a more positive check. It has tabs that permit you to check each pin in-circuit.

You can also insert thin wires into the cathode pin holes. A no-voltage reading could indicate that the wire has not made good contact, but good voltage indicates a good socket and a bad CRT. To make doubly sure, make a continuity check of the socket.

Check for a short between ground and the pin holes that are not supposed to be grounded. Many sockets have internal spark gaps that can short to ground.

13-10-3 Color Purity and Convergence

If you have discolored areas in the display, you have poor *purity.* This happens when the electron beams (Fig. 13-11) do not strike the correct targets on the inner surface of the screen. Each target or *phosphor dot* is made of a substance that gives off its own color when struck by an electron beam. There are red dots, blue dots, and green dots. They are arranged in tightly packed groups of three. These groups cover the entire face of the CRT. The beams from the three guns must converge on one of these groups, to produce a single dot made up of three color dots. For instance, to make a single white dot, each of the three phosphors must light up equally. To have a pure raster, the beam from each gun must hit its own target; otherwise, discoloration takes place.

Purity can be affected by stray magnetic fields. If the metal parts of a chassis become magnetized, the purity will suffer. There is an automatic demagnetizing coil that demagnetizes the field around a CRT every time the unit is turned on. If this demagnetizing or *degaussing* coil fails, the purity may be affected.

In some cases this built-in coil may not do the job even though it is not defective. In this case you will have to use a degaussing magnet and degauss the CRT by hand. This magnet is a coil, 14 inches in diameter, that works on 120 V ac. To degauss the tube, turn on the display tube circuit and plug in the degaussing coil. Then move the coil around the face of the tube with its plane parallel to the glass surface of the screen. Keep it about an inch away from the glass and move it in small circles as if you are wiping the CRT. Do not make abrupt motions. Move smoothly. You will see beautiful patterns develop on the screen. In some cases you will have to move the coil around the top, bottom, and sides or the forward part of the tube.

When you are finished, do not turn off the degaussing coil right away. Move the coil away from the tube before turning it off. Move the coil slowly in large circular wiping motions as you move it away.

If hand degaussing fails, you will have to adjust the purity magnets on the neck of the CRT. The procedure for adjusting the purity magnets involves turning down the two bias controls so that only one color, green for instance, remains. Consult the service manual for which color to use. It is best to have no signal. Some models have back-panel switch for this purpose. The set must be on for 15 minutes before adjusting the purity.

Face the CRT north or south for best results. Then position the yoke in accordance with the service instructions. The magnets are then adjusted to obtain a vertical green area in the center of the screen. Next reposition the yoke to obtain a pure, full, green raster. The bias controls are then readjusted to obtain a white raster. Consult your manual for more details or variations on this procedure.

When the edges of the objects in a color display are fringed with color, you have what is called poor *convergence*. This occurs when the beams do not come together at the target. There are two types of convergence adjustments. One is called *static convergence* and the other is called *edge convergence*.

Static convergence is accomplished by adjusting a set of magnets on the neck of the CRT. Each magnet is positioned near a gun in the neck. For instance, the green static-convergence magnet is next to the green gun. The others are placed similarly next to their respective guns. The magnets are individually adjusted to bring together the color dots or lines in the center of the display to produce white dots and lines in the center of the picture. Dot generators are available to help beginners make the adjustments.

Edge convergence brings together the color dots and lines at the edges of the picture to produce white dots and lines. Formerly, three electromagnets, one near each gun, were used for this purpose. Electronic convergence circuits taking their cues from the vertical and horizontal circuits were designed to control the electromagnets throughout the vertical- and horizontal-deflection scan times. This was necessary because the beams had to converge at a point much farther from the edges of the raster than the center. Thus the individual beams had to have different angles with respect to each other at different points on the screen.

The electromagnets were adjusted by a panel of 12 variable resistors and coils. Modern "edge" convergence is now accomplished by moving the yoke around. Consult your service manual if the display needs edge convergence. You can use a pattern generator to adjust the convergence on the edges.

Look at the vendor's demonstration models to see how close you have to adjust the convergence. You will find that none are perfect. The human eye does not have acute color vision at the average viewing distance.

13-11 COMPUTER MONITORS

The deflection and high-voltage circuits for computer monitors are basically the same as the circuits we have seen in this chapter and Chapter 11. The sync circuits are the same. You have the sync amplifier, vertical integrator, and horizontal AFC circuits. The sync and blanking pulses are basically the same. The only thing that looks different is the scope waveform of the video informaton between the sync and blanking pulses. You may not see any video—only the sync and blanking pulses. Unlike TV or VCR displays, the computer display has a lot of blank spaces. In the intervals that you do have video signals, they will be of even height.

Horizontal sync symptoms will appear different because most of the display is black. You will not be able to see the sync and blanking bars. Adjust the hold controls of a good unit so that you will know what a sync problem looks like.

The computer color monitor is similar to the color display tube circuits. It has the three color output amplifiers, the video amplifiers, and the same deflection and sync circuits. It may have an extra high-frequency dc power supply that most TV systems do not have.

13-12 OSCILLOSCOPE CRTS

The cathode ray tube for the oscilloscope uses electrostatic rather than electromagnetic deflection (see Fig. 13-14). At the forward end of the tube, we see two pairs of deflection plates. One is for horizontal deflection, the other for vertical deflection. In the back end we see the grids. They represent the gun structure. The screen grid is called the accelerating anode or the first anode. After it comes the focus grid. Then comes the astigmatism grid.

The horizontal-deflection plates are connected to a horizontal-deflection circuit that places alternately positive and negative voltages on the plates. The electron beam is attracted toward positive voltages and repelled by negative voltages. These voltages deflect the beam horizontally across the screen.

The vertical-deflection plates receive deflection signals that originate from the scope probes. These voltages are amplified before they reach the deflection plates. The deflection-amplifier circuitry is very similar to the amplifier circuits we have already seen. You might find differential amplifiers in some makes. You may see complementary-symmetry circuits in the preamplifier stages in addition to the driver and ouput stages that we have seen in Chapter 11.

There is one big difference with respect to gun voltages. You will find a high negative voltage on the cathode and a low voltage on the accelerating anode. The

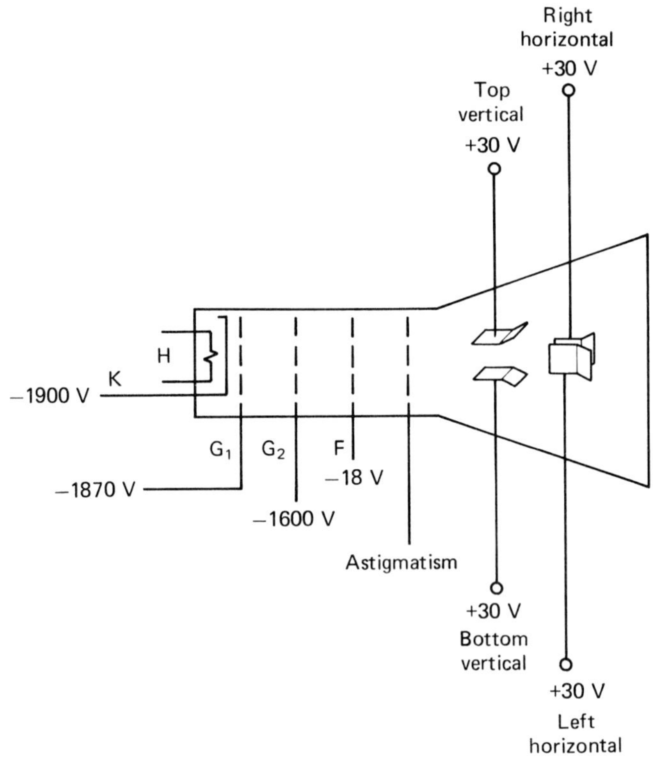

Figure 13-14 The oscilloscope CRT has horizontal and vertical electrostatic deflection plates.

cathode has typically −2000 V with respect to chassis ground. The accelerating anode has only about 10 V when measured from ground, but if you measure it from the cathode, it has 2010 V positive—enough to accelerate the electrons toward the face of the CRT.

There is good reason for the high negative voltage on the cathode. This allows the circuits feeding the deflection plates to operate at a low voltage. This permits all the amplifiers and multivibrator circuits to operate at low voltages. Also, the external input terminals may have to be connected directly to the deflection plates. We do not want any high voltage at the probes.

If you look at the schematic of an oscilloscope, you will see basic single transistor amplifiers, differential amplifiers, complementary-symmetry amplifiers, multivibrators, and a regulated power supply. These are all very familiar circuits. Even the high-voltage circuit is the same. Only the names may be different.

If you cannot get brightness—they call it intensity—check the cathode and grid voltages. If you cannot get horizontal deflection, follow the horizontal-deflection terminal leads back to the horizontal-deflection amplifiers. They may call it the *x* amplifiers. If you cannot get any signal, check the probe lead first, then check the vertical (or *y*) amplifiers. There is even a *z* amplifier for special applications that is used to vary the intensity of the signal. Finally, what would you do if you cannot get focus? Follow the lead from the focus terminal, of course. Then check out the focus-control circuit.

The most common problem with a scope is an open ground lead. The symptom is excessive signal voltage. The next most common problem is an open probe cable. An open shield connection has the same symptom as an open ground-lead connection. An open inner wire leads to a loss of signal. The fastest preliminary check for signal is to touch the ungrounded probe tip with your finger. If you get no response, the probe connections are probably bad. Wiggle the lead to check it.

Another common fault is a burned-out input circuit, due to excessive voltage being placed on it. The input connectors go bad quite often, due to constant usage. Other than these faults, the most likely circuits to go bad are the high-voltage and power-supply circuits (see Chapter 10 and Section 11-7).

WHAT YOU LEARNED TO DO

In this chapter you learned the procedures and tips for troubleshooting monochrome and color display tube circuits. You also learned to check out the oscilloscope circuits. This information will enable you to:

1. Trace dc paths and signal paths from the CRT pins back to their sources.
2. Be familiar with the typical monochrome and color CRT pin voltages.
3. Read vertical and horizontal sync symptoms and troubleshoot the deflection circuits.
4. Troubleshoot the color-sync and color-signal problems.
5. Distinguish between color-signal and dc-color symptoms and identify the circuits associated with these symptoms.
6. Check a color CRT for possible replacement.
7. Check the computer monochrome and color monitors.
8. Check the oscilloscope circuits.

QUESTIONS

1. What happens to the raster of a monochrome CRT if the cathode voltage is too high? What would the result be if the grid voltage is too high?
2. How can you check for high voltage?
3. What circuit could be bad if you have a raster with no display?
4. What section of the CRT is responsible for deflecting the electron beams?
5. Describe the functions of the yoke. Where is it located?

6. Where do the deflection voltages originate, and which describe the circuits that output them?
7. Name the circuits responsible for keeping the deflection voltages working in time with the video voltage.
8. Describe the symptoms for poor vertical sync.
9. Describe the symptoms for poor horizontal sync.
10. What defective parts can cause poor focus?
11. What circuits can provide dc voltages for the screen and focus grids?
12. What circuit is responsible for extracting the sync signal?
13. Where do the vertical-sync and horizontal-sync signals split off from each other?
14. How can you check out the vertical-sync signal by observing the display?
15. How can you check out the horizontal-sync signal by observing the display?
16. If you have poor horizontal sync, how can you tell whether the fault is in the AFC circuit or the oscillator circuit?
17. What color would predominate if you turn down the red cathode control? The green cathode? The blue cathode?
18. What circuits would you check if you have a red raster?
19. What circuits would you check if you have weak color and good black-and-white?
20. What circuits would you check if the color doesn't look right but the black-and-white looks good when you turn down the color control?
21. What are the differences you will find between a TV picture tube, a computer monitor tube, and an oscilloscope CRT?

PRACTICAL TROUBLESHOOTING EXPERIENCE

1. Try to roll up the picture on your TV by turning the vertical hold control. Would you say that your set has good hold?
2. Roll down the picture and adjust the contrast control. Does your set have a good vertical sync signal?
3. Can you turn the user horizontal hold control to knock the picture out of horizontal sync? If so, how many lines do you see just before you pull in the picture? Does your set have good horizontal sync? If you cannot get your picture out of sync, try turning your fine tuning until the picture goes out of sync.
4. Turn down the color-intensity control. Do you have a good black-and-white raster? If it is slightly green, what touch-up adjustment can you make? If it is slightly purple, what touch-up adjustment can you make? If it is slightly yellow, what touch-up adjustment can you make? (*Note:* It takes a lot of experience to make more than touch-up adjustments on these controls. Consult the service manuals before making extensive adjustments.)

5. Tape a small neon lamp onto the end of a 12-inch plastic stick. Turn on the set and hold the lamp several inches away from the flyback transformer. Do not let your hands get close to the transformer. Note how far away you can hold the lamp before the lamp goes out. Hold the lamp near or against the horizontal transistor. Does the neon lamp light up?
6. Locate the points on the PC board to which the CRT socket leads are connected. Usually, they are poked through the pin holes in the boards rather than to tie points. The board may have the circuit connecting lines drawn or printed on the component side. If not, consult a schematic to find the parts to which they are connected. Check the dc voltages on each lead to the picture tube except the focus and the heaters.

   Monitor the CRT cathode leads and the brightness control. Also check the grid voltage. What is the amount of cathode-grid voltage drop when the picture has maximum brightness? What is it at minimum brightness? Are all the cathode voltages exactly the same when you have a good black-and-white raster? If not, why?
7. Scope the cathode voltages and turn the contrast controls for a normal display. What is the peak-to-peak voltage at normal contrast?
8. Measure the vertical- and horizontal-oscillator base voltages. What measurements do you have?
9. Measure the sync-separator base voltage. What voltage do you have?

14

Linear IC Chips: the Pins Tell it All

EXPERIENCE REQUIRED

To make use of this chapter, you should know the basic rules for locating defective parts. You should be able to recognize an integrated-circuit chip when you see one. You should also know how to:

1. Check ac and dc signal paths.
2. Check shorted- and open-circuit blocks.
3. Check the input and output terminals of direct-coupled amplifiers, including the differential and complementary-symmetry amplifiers.
4. Solder and recognize a bad solder joint.
5. Inspect circuit boards.

14-1 CHIPS: THE BASICS STILL APPLY

Most electronic circuits today are encased in integrated-circuit (IC) chips. The old discrete transistor circuits are now limited to high-power, very-high-frequency, and other specialized applications. In some ways this makes troubleshooting a lot easier. The basics still apply. The overall game plan, of working from the outside, is even more applicable. Remember Fig. 1-1? There is one difference: We do not have to work our way into an integrated circuit. Everything is encased in little plastic chips, and we do not have to replace any of the inside parts.

Working on the bigger chips is similar to working around the outer connections of a PC board. The color-display circuit in Chapter 12 is the equivalent of a 14-pin IC chip. Some of these display circuits are mounted on a single board with many connector terminals. There are supply terminals, signal input terminals, control terminals, and output terminals to the CRT and speaker. Checking IC pins is like checking PC-board terminals.

In Section 5-5 (Fig. 5-16) we worked on the parts around a big chip. Now we can work on the chip itself.

14-2 LINEAR INTEGRATED CIRCUITS

There are two types of ICs: analog and digital. Analog circuits are also called linear circuits. That is the way they are listed in catalogs. They deal with voltages that faithfully follow the movement of natural processes such as radio and sound waves.

Digital circuits deal with discontinuous, on-and-off, high-and-low, pulse voltages. They often carry coded information. Their most important use is in computer circuits. We investigate digital chips in Chapter 15.

14-2-1 The Chip Names (Functions)

The first and possibly the most important thing you must know about a chip is its name or function. It may be possible to troubleshoot a problem simply by looking at a symptom and replacing the chip that has a name that relates to the symptom. Of course, this is usually not very wise, because the chip has so many pins—unsoldering and replacing or resoldering them can be very time consuming. You want to be sure the chip is bad before you replace it. This means that you will have to learn to trace the signal paths and check the voltages. Only then will the name of a chip and the symptoms associated with it really confirm that a chip is bad.

14-2-2 The Pins: Remember Their Names

Integrated-circuit chips are easily spotted; they have lots of pins (see Fig. 14-1). All the pins look alike, but each has a different job. They can be divided into three types: signal pins, control pins, and supply voltage pins. You already know some things about these functions.

1. Signal pins pass information to be processed—audio and video signals, for instance. There are two types of signal pins, input and output. It is very important to know which are the input pins and which are the output pins (see Fig. 14-1). Check these pins to find out where the signal is lost.

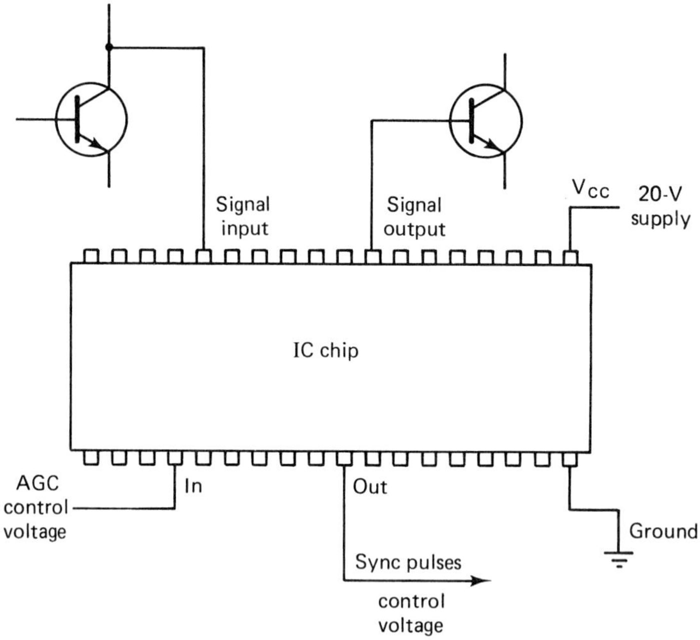

Figure 14-1 One way to find the input and output pins.

2. Control pins have voltages that change the way a chip works on the signal. They can turn a circuit on or off. They can change the output level or change the timing or shape of the signal. The AGC, AFT, and sync voltages are good examples of control functions.

   Some control pins are input pins and some are output pins. A chip is controlled by an input control pin. It controls other circuits with its output control pin.

3. Supply-voltage pins provide the chips with power so they can do their work. They are often designated by the letters V_{CC} for the positive supply voltage to the collector, and C_{EE} (or $-V_{CC}$) for the negative supply voltage. Check these pins primarily if you have no signal. However, do not neglect to check them if you have poor or weak signal output. And do not forget the ground pin.

We have already had experience with audio, video, sync, deflection, and color signal lines or paths in multitransistor circuits. We encountered AGC, AFT, AFC, and synchronization control circuits and lines. We learned about supply distribution lines in Chapter 5. We should have no problems with these lines in IC circuits. The problem is that there are so many of them. There are many more lines than parts. Fortunately, the problem becomes easier if you can trace the specific signal or control paths from chip to chip. You can also check the supply distribution lines to the chips. Then when you find the suspected chip, you can apply the following basic tip for troubleshooting IC chips.

TIP 14-1

If all the input signal and control voltages in a linear IC except the output are normal, as shown in the schematics or service information, check the output load circuit. If the output load circuit is good the chip is bad.

The second very important tip applies primarily to very-large-scale (VLSI) chips. In many VLSI circuits the external *discrete parts* count is very small. The following tip applies very often to smaller chips, too.

TIP 14-2

When you find a bad pin voltage, check the parts connected to the pin first. Before removing the IC chip, remove the parts in the circuits connected to the suspected pin. In some VLSI chip circuits, the parts count around the chips can be very low. In these cases it may be a lot easier to remove them all if necessary.

14-2-3 The Lines

If the names of the pins and lines are not shown on the schematic, you can still find them by following the lines and noting the names and functions of the circuits they begin or end at. The lines can be named after the source circuit function or after the destination circuit requirements. You can label these lines yourself.

Sometimes it is not easy to tell whether the pin is an output pin or an input pin. Again, you have to follow the line connected to it. If the line is connected to a transistor base, it is an output pin. If it is connected to a collector, it is an input pin (see Fig. 14-1).

More likely than not, your line is going to end up at another IC pin. There should be a name on the chip. You might have to look at the overall direction of all the signal paths in the schematic, which is usually from left to right. However, as modern circuits develop more and more inputs, outputs, and feedback lines, they cannot all be drawn from left to right. So look for the input and output connections all around the edges of the schematic. These will give you a clue as to the signal input, output, and control functions of the pins. Fortunately, many schematics for larger systems, such as VCRs do identify the pins and lines.

If you cannot identify the pin functions, you will have to consult an IC manual. Save all the literature you can find. You might even have to buy expensive manuals that list all the ICs.

If you work a lot on one specific model, it is worth memorizing the pin numbers and voltages so that you do not have to keep looking back and forth between the schematic and the chassis. It is very easy for your probe to slip off a pin and short something out while you are turning your head to look at the schematic.

> TIP 14-3
>
> Learn the names and functions of the IC pins and the lines that are connected to them. This will enable you to follow up on the symptoms and trace the signal and dc paths.
>
> Mark your schematic with the names of these lines. You can name them after the source functions or the destination circuit requirements. Draw arrows to indicate inputs and outputs.

14-2-4 Line Continuity, Connections, and Soldering

With the advent of chip technology, not only do pins and lines assume greater importance, so do their connections and continuity. The chances of these being defective have increased tremendously. One look at even a single-chip system would make this quite obvious. You learned how conductors were checked in Chapter 4. Many digital meters make a beeping sound to indicate continuity.

Soldering becomes more important. Desoldering chips and soldering in replacements requires a high degree of skill. Boards with sloppy soldering and loose copper traces lose their trade-in value. In-warranty boards with clean soldering are acceptable. Make doubly sure that the chip is bad before you replace it.

14-3 THE BASIC TROUBLESHOOTING PROCEDURE FOR IC CIRCUITS

There are four basic steps for finding a defective IC chip:

Step 1. Check the symptoms at the system output. Sometimes this may lead you directly to the faulty chip.

Step 2. Signal-trace the signal to the bad chip with a scope. If you have a signal at the input pin but none at the output pin, the chip could be bad.

Step 3. Check the supply and ground voltages.

Step 4. Check the other pins and external circuits that are connected to them with scope and meter.

Step 4 is needed to confirm the suspicions aroused by the preceding steps. This is the most difficult step. You must decide whether or not you want to unsolder the chip. This can be a big job. Sometimes it is not worth your time to unsolder a big chip. The following tips may be of help.

Generally speaking, you can use the fundamental rules and tips in Chapter 3, which dealt with circuit blocks. A shorted block will pull in the voltage and an open block will push it out. For instance, if a high-side pin voltage is low, the chip may be shorted or there may be an open above it. Figure 14-2 illustrates what might happen when a chip partially shorts. We are not saying that all bad chips act like this; we are saying only that if you *do* find a bad voltage in a situation like this, the chip could be bad.

TIP 14-4

When checking linear IC chips, treat them as block circuits that can pull the voltages when they short and push them when they open.

Check the resistance from pin to supply or ground to determine whether the trouble is in the chip or outside it.

The following tips are applications of the fundamentals you learned for checking specific parts in Chapter 5. Look for these easy-to-check pin connections. You might do these first as a matter of routine.

TIP 14-5

Routinely look for the connections that are easy to check: Check the voltage drops across the resistors connected to these pins.

Bypass capacitors connected to the chip pins should have little or no signal on them.

A coil between two pins of a linear IC chip should have the same dc voltage on both ends.

Any linear-IC pin not connected directly to ground that has no signal or dc voltage on it should be investigated. (Some pins have no internal or external connections.)

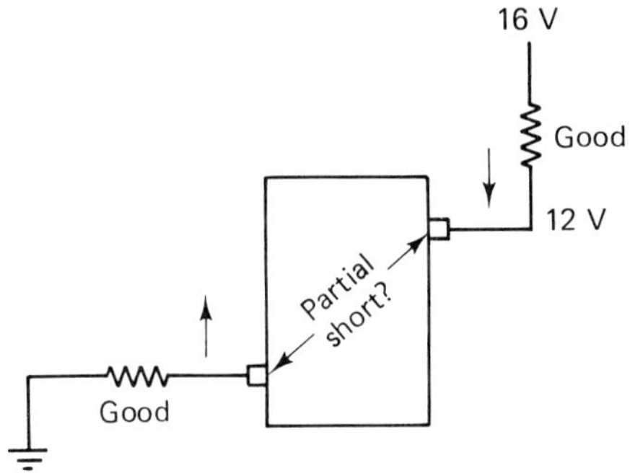

Figure 14-2 A shorted internal chip circuit will pull the voltages above it down and the voltages below it up. The chip can be treated as a circuit block.

The following tips are based on what we learned about amplifiers, multiple-stage circuits, and control circuits.

TIP 14-6

Note whether the IC output is supposed to follow or invert the input signal. If it does not do what it is supposed to do, the chip could be bad. Check the load, V_{CC}, and the control circuits.

TIP 14-7

If a linear-IC input pin has a normally low dc voltage, about 2 V or less, short it to ground and see if the output responds. If it does, it could be good; look elsewhere before coming back to it.

If the dc input voltage is too low, you can short it to supply with a 10-kΩ resistor and check the output response.

TIP 14-8

To check a linear IC, introduce substitute voltages into the *control* input pins to see if the operation can be restored. See what happens to the other pin voltages. Use a schematic or specifications manual to find the normal voltages.

TIP 14-9

Signal-trace the signals to the *control pin* of a linear IC with a scope. Find out where you lose it. Also look for weak or distorted waveforms.

TIP 14-10

If you have socketed IC chips, remove them and check the voltages. If a defective voltage returns to normal or beyond normal, the chip could be shorted.

A resistor should have no voltage drop if it is the only part connected to a socket terminal other than a capacitor.

A capacitor should have no voltage on its socket side if it is the only part connected to the terminal other than another capacitor.

A coil should have the same voltage on both ends.

We have covered some basic overall tips for troubleshooting chip circuits. Now we have to go into some of the more specific details.

14-4 TROUBLESHOOTING THE OPERATIONAL AMPLIFIER

To troubleshoot linear integrated circuits, we should familiarize ourselves with a very basic linear IC—the *operational amplifier* (op amp). It is used in a wide variety of circuits. The troubleshooter must be able to recognize these circuits and their functions. Although the op amp is now contained in larger IC chips, a familiarity with its inputs and outputs, and its functions, will speed up your work on the larger chips. You will find little op-amp chips occasionally among the larger ones. Sometimes you will find two or four of them packaged in one chip. This single chip could serve different circuits in various other regions of the board.

14-4-1 A Basic Operational Amplifier

Figure 14-3 shows the symbol for the op amp. The triangle indicates the direction of the signal. The input circuit of the op amp is a differential amplifier (Section 8-6-2). This gives it a very high input resistance or impedance; it is not likely to load down the preceding stage excessively. The output circuit has a low impedance, which is not likely to be loaded down by the following stage.

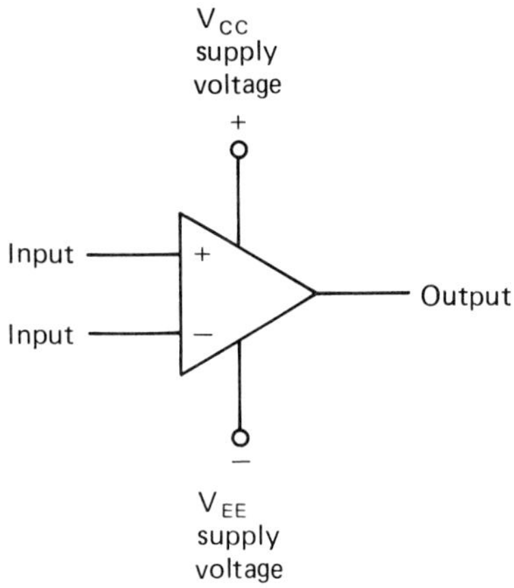

Figure 14-3 Symbol for an operational amplifier.

It has positive (V_{CC}) and negative (V_{EE} or $-V_{CC}$) supply terminals. With certain external circuit modifications, you may find the negative terminal tied to ground. There are also op amps that do not have negative supply-voltage terminals.

The basic op amp has two input pins and one output. Inside the chip, the inputs go to the bases of a differential amplifier. Do not confuse the plus (+) and minus (−) signs here with plus and minus dc voltages. The plus (+) sign at an input indicates that output follows this input. It is called the noninverting input. The minus (−) sign indicates that the output inverts the input. It is called the inverting input.

14-4-2 Inverting and NonInverting Op-Amp Circuits

Figure 14-4(a) shows a typical hookup for an inverting op-amp circuit. The gain of this amplifier can be controlled by the resistances of R_1 and the feedback resistor R_F. The higher the resistance of R_F as compared to R_1, the bigger the gain. ($A_V = R_F/R_1$, where A_V = voltage gain.)

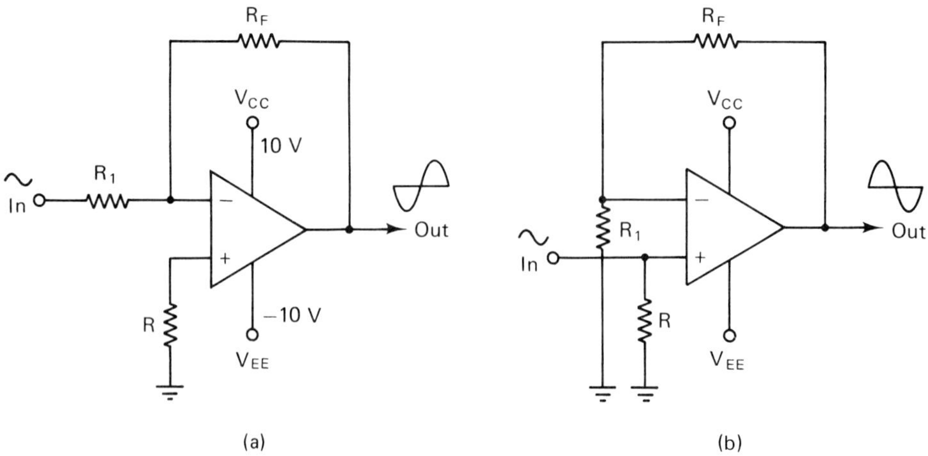

Figure 14-4 (a) Inverting op-amp circuit. (b) Noninverting op-amp circuit.

Figure 14-4(b) shows a hookup for a noninverting op-amp circuit. It has a very high input impedance. The gain of this amplifier is determined by R_1 and R_F. The higher R_F is compared to R_1, the bigger the gain. ($A_V = 1 + R_F/R_1$.)

If the input resistor, R_1, and the feedback resistor, R_F, have equal resistances, there will be no voltage gain. However, the output wave will be the inverse of the input wave, thus making it a *voltage inverter.*

The following tips provide some basic information concerning inverting and noninverting amplifiers:

TIP 14-11

An inverting op-amp circuit with negative and positive supply voltages should have 0 V dc with respect to ground on both input terminals.

If the source signal on the input resistor, R_1, is an ac signal or a pulse that has symmetrical top and bottom waveshapes, the output signal should average 0 V dc.

A noninverting op-amp circuit should have slightly differing voltages on the input terminals. If the input signal is symmetrically equal top and bottom, the output signal should average 0 V dc.

TIP 14-12

If you have a coupling capacitor at the input, or if the voltage is low (about 2 V), you can short the source side of the input resistor, R_1, to ground to test the basic inverting or noninverting op-amp circuit. You should have zero volts at the output.

TIP 14-13

An increased-value resistor, R_1, that is connected to the inverting terminal will reduce the gain of either the inverting or noninverting op-amp circuit. An increased-value feedback resistor, R_F, will increase the gain. An increased-value ground resistor, R_3, that is connected to the noninverting terminal of the op amp may increase and distort the output signal.

Offset Voltage. One reason for the resistor R_2, the *compensation resistor*, is to make up for a less than perfect input–output relation. Ideally, the output should be 0 V when you have 0 V across the input terminals. In practice, there is a difference. An *offset voltage* is required to reduce this difference. R_2 helps compensate for the discrepancy. In some applications, R_2 is connected to a variable resistor that can be adjusted to provide the right offset voltage. Some op-amp ICs have special terminals for introducing offset voltages. Some op amps have internal circuits that automatically provide offset voltages.

Frequency Compensation. The basic operational amplifiers have limited frequency response characteristics. Above 10 kHz its response begins to fall off. Negative feedback via R_F in an op-amp circuit helps to extend response to nearly 100 kHz, but at higher frequencies the feedback path increases the tendency for unwanted oscillations to occur. To overcome this, some op amps have compensation terminals that require external components. A small capacitor is usually connected to these terminals. Sometimes a resistor and capacitor combination is seen. Most op amps are internally compensated.

14-4-3 Op-Amp Applications

There is no telling where you will come across an op amp that is doing other than amplifying. The op amp has many uses. You should be able to identify the basic ones and check them out.

Figure 14-5(a) shows only a conductor between the output and the inverting input. It is a *voltage follower.* The output voltage follows and equals the input voltage at all times. Its purpose is to keep the load circuit from affecting the source voltage.

Figure 14-5(b) shows a summing amplifier or mixer. A mixer, in this case, is a circuit that enables two or more signal sources to feed a single line without disturbing each other. If the three input resistors are equal to each other and also to the feedback resistor, the output will equal the sum of the three input signals. Since the inverting input of the op-amp chip always has 0 V on it, the three inputs are isolated from each other. This allows the circuit to be used as a mixer. In this case the feedback resistance can be greater for more amplification.

Interesting Sidelight. Use Ohm's law to figure out how the op-amp adder works. Since sum of the three currents flows through R_F,

$$V_f = I_1R_F + I_2R_F + I_3R_F \quad \text{or} \quad V_1 + V_2 + V_3$$

where $R_F = R_1 = R_2 = R_3$ and I_1, I_2, I_3, and V_1, V_2, V_3 are the currents and voltages through the respective input resistors. Since the input side of $R_F = 0$ V, the output side must equal V_F.

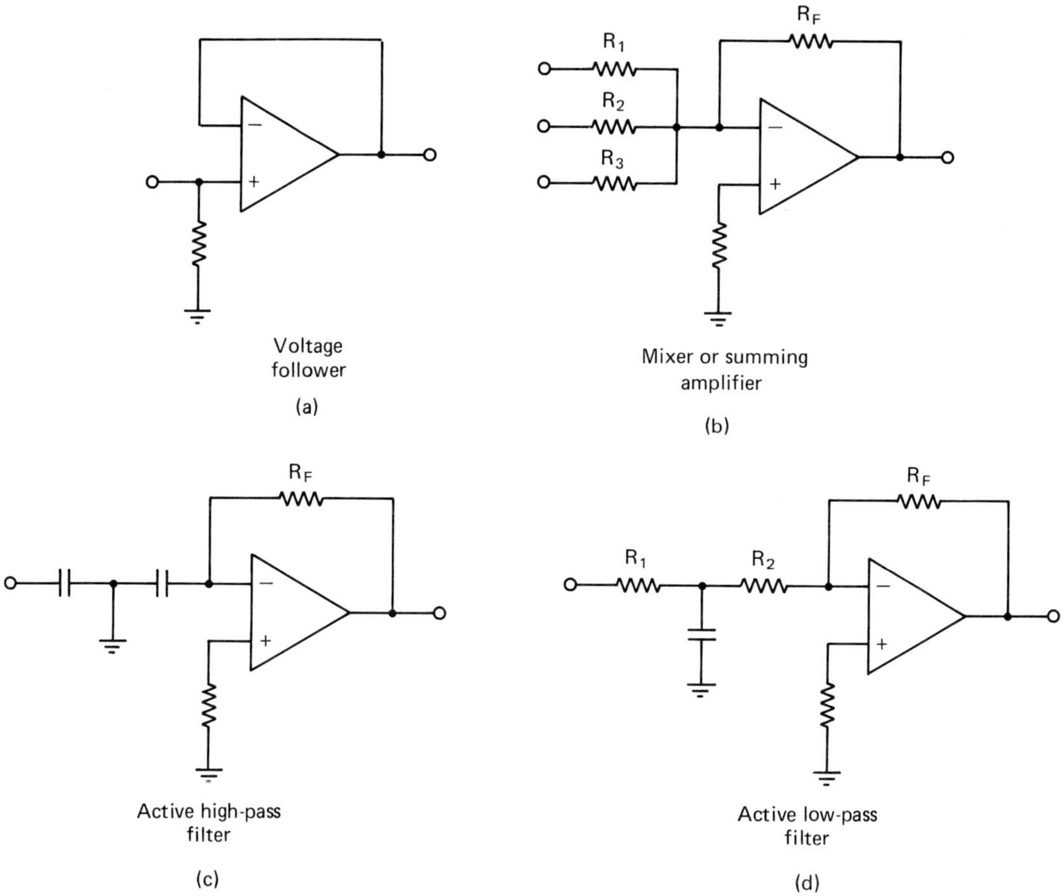

Figure 14-5 Op-amp circuits. (a) Voltage follower. (b) The summing amplifier adds the input voltages. (c) High-pass filter. (d) Low-pass filter.

Figure 14-5(c) shows an active high-pass filter. The term *active* refers to transistor and IC devices, as opposed to *passive,* which refers to *RLC* circuit parts. The impedance of the input circuit of this op-amp circuit will decrease at the high frequencies. Since the inverting-amplifier gain goes up when the input impedance goes down in comparison to the feedback resistance (or impedance), the gain will go up at the high frequencies. Conversely, it will fall at the lower frequencies.

Figure 14-5(d) shows an active low-pass filter. At the low frequencies a greater current will flow through the input resistors, since it will not be bypassed by the capacitor. This large current cannot enter the op amp because the impedance there is too high. Therefore, it must pass through this 0-V point and go on through the feedback resistor, so you must have a bigger voltage at the output end. (Ohm's law associates a bigger voltage with a bigger current.) At the higher frequencies the capacitor will bypass the current, and the smaller current must have a smaller voltage at the output end.

14-4-4 Pulse Circuit Applications

Figure 14-6(a) shows a *comparator* circuit that converts various types of input signals into rectangular pulses. It does this by detecting the input-signal voltage level. When the input reaches 0 V going positive, the output voltage switches rapidly negative. Just the opposite happens when the input voltage hits 0 V going negative. We call this circuit a comparator because the input voltage is compared to 0 V. The 0 V point is called the *reference voltage*.

Figure 14-6(b) shows a Schmitt trigger circuit. It is similar to the comparator, but it triggers at two points instead of one. The dots on the input wave show the trigger points. Note that they trigger going away from zero, not when going toward it. The two trigger points are determined by the values of the R_1 and R_F. The higher the resistance of R_F compared to the sum R_1 and R_F, the nearer to 0 V the points will be. (*Optional*: The equations for upper and lower trigger points are UTP $= +V_{max}$ $[R_F/(R_1 + R_F)]$ and LTP $= -V_{max}$ $[R_F/(R_1 + R_F)]$.)

The Schmitt trigger is immune to very short-duration low-level noise pulses. The comparator can be triggered by a low-level noise pulse just before and after the

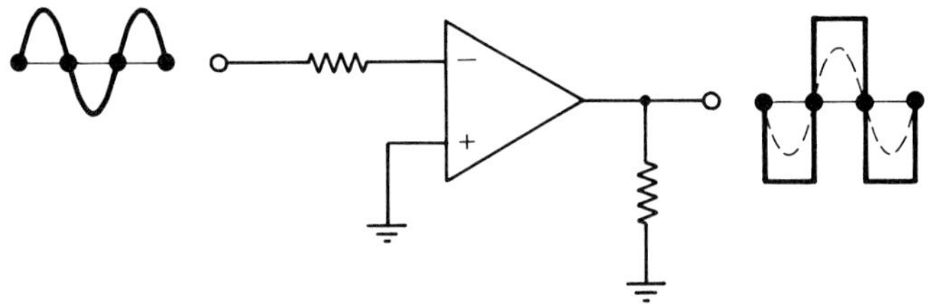

(a)

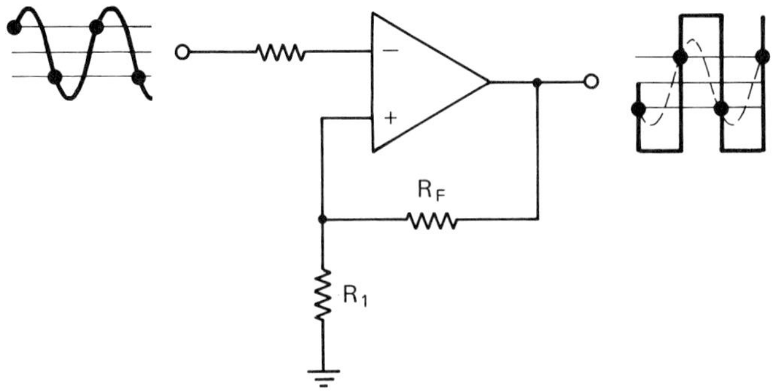

(b)

Figure 14-6 (a) Comparator. (b) Schmitt trigger.

signal reaches the 0-V point. The Schmitt circuit may be triggered just before it reaches the trigger point but not after it because the next trigger point is much too high or low.

Now that we have seen some basic circuits for handling various types of signals for various purposes, let us look at an IC that generates these signals.

14-5 THE 555 TIMER

14-5-1 A Basic Timer Circuit

Figure 14-7 shows a simplified diagram of a 555 IC circuit. It is connected as a basic timer. Let's say that it is for a photo lamp that is required to stay on for a certain length of time. Basically, when pin 2 is triggered momentarily, the output voltage will go high for a period of time that is determined by the variable resistor, *R*, and capacitor *C*.

Let's investigate what each pin does. Pins 8 and 1 are the supply and ground pins, respectively.

Pin 3. The *output* pin normally has a low or "off" voltage. When pin 2 is triggered by a low voltage, it goes high or "on," and it stays on for a period determined by *C* and *R*. The output voltage is fed to a load circuit consisting of a power transistor switch (Section 7-9-1) which turns the photo lamp on.

Pin 4. The *reset* pin is activated by a high positive voltage. This voltage maintains the initial off state of the chip until it is triggered. In this initial state, the output voltage is low or off. The reset voltage also grounds pin 7 internally to keep capacitor *C* discharged.

Pin 2. The *trigger* pin normally has about one-third of the supply voltage on it.

a. A momentary low trigger voltage from a switch or a negative pulse turns the output high.

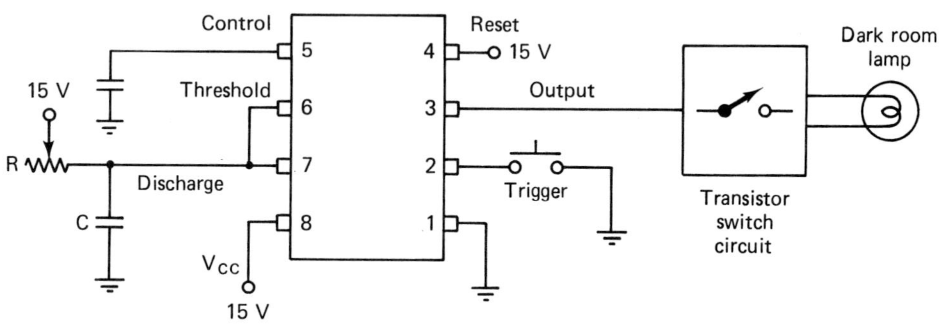

Figure 14-7 Timer chip. Resistor *R* and capacitor *C* determine the length of time the lamp stays on after triggering.

b. At the same time the low trigger voltage "ungrounds" pin 7 so that the capacitor can be charged up. The capacitor charge is placed on pin 6, which turns the output off when it reaches a certain voltage.

Pin 6. The *threshold* pin controls the amount of time the output stays high after triggering. The output turns off when the capacitor voltage reaches two-thirds of the supply voltage, V_{CC}. Normally, it is held down by pin 7.

Pin 7. The *discharge* pin keeps the capacitor discharged when it is grounded or "held low" by the reset voltage on pin 4. It is "ungrounded" by a low trigger voltage on pin 2. In this circuit its voltage is placed on pin 6, which can turn the output off when it is high.

Pin 5. The *control* pin is not being used in this circuit. A high positive voltage on it will turn the output on. A low voltage will turn the output off. It normally has two-thirds the supply voltage on it. The capacitor on the pin bypasses interference voltages, which may trigger the output on or off.

Understanding the inner circuitry of the 555 requires an understanding of the "flip-flop" circuit that is taken up in Chapter 15. The threshold and trigger action makes use of the comparator circuit in the preceding section. The threshold voltage is compared to a voltage that is two-thirds of the supply voltage. It turns the output off when its voltage equals $\frac{2}{3} V_{CC}$. The trigger voltage is compared to $\frac{1}{3} V_{CC}$. It turns the output off when it goes below that voltage.

Note on *RC* Charging Time. Here we have another use for the *RC* charge and discharge circuit. A higher capacitance or resistance takes longer to charge or discharge. We saw this in the multivibrator circuit. A larger capacitor and resistor put out a lower frequency. The charge time is measured in *time constants* or *RC* ($R \times C$) seconds. It takes about 5 *RC* seconds to charge a capacitor to nearly the full voltage. At first the charging voltage goes up very rapidly and begins to slow down as it reaches the full charge. Theoretically, it would take forever to reach a full charge.

For instance, in the timer circuit above it only takes about 1.1 *RC* second to reach the threshold voltage, which is two-thirds of the full supply voltage. As an example, a 10-μF capacitor and a 1-MΩ resistor would leave the lamp on for about 11 seconds.

Checking the 555 Timer. The 555 can be checked for normal voltages while in the initial inactive state. Substitute voltages can be placed on various pins. If the output does not respond properly, the chip could be bad (see the chart in Fig. 14-7). Check the voltages on the other inputs before you draw any conclusions. Check the output load circuit, too. (Do not forget to check the lamp—and the supply voltage, too!)

Pin 7. The discharge pin should be low when the circuit is inactive. It is an output pin, so do not try to put a substitute voltage on it. In fact, it might cause damage. Disconnecting pin 7 should allow the capacitor to charge up. Otherwise, the capacitor, resistor, or pin 6 could be bad.

Pin 6. The threshold pin should have no voltage on it when it is disconnected from the circuit. It must be disconnected from pin 7 before a substitute voltage can be placed on it. After it is disconnected, a jumper from the supply voltage point should turn the output off if the trigger voltage is held low. If the trigger voltage is then brought up, the output should remain low. (This shows that the internal flip-flop circuit is working.)

Pin 5. The control pin can be grounded to see if the output can be turned off. It can be shorted to the supply voltage point to see if the output can be turned on.

Pin 2. Grounding the trigger, pin 2, should turn the output on if the threshold voltage is low. Shorting the trigger pin to the supply should not turn it off immediately. The time delay checks the capacitor–resistor circuit.

Pin 4. This pin should not be grounded to test it. If it is disconnected from the supply, nothing predictable will happen. It works only when the circuit is first turned on and when the output goes out after it has been on.

Pin 3. If the output cannot be turned on, either the input circuits, the chip, or the load circuit could be bad. Check the load circuit as in Section 7-9-2. Check the chip and other circuits as shown above. If the output cannot be turned off, ground the pin momentarily to check the load circuit. If the lamp does not go out now, the load circuit is bad. Check the other circuits as above.

14-5-2 A 555 Multivibrator

The 555 chip can also be used as a pulse generator. In Fig. 14-8, the trigger pin is fed by the capacitor along with the threshold pin. The discharge pin discharges the capacitor through R_2 when its voltage goes low. When it goes high, the capacitor is

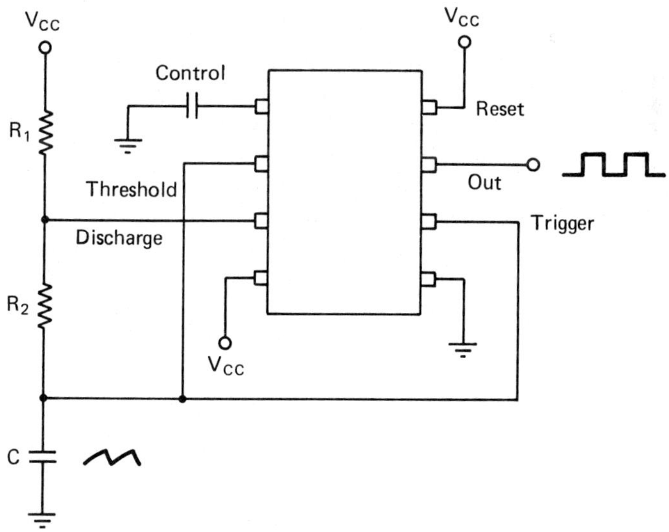

Figure 14-8 This timer chip is connected as a pulse generator.

allowed to charge through R_2 and R_1. In other words, the discharge time is shorter than the charge time.

When the circuit voltage, V_{CC}, is first turned on, reset voltage will put a high voltage on the discharge pin. This will allow the capacitor to charge. At first the capacitor voltage is low, and both the trigger and threshold pins would be below their comparator trigger voltages. This would place a high voltage at the output. Also, this keeps the high voltage on the discharge pin. This will allow the capacitor to keep charging and increasing the voltage on the trigger and threshold pins. See the capacitor waveform in Fig. 14-8.

When the threshold and trigger voltages are finally reached, the output will go low and the dicharge pin will go low. The low on the discharge pin will cut off the voltage from V_{CC} and discharge the capacitor through R_2. When the capacitor voltage goes below the trigger point, the output will go high again, and a new cycle will begin.

The on-time duration of the pulse is determined by the two resistances and the capacitor. It is equal to $0.7C(R_1 + R_2)$. The off-time duration of the pulse is equal to $0.7CR_2$.

14-6 SOME REPRESENTATIVE IC CHIPS

Figure 14-9 gives you examples of some of the wide variety of IC chips that are available. Figure 14-9(a) shows a four-terminal adjustable voltage regulator that includes a built-in thermal shutdown circuit. If the load shorts, the current will shut down. Compare the pin functions with the corresponding points in the transistor regulators in Fig. 10-2.

Figure 14-9(b) shows a switching voltage regulator (see Section 10-4-2 on switching regulators). The inductor and capacitor at the output are just about the only main components that are not built in. C_1 determines the average frequency of the built-in pulse generator. Pin 1 is the input to the transistor switch. R_1 is for overload protection. A short in the load will reduce the voltage on pin 2, and this will reduce the output voltage and current to a safe level. R_2 is for adjusting the output voltage. Compare the pin functions with the corresponding points in the discrete switching regulator shown in Figs. 10-6 and 10-7.

Figure 14-9(c) shows an FM detector. It requires an input IF transformer and a FM detector coil L. C_1 and C_2 are bypass capacitors. C_3 is a deemphasis capacitor. It reduces the high frequencies which were preemphasized at the transmitter. The coil L is not center tapped. It is called a quadrature coil. This chip has many industrial and communications applications; it is used wherever frequency detection is needed.

Figure 14-9(d) shows a phase-locked-loop (PLL) IC that can be used for FM detection. The output voltage of this IC changes when the input frequency or phase deviates from the frequency of an internal voltage-controlled oscillator (VCO). A portion of the output voltage is fed back to a phase-detector circuit that tends to lock or keep the VCO phase in time with the input phase. R_1 and C_1 determines the center

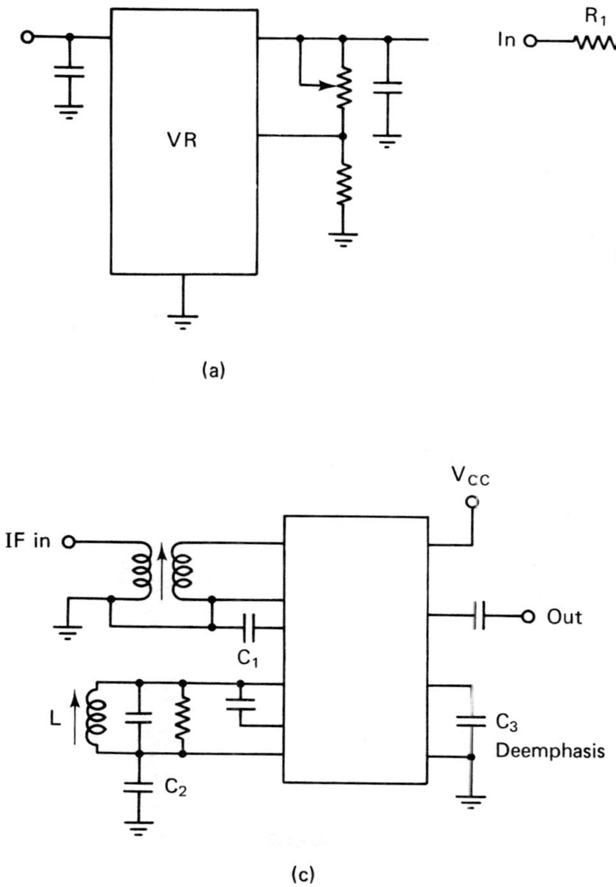

Figure 14-9 Variety of IC chip circuits.

frequency of VCO. Pins 1 and 2 are the inputs to a differential amplifier. Pins 5 and 6 are outputs that go to an op amp (not shown). PLLs has many uses in communications systems. At the lower frequencies there are PLL ICs that are used for tone detectors.

Troubleshooting Representative IC Circuits. Troubleshooting the ICs above involves checking the input, output, control, and supply voltages. In the four-terminal adjustable voltage regulator in Fig. 14-9(a), the control circuit resistors can be checked out like their counterparts in the voltage-regulator circuit in Section 10-2-2. With this regulator there is another check: When the power goes out, turn off the ac input power. Wait a minute or so and turn the power back on. If the equipment shows signs of life, the regulator chip is probably okay, and the trouble is in the load circuits.

The output and control circuits for the switching regulator in Fig. 14-9(b) can be checked out like the switching regulator in Section 10-4-2. The oscillator capacitor, C_1, should have a pulse. The width of this pulse and that at the output should vary as you adjust the output voltage.

The input for the FM detector circuit in Fig. 14-9(c) is a differential amplifier. The specifications manual may show that this circuit should have a hi–lo resistance reading from pin 1 to ground. Pin 1 should have no signal on it when the circuit is operating. The dc voltage on pins 1 and 2 should be equal. The dc voltage at the output should be about one-half the supply voltage. An excessive high-frequency response would indicate an open deemphasis capacitor. Both the transformer secondary and the detector coil should have zero ohms across them. In the case of FM radio or TV sound detection, the quadrature coil can be easily tuned for maximum sound level and clarity.

TIP 14-14

Some signal processing ICs have hi–lo resistance readings between input and ground, or between output and ground. Check the IC specifications manual.

If the output of an IC is supposed to be an undistorted sine wave or a pulse with symmetrically equal alternations, the dc voltage at the output should split the supply voltage or voltages into approximately equal halves.

The PLL circuit in Fig. 14-9(d) has an input that can be checked like an op amp. See the tip above. Pin 3 should be pulsing. The output pins should have a varying voltage. The amplitude of the input signal should not vary.

14-7 CHECKING MULTISTAGE IC CHIPS

14-7-1 Checking an FM-Detector and Audio-Output Chip

Figure 14-10 shows a complete FM sound IC from the IF input to speaker output. Let's see what we can do to check it out. If the sound is bad, tune the IF coil T_1 for maximum sound (do not do this if there is more than one IF coil), then tune the detector coil L_1 for best sound. If the sound is still bad, weak, or dead, you can start checking voltages. Make sure that you take all the voltages before making a decision. The following checks should help reveal whether the chip is bad or not.

Pin 9. Check the supply voltage for 20 V dc. If the dc voltage here is too high, the chip or its ground connection is probably open. Confirm this by touching the chip for coldness. (Supply resistor R_4 is not very likely to be shorted.)

If the pin 9 dc voltage is too low but not 0 V, the chip or the bypass capacitor C_5 has probably shorted partially. (R_4 is a wirewound resistor and will open completely.) If pin 9 has 0 V, the supply resistor R_4 could be open or C_5 could be shorted.

Pin 7. The ground pin should have zero volts. Otherwise, its connection could be open. Examine the pin connection carefully before removal. Use a magnifying glass if necessary. Resolder the connection if it is obviously bad. If it looks good, carefully press against the pin with a plastic prod. If you get sound, the connection is bad. Clean and resolder the connection. Make sure that

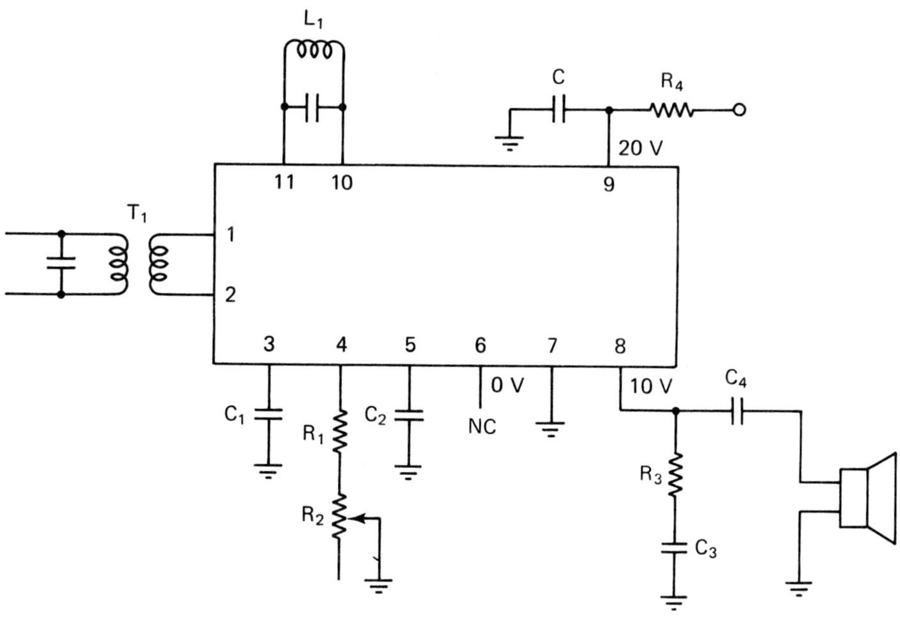

Figure 14-10 FM sound chip.

you obtain a thin shiny coating on both the lead and the foil. At least clean the visible part of both. (Section 16-3-8 has more on soldering intermittent connections.)

Pin 8. The audio output pin should have about half the supply voltage. (We assume that the speaker and output capacitor have been checked.)

If the pin 8 voltage is too high, the chip may be shorted internally to the supply or open internally to ground. Also check the input pins and other pins, too. We assume that the supply pin 9 has been checked already.

Also, if pin 8 is too high, the ground pin can be open. We took care of this when we checked ground pin 7.

If pin 8 is too low, the chip may be open internally on the supply side, or shorted internally to ground. Also check the input pins and other pins, too. Check ground pin 7 if you have not done so.

Also, if pin 8 is too low, capacitors C_4 and C_3 could be shorted. Capacitor C_4 should have the same voltage as pin 8. If lower, it is shorted. C_3 keeps the tone mellow by reducing the high frequencies. If it is open, the dc voltage will not change, but the tone may be too high.

Jump capacitor C_3 with a substitute to see if you get a better tone. If the speaker has dc voltage on it, capacitor C_4 is shorted. If you have no sound, jump capacitor C_4 with a good one and see if you get sound.

Pin 1. Check the dc voltages on pins 1 and 2. They should match the voltages on the schematic and should equal each other. If they do not, the transformer secondary coil is open. If the coil is open, short pins 1 and 2 and see if the output dc voltage returns to normal. Also check to see if the input voltages return to normal. If they do, the chip is okay.

The input circuit is probably a differential amplifier. If so, the signal voltage between pins 1 and 2 should be no more than a fraction of a volt. If you have a large dc voltage drop between the two pins, the coil is open and the chip is possibly bad.

If the coil is open, both inputs should have hi–lo ohmmeter readings to ground. These two readings should not differ greatly. They should also have a hi–hi reading between them.

An open or shorted T_1 primary will not affect any of the dc voltages on the chip. Neither would a shorted secondary.

A shorted C_1 will bring down the voltages at pins 1 and 2. An open C_1 may place a signal on pin 2, and the difference voltage may be affected. The output signal will be weakened. Scope pin 2 to see that there is no signal on it.

A primary-to-secondary short may burn the input circuit or place a higher-than-normal voltage on them, provided that there is a direct connection to the IF collector or other dc source.

Pin 10. Check the dc voltages on pins 10 and 11. Pins 10 and 11 should be equal, or else the coil L_1 is open. Scope the pins for a signal.

Pin 4. Check the voltage at pin 4. If there is an incorrect voltage on it, see if you can make it vary by adjusting R_2. If you can, R_1 and R_2 are probably okay. If not, check the resistance of R_1 and R_2.

Pin 3. Check the dc voltages on capacitors C_2 and C_3 at pins 3 and 5 if the schematic gives you the pin voltages or if the IC manual indicates the presence of dc voltages. If the voltages are lower than normal or seem to be, the capacitors could be partially shorted. If higher than normal, something could be wrong with the chip.

You can jump a capacitor with a substitute. If the operation returns to normal, the chip is good.

If you suspect a shorted capacitor, disconnect one of its leads and try a substitute.

TIP 14-15

If a chip has a history of repeated failure in a particular circuit, it may be a good idea to install a socket into which you can insert new ICs in the future.

14-7-2 Schematic of a TV Sound Section

Figure 14-11 shows the sound section of an RCA CTC-123 color TV schematic with the actual dc voltages on each pin. The pin functions are also given. This should be of great help.

Pin 1 has the supply voltage of 23.2 V, which is taken from the supply line through R_{208} and filtered. A second line at pin 1 goes to a protection circuit (not shown). It guards against shorts in the power-amplifier circuit of this chip.

Ceramic filter, CF_{201}, passes the sound IF signal to pins 4 and 5. There should be no dc voltages on any of its terminals. If C_{213} shorts, the dc voltage at pins 15 and 14 should drop. If it opens, the signal may not be amplified.

L_{201} at pins 10 and 11 may be adjusted for best sound. C_{206} couples the attenuated audio signal from pin 9 to the audio power-amplifier input at pin 7. The amplitude of the signal at this point should be controllable—by adjusting the volume control that connects to the internal attenuator. The schematic shows that this control can change the dc voltage on pin 12 from 0 V to 6.3 V.

14-7-3 A Very-Large-Scale (VLSI) Chip

Figure 14-12 shows part of a 64-pin IC. Only a tuner and a few other parts plus a picture tube need be added to make a complete color television receiver. All the signal inputs and outputs on the chip are shown. For simplicity, not shown are all the coils, capacitors, resistors, and controls required to make this chip work. Three pins are shown connected to a 3.58-MHz crystal. There is a main 18-V supply pin plus three more 11-V auxiliary supply pins.

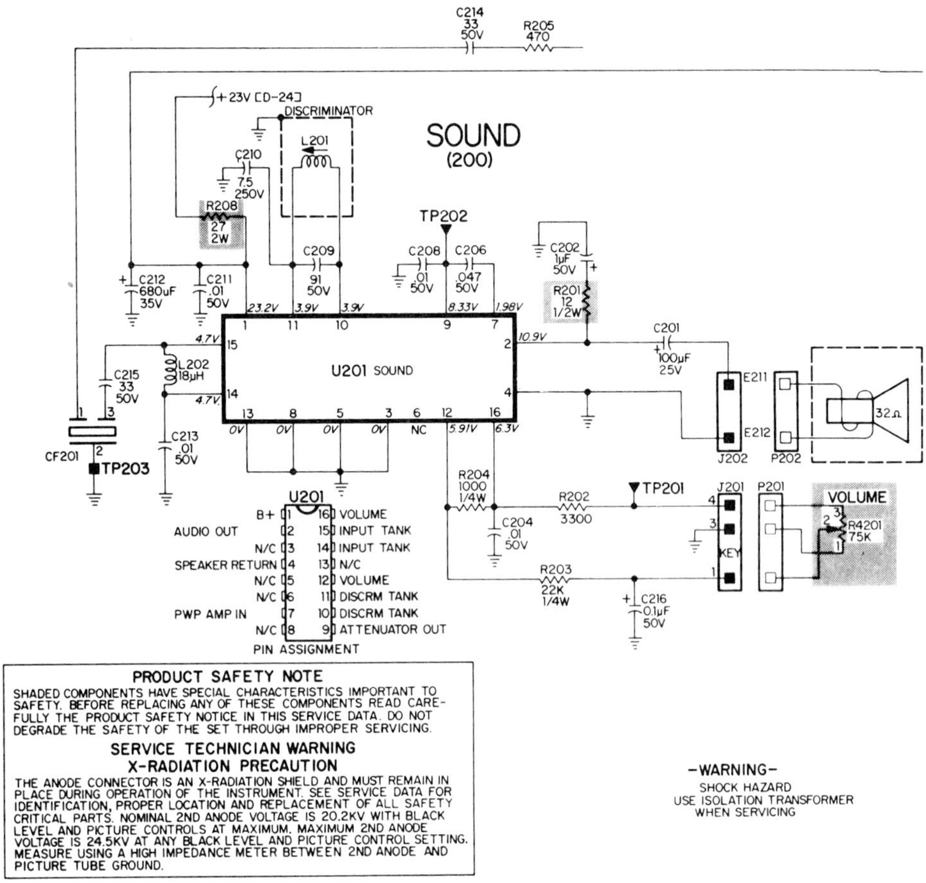

Figure 14-11 RCA TV sound circuit. (Reprinted courtesy of Thomson Consumer Electronics)

In color TV receivers, most circuits derive their supply voltages from the horizontal output circuit. If any of these circuits short out, the horizontal output will be seriously affected or even killed. Practically all circuits could lose their voltages, and the voltmeter would not do any good. We can solve this problem by using a substitute power supply.

If you have the supply voltage, start troubleshooting the IC and its output amplifiers in exacty the same way as with the basic color system. Start at the CRT and work your way back. The output amplifiers are the same. Now, instead of ending up at a demodulator, you end up at an IC pin.

When you find that the IC pin has no output, you naturally check the supply voltages first. Then you check the input pins responsible for the missing output. For instance, if the problem is luminance or synchronization, check the video input pin.

Then check the relevant variable-resistor control circuits for your particular problem. All the pins and controls are labeled. Varying the controls may give you an

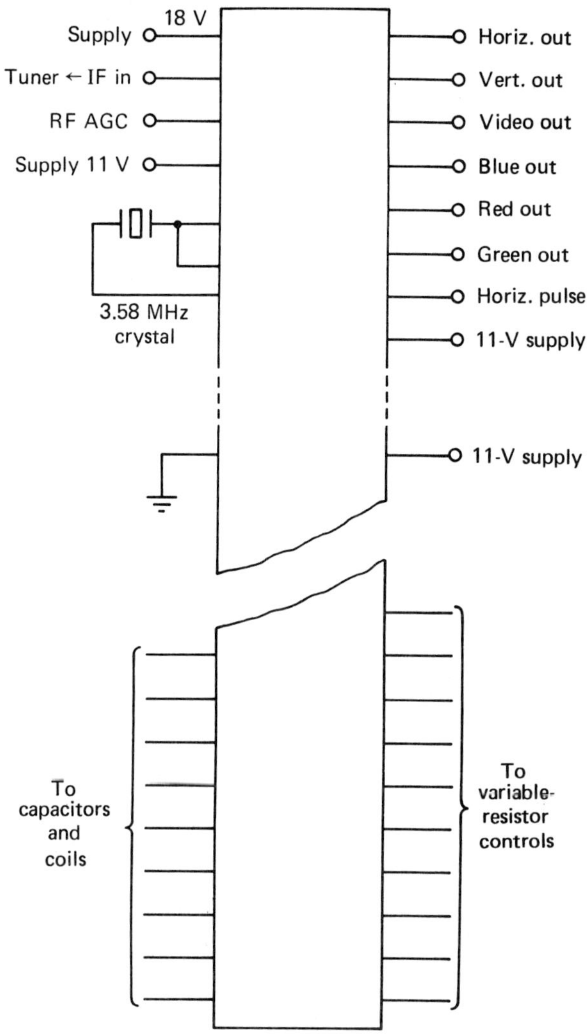

Figure 14-12 Very-large-scale (VLS) chip. It lacks only the input and output circuits of a color TV set.

idea of the nature of the trouble and where it is. You must be familiar with all the names and functions.

Next check the automatic feedback control inputs and outputs that may have something to do with your trouble. Introduce substitute voltages at these points. Even if operation is not restored, these voltages may give you important clues.

Check the relevant capacitors and resistances. Check the coils for signals, very low dc voltage drops, and very low resistances. Finally, check the parts you are not sure of out-of-circuit. The 3.8-MHz crystal is checked by substitution. The parts count on the board is very small.

The parts that relate to your trouble are even fewer, so it may be a good idea to pull them out and check them all out-of-circuit. The whole procedure up to now should not have taken more than a few minutes. So this may affect your decision as to whether you should do any more work on the chip and on the board. You may decide to exchange the board at this point if there is a manufacturer's exchange policy on it.

Let us stop for a brief summary of the major troubleshooting steps given above. First, you look at the symptoms and find the input or output devices that are not performing as expected. Then you find and check the pin or pins that work with these devices. You make use of the fundamentals for finding shorts and opens not only on the pins, but also in the attached parts around the chip.

TIP 14-16

The general procedure for checking all types of chips, large or small, is as follows:

1. Check the symptoms at the system output.
2. Check the relevant input and output devices that are associated with the symptoms.
3. Check the output amplifers that drive the relevant system output devices.
4. Check the relevent IC signal input pins and circuits.
5. Check the IC supply-voltage pins and circuits.
6. Check all the relevant control pins and circuits.
7. Check the relevant automatic feedback control pins. Introduce substitute control voltages.
8. Check the peripheral capacitors, coils, and crystals associated with the relevant IC pins. Remove and substitute these parts if necessary.

TIP 14-17

Very-large-scale (VLSI) chips have very simple external circuitry. The parts count is very low. Check all the parts, connected to the bad pins. If necessary, remove the parts to check them before removing the chip.

14-7-4 IC Manuals

With the wide variety of integrated circuits in use today, you will have to save all the IC spec sheets you can get a hold of. In fact, you would do well to purchase one of those big manuals that cover just about all the chips made. Of particular interest to you are the signal paths inside the chips. They also give you the functions of each pin, so you will be able to work on them actively with substitute parts, jumpers, and voltages.

Another consideration is that many manufacturers do not have schematics available for their products, not even for their own and authorized service centers. Since the parts count on modern circuit boards are so low, you may get by without schematics. At any rate, IC specifications manuals would be of great help in this situation.

14-8 SURFACE-MOUNTED CHIPS

Working with surface-mounted chips or devices (SMDs) requires a lot of skill and patience. They are soldered directly to the copper traces on the board. There are no pin holes. They are very small and you will not find much space between the pins. Care must be taken when taking measurements on the pins. It is very easy to short them out. Use sharp, clean probes. Special soldering tools should be used. Patience and skill are extremely important here.

WHAT YOU LEARNED TO DO

In this chapter you learned:

1. To use the basic rules for shorts and opens when troubleshooting IC chips.
2. That checking the input, output, and control voltages is of greater importance when working on IC circuits.
3. To use your knowledge of the basic internal functions of the chip in order to check the pin voltages.
4. That you must be able to identify the pins by names.
5. To check basic op-amp circuits.
6. That the parts count around a VLSI chip is small, so that it may be best to check all these parts out-of-circuit, before changing the chip.
7. To make sure that all the lines and connections are good before removing a chip.
8. The importance of having IC manuals on hand.

QUESTIONS

1. Why is it important to know the names and initials of various types of electronic circuits and functions?

2. What three types of pin functions are there?
3. Name some control functions and their initials.
4. If a pin has a higher-than-normal voltage, what could be wrong above it or below it, according to the basic rules?
5. Compare the input pin voltages of an op amp.
6. How many dc volts would you expect at the output of a typical inverting op amp if it has equal positive and negative supply voltages, and its input terminals are shorted together?
7. What can the voltages on the pins that are connected to variable resistors on a VLS chip tell you?
8. In what ways do the IC control signals affect the outputs?
9. About how many dc volts would you usually expect on an output pin for a signal that has symmetrically equal alternations?

PRACTICAL TROUBLESHOOTING EXPERIENCE

Obtain a PC board with a basic op amp in it. Some chips have two or four op amps in a single package. You need a power supply with positive and negative outputs.

1. Check your power-supply connections by measuring the positive and negative (or ground) supply-pin voltages on your IC.
2. Check all the pin voltages.
3. Does the op-amp output divide the supply voltage or voltages equally? If not, are the input voltages equal?
4. If the input pin voltages of the op amp are not equal, say by about 0.9 V, what is wrong?
5. With a 10-kΩ resistor connected to the positive supply, insert a voltage at the input. If there is a coil connected across the inputs, will this affect the output voltage substantially? In this case, if the output voltage changes substantially, is everything okay or not? If not, what part could be bad? Is the input an inverting input or a noninverting input?

15

Checking Voltages and Resistances in Digital Circuits

EXPERIENCE REQUIRED

To understand the simple troubleshooting checks in this chapter you should know how shorts and opens are checked in both series and parallel circuits. You should know the basic ways in which electronic switches can produce high and low voltages. You should know how a switch that is "on" can produce a low voltage and how a switch that is "off" can produce a high voltage (see Figs. 5-14 and 8-16).

An understanding of synchronization would be helpful. You saw how multivibrators were synchronized in Section 13-5. The details may differ, but the idea of using timing pulses to ensure that different circuits work together is similar.

You should be aware of such widely known computer words as *software, hardware,* and *peripheral devices* (disk drives, monitors, and keyboards). It would be helpful to know about computer programming.

15-1 DIGITAL SIGNALS AND THEIR USES

Digital signals are on-and-off pulses. They have two voltage levels: high and low (hi and lo). These signals may be sustained hi or lo dc voltages as well as pulses. They can be obtained by switching voltages on and off.

Mechanical switches are adequate for turning on lamps and appliances. Electronic switches, however, are necessary for fast switching. Switches are the basic means for starting and stopping all kinds of operations. They are also used for sending coded information that closely controls these operations. Digital code consists of voltages that are switched on and off or hi and lo. Computer circuits have thousands of high-speed transistor switches.

15-1-1 Basic Digital Operations

In complex operations, digital circuits and signals are required to perform three basic tasks. They must:

1. Separately activate or deactivate many different devices or circuits.
2. Route signals between various devices and circuits.
3. Activate and deactivate devices and circuits at the right times.

Because of the step-by-step nature of most operations, and because they must work simultaneously with many devices and circuits, and because of the limited number of signal lines that can be hooked up economically, digital signals must be able to:

1. Operate sequentially.
2. Share the same lines.
3. Work at high speeds.

The washing-machine cycle is a familar example of the first condition, sequential operation. Your telephone is a good example of the second condition—sharing the same line. You do not see a separate wire between your phone and all the other phones in this world. Since the signals are on shared lines, digital circuits may have to meet the third requirement—high speed. Each signal needs to wait only a microsecond, perhaps even less, before it gets on a line.

15-1-2 What to Look For

There is a big difference between troubleshooting analog and digital circuits. In analog circuits, you have series circuits with various dc voltage levels. This made it easier to isolate the faulty parts. In digital circuits you have only two basic dc operating voltage levels and a third level called ''floating'' (also called ''open'' or ''high impedance'').

Furthermore, in the typical digital system all the chips have the same supply voltage, 5 V for TTL chips. They are wired in parallel. This makes it harder to find the bad section or chip.

In digital circuits, more attention has to be given to the routing of signals and to the sequence in which they activate various circuits and operations. These signal voltages appear on the pins of various chips. To check them intelligently, you must know the pin functions and how the chips are working together.

15-1-3 A Look at a Digital Circuit Operation

To understand digital circuit operations, let us look at a simplified version of the loading and unloading sequence of a VCR cassette. (The same circuit can be used for huge carriages in industrial operations.) In Fig. 15-1, SW_1 is the forward switch. It turns the motor forward. SW_2 is a stop switch. It stops the motor when the cassette hits bottom. SW_3 is the reverse switch. It reverses the motor. SW_4 is another stop switch. It stops the motor when the cassette reaches the top. The forward driver is the amplifier that drives the motor forward. The reverse driver does the opposite.

At first glance the operation of these switches seems rather simple. However, as is true generally in digital circuits, things are not so simple once you get down into the actual circuits.

> *Note:* To understand the VCR cassette loading sequence, we must remember that once a switch initiates an action, it has no further control over it.

What happens is that the digital circuit (DC_1) takes over and prevents any conflict that may arise between the switches. You cannot have two switches on and in control at the same time. These two actions, the releasing of control and the taking over of control, are very important in digital electronics. Always look for this. In the following sequence, note the voltages on all the switches at each step.

Step 1. The cassette is pushed in, and the forward switch SW_1 turns on. This switch remains in the on position as long as the cassette is in the carriage.

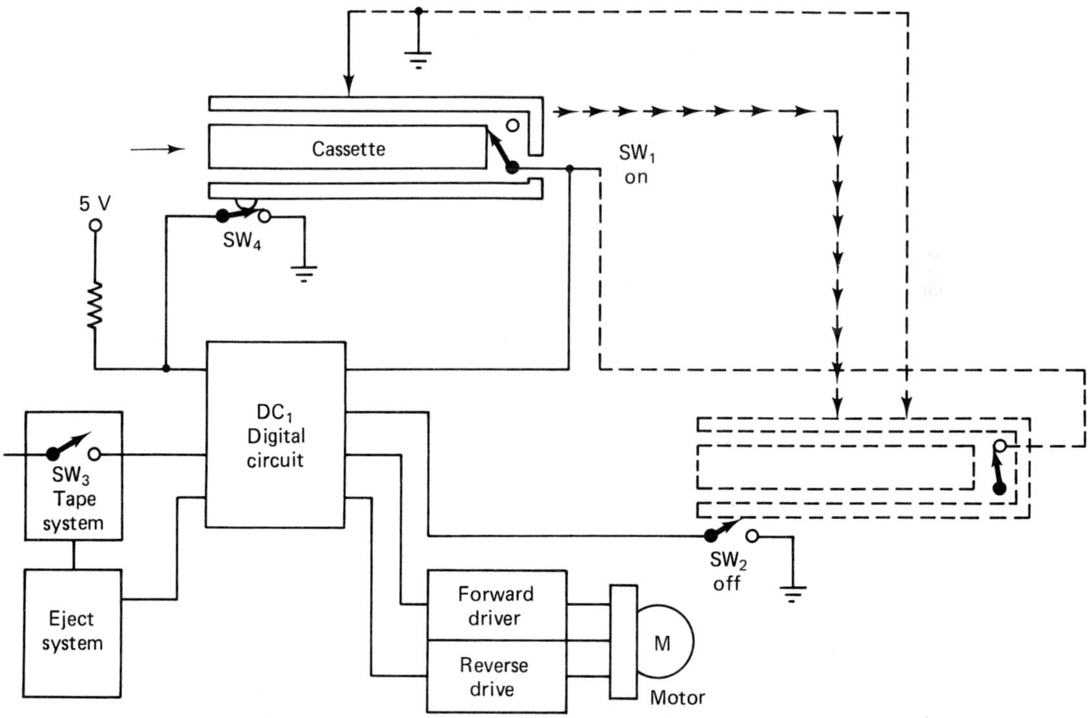

Figure 15-1 Diagram for a cassette loading system. The switches play a key role.

Step 2. SW_1 activates digital circuit, DC_1, which makes the forward driver circuit start the motor forward. At the same time, DC_1 must keep the reverse driver off.

Step 3. The motor turns forward and moves the cassette carriage down to activate switch SW_2, which makes DC_1 stop the forward driver. SW_1 and SW_2 are both on, but SW_1 had given up control to DC_1.

Step 4. After the motor has stopped, the user pushes the play button and control passes to the play circuit. During play, digital circuit DC_1 keeps the cassette load motor stopped.

Step 5. At the end of play, switch SW_3 in the tape system activates DC_1 to start the eject cycle.

Step 6. Digital circuit, DC_1, turns on the reverse driver. Note that the stop switch, SW_2, has given up control.

Step 7. The cassette carriage reaches the end of its travel, and switch SW_4 turns on.

Step 8. SW_4 causes digital circuit, D_1, to deactivate the reverse driver, and the motor stops.

Step 9. The user pulls out the cassette and SW_1 is deactivated, but nothing happens, as would be expected.

The sequence above tells us several things about the digital circuit DC_1.

1. It has several unique internal routes for conveying the information from the four switches to the two motor drivers.
2. After a switch activates DC_1, the switch loses control of the operation. DC_1 takes over.
3. The sequence of tasks is determined by the switches and the mechanical operations.

Let us look at each of these digital-circuit functions. We check the routing circuitry in DC_1 first. The basic routing circuits are called *logic gates.* They are enclosed in *logic chips.*

15-2 DIGITAL LOGIC CHIPS

The following is a short rundown of the basic digital logic chips. The simpler ones come in packages with four or more gates to a chip. They do the basic routing operations. These chip circuits can be hooked together in accordance with engineering logic equations, hence the name *logic circuits.* The most popular types are called "transistor-transistor logic" (TTL) chips. To keep things short, we focus mainly on TTL chips in this chapter.

As troubleshooters, we concentrate on the inputs and outputs of the gates. Of course, we cannot overlook their supply voltages. However, to keep things short, we skip the supply voltages and assume they will not be forgotten. Anyway, they are not always shown on logic-chip diagrams.

15-2-1 The Basic Logic Chips

The TTL chips work off 5-V dc supply lines. They input and output hi and lo voltages. The hi voltage on a line is greater than 2.4 V, and the lo is below 0.8 V.

The AND gate in Fig. 15-2(a) routes three signals through one output. You will find several gates encased in one chip. All inputs to the AND gate must go hi, before the output goes hi. In addition to the three-input gate shown here, you'll find some with more inputs and some with just two. Just one lo input is enough to output a lo.

The OR gate in Fig. 15-2(b) routes any of several signals through one output (in this case, three). Any hi input will cause the output to go hi. A lo has no effect unless it replaces a hi to produce all lo's at the inputs.

The NAND gate in Fig. 15-2(c) does the opposite of the AND gate; you must have all his at the input to output a lo. Any lo input will produce a hi output. The round bubble at the output indicates that it inverts the AND gate. One way to remember what the NAND gate does is to recall what the AND gate does, then invert it.

The NOR gate in Fig. 15-2(d) does the opposite of the OR gate. Any hi input will output a lo. It takes all los to output a hi.

The buffer in Fig. 15-2(e) protects the input and output from each other. It also provides more drive current. The inverter in Fig. 15-2(f) changes a hi input to a lo output, and a lo to a hi. It is often called the NOT gate. It is shaped like a buffer with a bubble-like nose for the output.

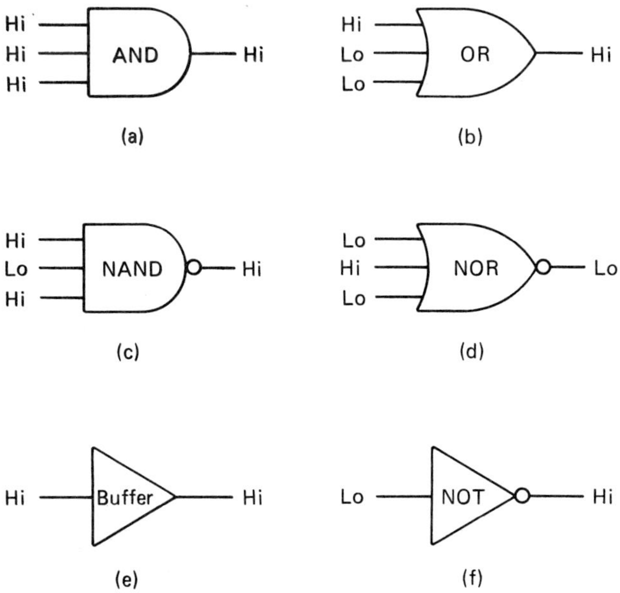

Figure 15-2 Basic logic gates.

Logic gates are basically used to route signals. They get the desired hi or lo voltages to the right places and keep the wrong ones out. We will find out how the OR and NOT gates are used when we return to cassette loader in Section 15-3 (Fig. 15-8).

15-2-2 The Decoder Chip

To know what a decoder is, you have to know what is meant by ''code.'' Let's take an example in which the code voltages are sent over two parallel lines. If, for instance, one line has a hi on it and the other has a lo (hi–lo), the signals could activate a circuit that turns on a green traffic light. If both lines have hi's (hi–hi) the red light turns on. There are four possible code combinations: lo–lo, lo–hi, hi–lo, and hi–hi. Each of these codes activates one of the four circuits. Lo–lo could ''mean'' yellow, and lo–hi could cause a flashing red light.

Instead of writing his and los, we can use the symbols 1 and 0. The 1 stands for hi. The 0 stands for lo. Figure 15-3(a) shows that in our cassette-loader example 00 stands for ''forward,'' 01 stands for ''reverse,'' 10 stands for ''slow,'' and 11 means ''stop.'' This is called a 2-bit code. Each digit (1 or 0) is called a *bit*.

In Fig. 15-3(b), the decoder has two input lines, A_0 and A_1. A_0 inputs the first bit, and A_1 inputs the second bit. There are four outputs. They are named here respectively, from top to bottom, ''forward,'' ''reverse,'' ''slow,'' and ''stop.''

When the two inputs, A_0 and A_1, receive the bits 00, the output line ''start'' is activated. If 01 comes in, ''reverse'' is activated. If 10 comes in, ''slow'' is acti-

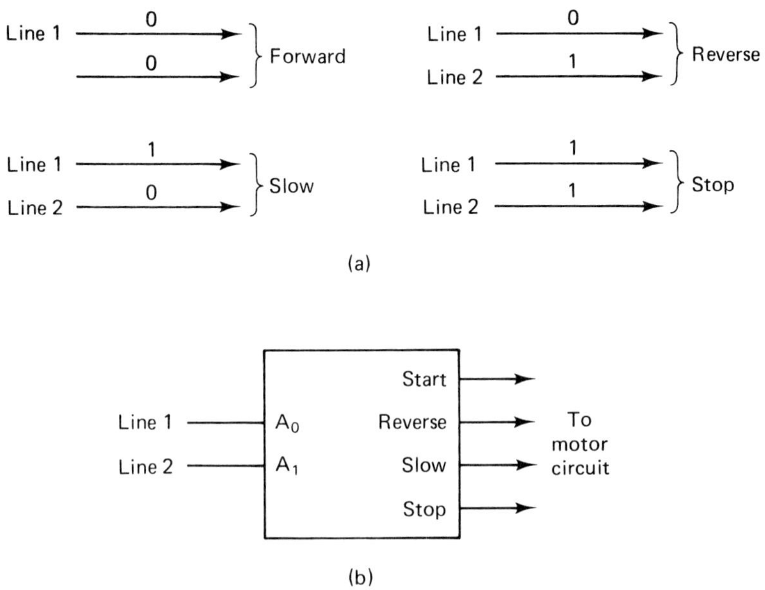

Figure 15-3 (a) Example of a two-bit code with four words. (b)Decoder that activates one of four lines, depending on the code that is received on the two input lines.

vated. If 11 comes in, "stop" is activated. The following table summarizes the action.

Input Code	Decoded Output Line
00	1—Forward
01	2—Reverse
10	3—Slow
11	4—Stop

This chip is called a decoder because the selected line gives user "meaning" to the input code. That is, if the code selects the line that starts the motor, the user decodes it as "start."

From two lines we could select four output lines. Further study will reveal that if we had four inputs, we would have 16 outputs. We could assign 16 meanings or "words" to the outputs. Actually, a "meaning" is the name of the circuit or circuit function that is selected.

Note: The use of 1 and 0 is very convenient, because they permit the use of *binary numbers.* You should study this number system if you are going to do a lot of work in digital circuits. Just to give you an idea of what it is like: 00 = 0, 01 = 1, 10 = 2, 11 = 3, 100 = 4, and 101 = 5. Going beyond this involves a knowledge of place values for each digit. In principle, you already know something about place values. In the number 555, the first-place value on the left is 100, to give you 500. The second is 10 for 50, and the last is 1, which gives you 5. The binary system has its own place values. You should study it if you are going to do more work in the digital and computer field.

15-2-3 Input–Output Tables (Truth Tables)

Integrated-circuit manuals have charts that list all possible combinations of input and output voltages. They are called *truth tables* because designers spent a lot of time with logic equations. Actually, they are input and output tables.

Figure 15-4(a) shows the truth table for an AND gate. H stands for high; L stands for low. Sometimes, the highs are listed as 1's and lows as 0's [see Fig. 15-4(b)].

Figure 15-4(c) shows the truth table for a 1-of-4 decoder. The lines over the letters indicate that they are normally hi and drop to lo when selected.

15-2-4 The Multiplexer

The multiplexer selects signals from many lines and transmits them over one line. Figure 15-5 shows a 4-to-1 multiplexer. There are four inputs, *A, B, C,* and *D*. The output is *Z*. If S_1 and S_2 have 0 and 0, or if S_1–S_2 is 00, the output is taken from *A*. If S_1–S_2 is 01, the output is taken from *B*. If S_1–S_2 is 10, the output is taken from *C*. If S_1–S_2 is 11, the output is taken from *D*.

In 1	In 2	Out
L	L	L
L	H	L
H	L	L
H	H	H

Truth table
AND gate

(a)

In 1	In 2	Out
0	0	0
0	1	0
1	0	0
1	1	1

Truth table
AND gate

(b)

In 1	In 2	$\overline{O_1}$	$\overline{O_2}$	$\overline{O_3}$	$\overline{O_4}$
L	L	L	H	H	H
L	H	H	L	H	H
H	L	H	H	L	H
H	H	H	H	H	L

(c)

Figure 15-4 Truth tables. (a) and (b) are both for the AND gate. (c) Decoder truth table.

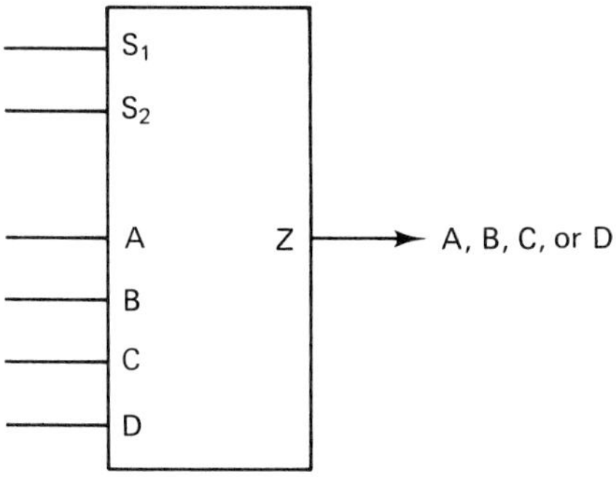

Figure 15-5 The multiplexer places one of four inputs on the output line, depending on the code it receives on S_1 and S_2.

We can say that the S_1–S_2 bits "decide" when and from where Z gets its signal. It also decides when Z gets a change of signal. Timing is of utmost importance in many digital circuits. There may be more than one receiver on line Z. They may be timed (synchronized) with S_1–S_2, so that they can be activated to receive a signal only at certain times. The tri-state buffer discussed next is one way of activating a receiver.

15-2-5 The Tri-state Buffer

The tri-state buffer chip in Fig. 15-6 transfers voltage levels from input to output only when enabled. When the chip is not enabled, the input voltage is cut off from the output line. The input-to-output path is open. This allows another circuit to use the line. Thus many circuits with tri-state outputs can share the same line if they can take turns being enabled. The tri-state buffer is so called because it has three output states: hi, lo, and open (or "floating").

The third state needs elaboration. An AND gate, for instance, has either a hi or lo output on the line at all times. It cannot share the same output line with other outputs. If it did, you could have conflicting voltages on the line. The tri-state buffer keeps its input voltage isolated from the output line until it is needed. It "dumps" the voltage on the line only when an enable voltage is applied to its *E* pin.

Most of the big chips have built-in tri-state buffers. It is very important to check the voltages on their enable pins. They tell you which output pin is connected to its input and which is not. This is one of the key checks in troubleshooting digital circuits.

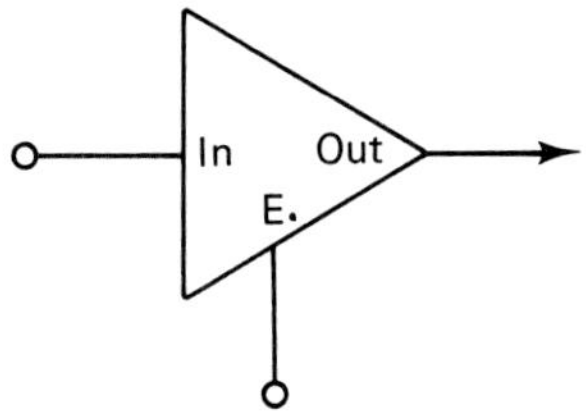

Figure 15-6 The tri-state buffer outputs the input voltage only when activated by a voltage on the enable pin. When not enabled, the output circuit is completely isolated from the input circuit. The output is freed for voltages put on the line by other chips.

A tri-state buffer can also be used to protect a receiver chip from unwanted signals on the input line. When the desired signal is on the line, the tri-state buffer can be enabled to put the receiver on the input line.

> TIP 15-1
>
> When checking a digital chip, the enable pin will tell you whether the output-pin voltage is coming from that chip or another chip on the same line.

> TIP 15-2
>
> When checking a pin voltage on a digital chip without enable pins, you can be sure the output is its own and not some other chip on the line—unless something is wrong.

Obviously, the enable pin needs to be controlled by some method of sequencing or timing so that the buffer will output at the right time. We take this up soon.

15-2-6 The Type *D* Flip-Flop

Figure 15-7 shows a type *D* flip-flop. It can pass control from one circuit to another. The input terminal is the *D* pin. The letter *D* is for *data*. Digital signal voltages are

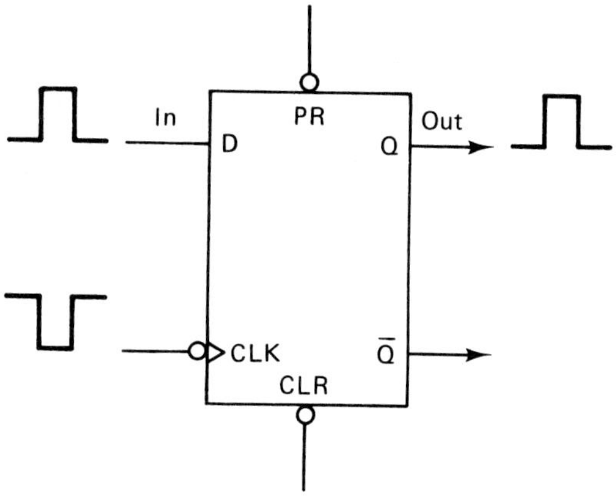

Figure 15-7 This type *D* flip-flop chip outputs the input voltage only on the down-going edge of the clock pulse. At all other times, the input is isolated from the output. Unlike the tri-state output, the output is not free to be changed by other chips on the output line.

called data. The two outputs are the *Q* pin ,which follows the action of the *D* pin when "clocked in" as we shall see, and the $\overline{Q}$ pin, which has the opposite voltage or logic level of the *Q* pin. When the *Q* pin has lo, the $\overline{Q}$ pin has hi, and the opposite happens when *Q* is hi.

This circuit can input one data pulse and shut off the next pulse. It does this with the help of a pulse at the clock pin. The pulse has upgoing and downgoing "edges." The leading edge of this downgoing clock pulse "latches" in the data input voltage. We call this *edge triggering*. The small triangle symbol at the clock input indicates an edge-triggered input. The bubble indicates that it is activated by the downgoing edge of the pulse. The clock pulse is also called a *strobe* pulse.

The flip-flop is the most important electronic circuit of the modern era; it stores a voltage. When you latch a voltage from *D* to *Q*, it will remain on *Q* as long as you please. No changes on *D* can change *Q* until another down-going clock pulse is received. This is memory.

We repeat: On this chip, only a downgoing pulse on the CLK pin can get a voltage through from *D* to *Q*. Nothing can get through at any other time.

In addition to the clock, you will find two more control inputs to the *D* flip-flop: They are called the clear (CLR) and preset (PR) pins. These pins have priority over the *D* and clock pins. When the CLR pin is activated, a lo is placed on the *Q*-pin output. The PR pin puts a hi on the *Q*-pin output when it is activated.

TIP 15-3

When troubleshooting digital chips, check that the outputs are correct for the inputs observed. This means checking the enable, clock, and preset and clear pins, as well as the input and output pins.

15-2-7 Registers

Although an investigation of the various types and uses of flip-flops is beyond our purposes, we should be aware of flip-flops that are grouped together as registers. They perform a variety of very important functions. Most often, they are used as memory locations, that is, they provide a place for storing voltages. They store whole *words* of information consisting of many hi and lo voltages called *bits*. Each bit is stored in one flip-flop. Eight bits makes one *byte*. Some memory locations can hold one byte of information per register. Many store more than one byte.

A shift register is a group of flip-flops that can input parallel code and shift it out serially, or vice versa. Parallel information is faster, because it works with many signals simultaneously on parallel lines. Serial information is often more convenient and less expensive, because it requires only one line. This is important for long-distance transmission.

A counter is a series of flip-flops that will count pulses. It will output a signal or control pulse after a certain count. It can activate circuits that must join the action at periodic intervals. This is good for sequencing operations. It is also used to obtain slower pulse rates.

15-2-8 CMOS Chips

CMOS chips find wide usage because of their low power dissipation. They can be used for lightweight battery-powered equipment. While TTL chips cannot be operated very much above 5 V, CMOS chips can use voltages from 3 to 18 V. Because of this range, their hi and lo voltages must be specified as percentages of the supply voltage, V_{CC}. Hi is any voltage above 70% of V_{CC}, and lo is any voltage below 30% of V_{CC}.

Preventing Damage by Static Electricity. Several precautions against static electricity must be taken when working on digital circuits—especially with CMOS chips. They are highly subsceptible to damage by low voltages. You must discharge the static electricity that you may be carrying on your body before you work on the circuit. You do this by touching a grounded chassis or the metallic part of the cover of an electrical outlet. When directly handling CMOS chips or cards containing them, you can wear a conductive wrist band attached to a long wire connected to ground. Make sure that the power is off when you are inserting or soldering CMOS chips. Your chip insertion tools should be similarly grounded. Your soldering iron should be the three-wire type that has the extra ground lead. Do not touch or keep static-producing synthetic materials such as Styrofoam or nylon near the work area. Styrofoam is great for generating static electricity.

15-3 A DIGITAL CONTROL CIRCUIT

The VCR example in this section should be studied carefully. The same circuit can be used in many industrial operations. All you need are the big heavy-duty electrical and mechanical components associated with them. We have here an example of the second of three digital circuit functions mentioned in Section 15-1-3: the passing of control from one circuit to another.

VCR control circuits are actually enclosed in very-large-scale (VLS) IC chips, but let us look at one that is made of small chips. These chips are still found here and there on big-chip boards. Let's look at a key action in the cassette loading operation (see Fig. 15-8).

If you recall, in Fig. 15-1 the forward switch, SW_1, remains on, even when the motor is stopped or reversed. Let's see how this conflict is resolved with respect to the

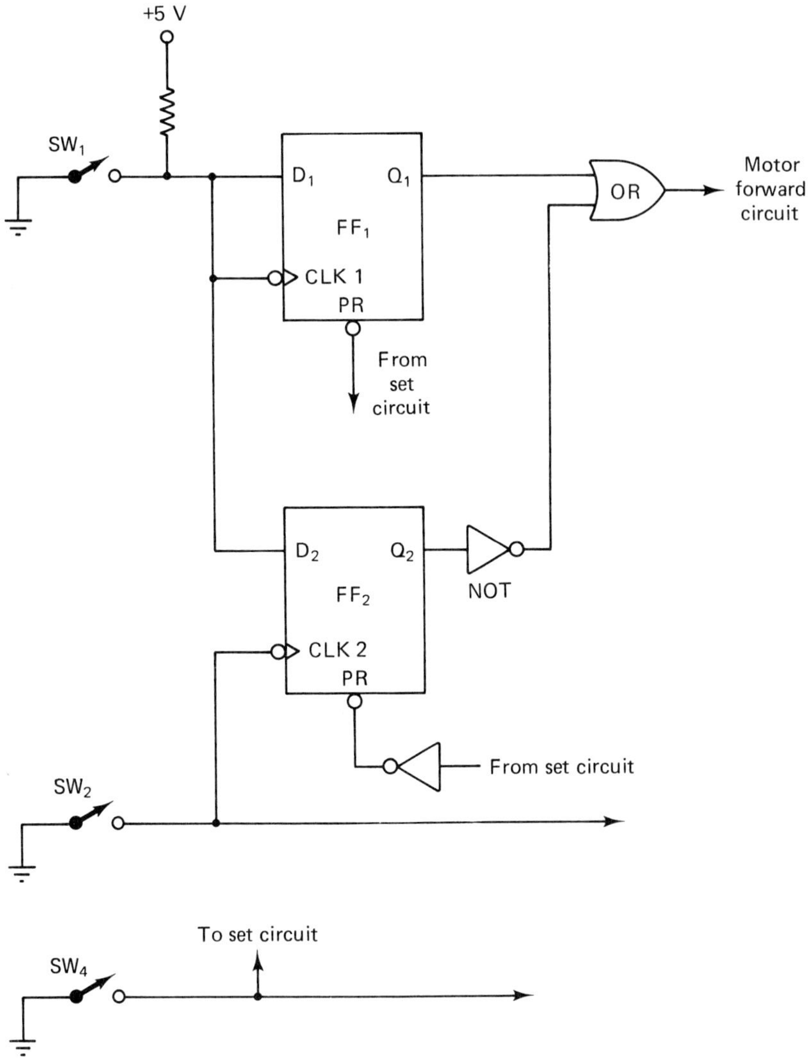

Figure 15-8 Digital control circuit for the VCR motor-drive amplifier. The flip-flops give control to the switches only at the instance they switch on (go lo). At all other times before or after, the switches have no control.

action of stop switch SW_2. The illustration shows two flip-flops, an OR gate, and two inverters.

Let us follow the action of this circuit. At turn-on, the system puts los on the preset pins of the flip-flops. This presets Q_1 and Q_2 to hi's. At the same time, both D_1 and D_2 are pulled hi by R_1. Nothing else happens in this circuitry until the clock CLK_1 goes down—and only at the instant its voltage goes down.

Since both Q_1 and Q_2 have been preset to hi, the OR gate has a hi from FF_1 and a lo from FF_2 through a NOT gate. This places a hi on the driver, thus keeping the motor stopped. (Remember, the driver circuit needs a lo to start the motor.)

> Q_1 is now hi, and Q_2 is also hi. OR gate input is hi–lo, and its output is hi. The motor is stopped.

When a cassette is loaded, SW_1 closes and D_1 goes lo. At the same time, the clock goes lo and transfers the lo at D_1 to Q_1. The OR gate now has two lo's and outputs a lo. This activates the forward-motor driver circuit and the cassette carriage travels down. Note that SW_1 no longer has control. Only CLK_1 can give it control, and only at the instant the SW_1 voltage is going down. Similarly, it cannot do anything to FF_2 and Q_2, because SW_2 does nothing to CLK_2.

> Q_1 is now lo, and Q_2 is still hi. The OR gate input is lo–lo, so its output is lo. The motor turns forward.

When the cassette carriage finally reaches bottom, SW_2 clocks CLK_2 with a downgoing pulse. This passes the lo on D_2 to Q_2. (SW_1 is still on, but not in control.) The inverter takes the lo from Q_2 and puts a hi on the OR gate. Since any hi on an OR gate will output a hi, the output of the OR gate changes from lo to hi. This stops the forward rotation of the motor.

> Q_1 is still lo, and Q_2 is lo. The OR gate input is lo–hi, and its output is hi. The motor stops.

Since the reverse circuit is similar to the forward circuit, we can skip over it to keep things short. The reverse circuit is activated when the end-of-tape is reached. The cassette carriage will then move away from SW_2 and output a hi. Since CLK_2 only responds to a low-going pulse, the hi has no effect on FF_2 and the rest of the forward drive circuit, which is now inactivated. Thus we have no conflict while the motor is reversed, and the cassette and carriage freely travel to the out position.

When it reaches the out position, the SW_4 switch stops the reverse motor. It also sends a lo to a circuit that outputs a single pulse that activates the PRESET pins of FF_1 and FF_2, so they are prepared for the next cassette insertion.

We have now seen one example of what a digital control circuit can do. Even though our example was fairly complex, there are circuits with even more complex actions. In big 42-pin chips, there is no way we can figure out what is going on inside them. We must rely heavily on the manufacturer's reference manuals. They not only give you the names of the output and input pins, they also give you the timing.

15-3-1 Checking the Digital Control Circuit

The first and most important step in troubleshooting electromechanical operations such as the VCR is to determine whether the trouble is in the electronic circuits or the mechanical system. Mechanical troubles can make the electronic circuit voltages look bad. Similarly, electronic troubles can make the mechanical system look bad.

For example, there is one important circuit that senses a drag on the motor. As soon as it senses an overload in the mechanical system, it sends a signal that brings the entire system to an orderly stop. When this happens, a good electronic system will appear to be bad, because it will not be able to start the motor.

For VCRs it is generally better to check the mechanical system first. Nine times out of 10 a good cleaning and change of belts will do the job. If not, you can isolate certain movements by hand and check their mechanical interactions with other parts. Only after you have made sure that the mechanics are okay can you go into the electronics.

In big industrial systems it may be better to check out the electronics first. The big relays (electromagnetic switches) that operate the motors can be checked independently. The motors can be also be run independently of the mechanical and electronic system.

You can start your electronic analysis of the VCR cassette loading system by moving the switches by hand and checking their resistances or voltages. Let's check out the VCR digital chip circuit in Fig. 15-9.

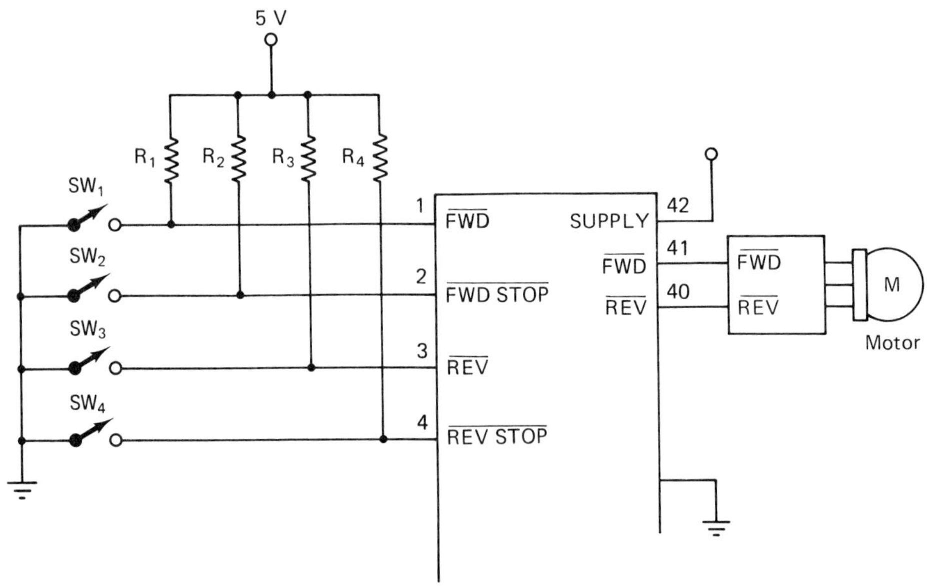

Figure 15-9 Troubleshooting this VCR chip involves a mechanical check of the switches, motor, gears, bearings, levers, and slides, as well as the electrical checks on the chip pins, amplifier, and motor.

Check 1. Check the motor. The top side should measure 12 V when you push the cassette in. Do not forget the lo side; it should measure 0 V. If both sides have 12 V, the motor will not turn.

Do not forget the ground connections; they could be open. The driver circuit was discussed in Chapter 10.

Check 2. Check the switch SW_1 circuit. Measure the voltage on pin 1. It should drop to zero when the cassette is pushed in. If it does not, check the switch. Push it with a prod while you measure the resistance of pin 1 to ground.

Check 3. Next check the output, pin 41. Its voltage should go down when the cassette is inserted. (Check the supply voltage.)

The checks above are the most obvious. Before you do anything else, study the service manual to see what other circuits might have anything to do with the load-motor operation. If the tape-threading and play functions are okay, the probability that the SW_1 and SW_2 circuits are bad is very high. Check the other actions. Check the eject acion. If eject action is okay, the probability that the SW_1 front-loading circuits are bad is even higher. Check the other pin voltages.

15-4 LOCATING BAD CHIPS

Working Your Way In from the Outside. The first steps in digital troubleshooting are similar to troubleshooting linear-IC circuits. You work your way in from the inputs, outputs, and supply terminals. In the case of the VCR, you checked the output voltages to the motor and worked back. Then you checked the input switches. It was understood, of course, that you checked the supply voltage as soon as you found a bad chip voltage.

In general, after tracing the lines to a suspected chip, the next task is to check it out. A chip is suspected if its output is not doing what it is supposed to be doing for a given set of input and control pin voltages. For instance, if a pin is not pulsing when it should, you should investigate it. Check it either against the schematic or against its inputs. The output load circuits should also be checked.

15-4-1 The Logic Probe and the Pulser

The digital *logic probe* is a handy tool for detecting hi and lo voltages (see Fig. 15-10). It lights up when you have a voltage. The lights indicate whether you have a hi, lo, pulse voltage, or an open (floating) line. Many schematics list these voltages on selected chip pins. Technicians themselves record these voltages on chip-location diagrams. They record these voltages when the system is first turned on and during other key operations.

Ordinary dc voltmeters interpret pulses as in-between voltages, neither hi nor lo. These are inconclusive; you cannot tell whether the line is pulsing or floating. On the other hand, voltmeters can spot borderline hi or lo voltages that will not show up

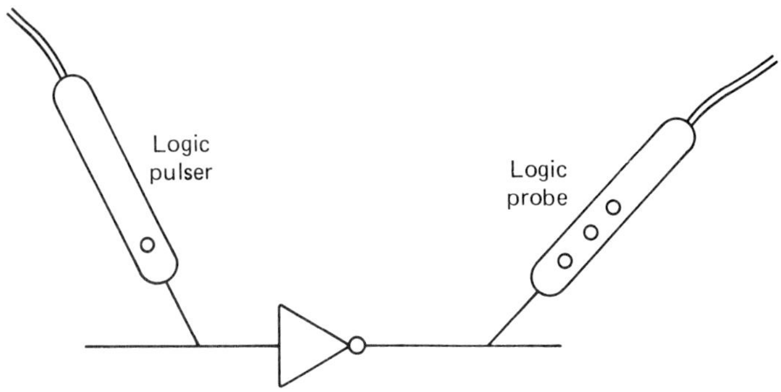

Figure 15-10 Logic pulser and logic probe.

on a logic probe. A logic probe may indicate a hi with a voltage as low as 2.4 V. Voltages in this range are acceptable, but they are prone to intermittent troubles. The scope is the best test instrument for this and other reasons.

The logic probe cannot give you the frequency or the timing of a pulse. The scope can do both. A dual-trace scope can show the timing between two pins. The scope gives the best results, but it is not as portable and handy as the logic probe.

The logic probe often has provisions for picking up a short-duration pulse. You can switch it to the memory position, and it will pick up a pulse and store it. The indicator light will stay lit after the pulse is gone.

The *logic pulser* shown in Fig. 15-10 generates pulses that can be picked up by the logic probe. The pulser is capable of overriding the regular signals on the line. It will generate a voltage that is the opposite of the voltage on the line. If the line is hi, the pulser will generate a low. Of course, it does not have enough power to overcome a line shorted to supply or ground, but it does have the ability to survive dead shorts to ground. This enables you to find a short to ground.

TIP 15-4

A logic pulser and logic probe can be used to check the continuity of a line. If the line is good, the probe will respond to the pulser.

They can also be used to check a chip input and output path. The logic probe on the output pin should respond to the pulses on the input pin, provided that the chip is enabled.

They can be used to find a line or pin that is shorted to ground. The logic probe will not pick up a signal from the pulser on a line that is shorted to ground.

Similarly, if a line that is stuck hi cannot be pulled down by a pulser, it is shorted to the supply line.

15-4-2 Checking Pin Voltages

Using the logic probe makes it easier to check out the pin voltages on suspected chips. The voltmeter or scope can also be used for checking borderline voltages. The following tips are helpful. We are dealing with TTL chips.

> TIP 15-5
>
> Every line and pin must have either a hi, lo, floating, or pulsing voltage. Hi is supposed to equal 2.4 to about 5 V. Lo should be from 0 to 0.8 V, without noise (for TTL only).

> TIP 15-6
>
> Opens are much more likely to appear than shorts. An open TTL output will allow a following TTL input to put about 1.5 V (with a digital meter) on the line if enabled. (When enabled, the other outputs are disabled.)

> TIP 15-7
>
> An open input pin on a noninverting TTL gate will produce a hi at its corresponding output. For an inverting TTL gate it will produce a lo.

> TIP 15-8
>
> High-side internal shorts at the input pin are not likely.

15-4-3 The Simplest Case: Two Pins on One Line

Let us assume that we have narrowed our search to a line between two chips [see Fig. 15-11(a)]. The line between the two chips has a bad voltage on it. There are three

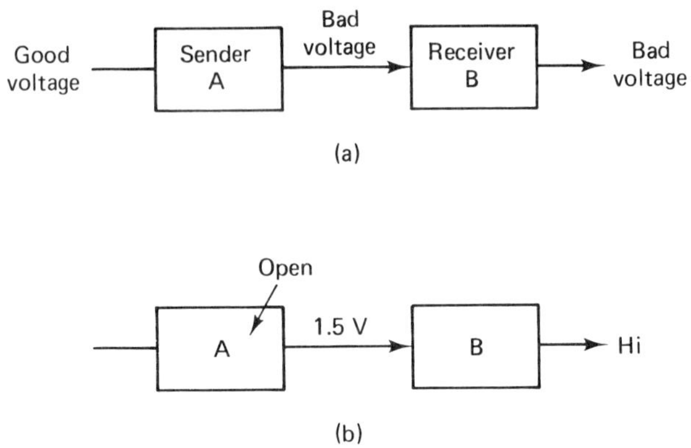

Figure 15-11 (a) In the simplest case a good input is followed by a bad output. Either chip B or its load is bad. (b) A floating voltage indicates an open sender or a chip that is not enabled.

types of bad voltages that we should be able to handle at this point: floating, stuck hi, and stuck lo. Floating occurs when the output of *A* has an internal open, and the line settles at about 1.5 V. If *B* is a TTL noninverting chip, as we are assuming, its output will be hi [see Fig. 15-11(b)]. Floating is normal if the sender chip is not enabled.

> TIP 15-9
>
> If a solitary line is floating between two chips and if the sender is enabled, the sender is most likely to be bad. The line voltage is definitely bad. The use of a pulser at the relevant input and a probe on the line should isolate the fault to one chip or the other. Removing the line connection and making resistance checks at either end will also isolate the fault. (See Tip 15-20 for checking resistances.)
>
> If the sender is not enabled and the line is floating, the chips are probably good.

> TIP 15-10
>
> A line can be removed by desoldering a pin at one end. Use a desoldering wick or a solder sucker to remove all the solder. The pin hole may be large enough to leave the pin free from contact. Both the pin and the line can now be tested independently. (Beware of multilayer boards. The pin may be soldered to several layers of foil.)

TIP 15-11

If a solitary line between two chips is stuck hi or lo, at least one of the two chips is bad. A logic probe and pulser may isolate the fault. Removing the line and making resistance checks on either end may isolate the fault to one of the two chips. (See Tip 15-20 for checking resistances.) The receiver chip has an internal diode connected from input pin to ground. A hi–lo resistance reading will check this out.

TIP 15-12

It is possible to ground a floating line to check a chip. Ground the line and check the output.

15-4-4 Checking Nonpulsing Voltages on Shared Lines

Most pins are connected to more than one other pin. In Fig. 15-12 we see several outputs and inputs sharing the same line. Such lines are typical. The following should help.

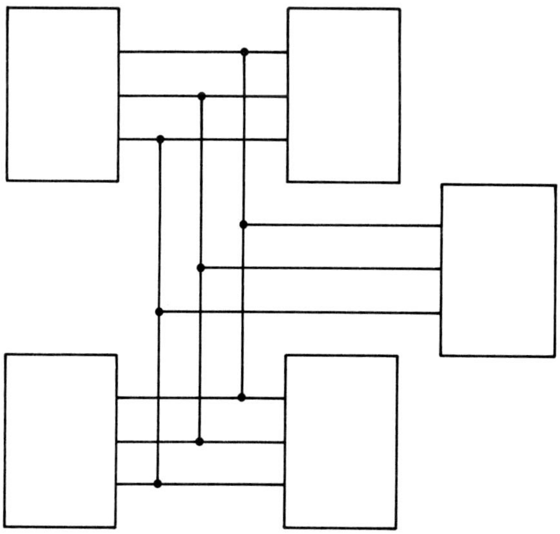

Figure 15-12 All the voltages and resistances to ground on a shared line should be equal.

TIP 15-13

All pins on shared lines should have the same voltage. A pin with a different voltage has an open line or connection.

TIP 15-14

If more than one chip is enabled on shared nonpulsing lines, pulse the suspected enable pin. If the output responds correctly, the enable pin or circuit is bad.

TIP 15-15

All pins are inputs if they are connected to the output pin of a chip that has no enable, select, or address pins.

TIP 15-16

With the TTL series chips, an internal open at the output terminal of an enabled chip will allow the lines to float above the lo threshhold, about 1.5 V or more. (All other output pins are disabled.)

TIP 15-17

A stuck hi at an output pin on a shared line indicates a probable defect in the sender or a short from line to the 5-V supply. A defect in the receiver is less likely.

TIP 15-18

If two adjacent lines have the same nonpulsing voltage on them, they could be shorted together. Check for a solder short.

15-4-5 Checking Pulses on Shared Lines

As we learned, several different pulsing signals can share the same lines. They are isolated from each other only by timing the enable pins or clock pins. The following tips should help you find the bad pin that is connected to several shared lines that are pulsing.

TIP 15-19

If shared IC circuit lines are pulsing, all pins connected to them should be pulsing; otherwise, the nonpulsing pin has an open line or connection.

TIP 15-20

If the shared IC circuit lines are not pulsing when they should be pulsing, look for the input or enable pin that is not pulsing.

TIP 15-21

With a nonpulsing input pin on a line buffer, the output pulse must be in time with the enable pulse. Otherwise, the chip is probably bad.

TIP 15-22

Generally speaking, two output pins on the same line should not have the same timing on their enable pins; if they do, look for the source of the conflict with a dual-trace scope.

TIP 15-23

Conflicting pulses on the same line can be viewed with a scope. You will see different-level pulses, where the hi and lo pulses conflict.

15-4-6 Resistance Checks

Aside from the usual difficulties encountered with resistance measurements, such as electolytic charging and warm-up problems, we have many more parallel lines in digital circuits. Despite this, resistance checks are often helpful for backing up voltage checks. They may not tell everything, but if they indicate a nearly dead short, it is definitely a short.

TIP 15-24

A shorted IC is more likely to have more than 1-Ω resistance. Shorted lines are more likely to have less than 1-Ω resistance between them.

Good lines should have high resistances to ground, at least 1 MΩ when measured with a digital ohmmeter in the low-voltage position. Take the higher of two reversed readings if you are using an analog meter.

15-4-7 Finding Shorted Chips on Shared Lines

If any of the tips above lead you to believe that a pin is shorted, remove the line connected to the suspected output pin and then check the voltages. Sometimes there are lines connected to the same pin on both sides of the board. Remove both sides if you think it is worthwhile to do so. You can use a pulser and logic probe at the input and output. If the sender chip is okay, one of the receiver chips is bad. Another evidence of a short is given in the following tip.

> **TIP 15-25**
>
> A shorted chip will often run warmer than normal.

15-4-8 Instruments for Finding Shorts

There are instruments for finding shorts without disconnecting any lines. The current tracer is capable of picking up the direction of current flow on a line. Since current always flows toward the short, it can be used to find the short.

The microvoltmeter is capable of picking up minute voltage differences on a line. The shorted side of the line always has the lower voltage. Read the voltage drop across two points on a line. The short is toward the common lead if the voltage is positive. The short is toward the red lead if the voltage is negative.

The short finder is an instrument that is specifically designed to find a short. It outputs an audible tone that gets higher as the probe tip approaches a short.

15-5 MICROPROCESSORS

15-5-1 A Basic Microprocessor System

If you recall, in Section 15-1-3 we mentioned three digital circuit functions in the VCR cassette-loading operation. It routed information, transferred control, and performed a sequence of tasks that were determined by the switches and the mechanical "circuit." A microprocessor circuit, as shown in Fig. 15-13, does all these things and more.

Besides sequential actions, the microprocessor controls many simultaneous actions. The bus path between the various sections consists of many lines. Many of these lines control simultaneous operations. The microprocessor sends out timing or *clock* pulses on the bus lines to make sure that all operations are working together without conflicts.

More important, the microprocessor sends out *coded signals* to initiate and control various actions. We saw a very simple example of code when we looked at the decoder in Fig. 15-3. It could select four different code "instructions" for the motor. Many of the big chips that work with the microprocessor have decoder circuits built

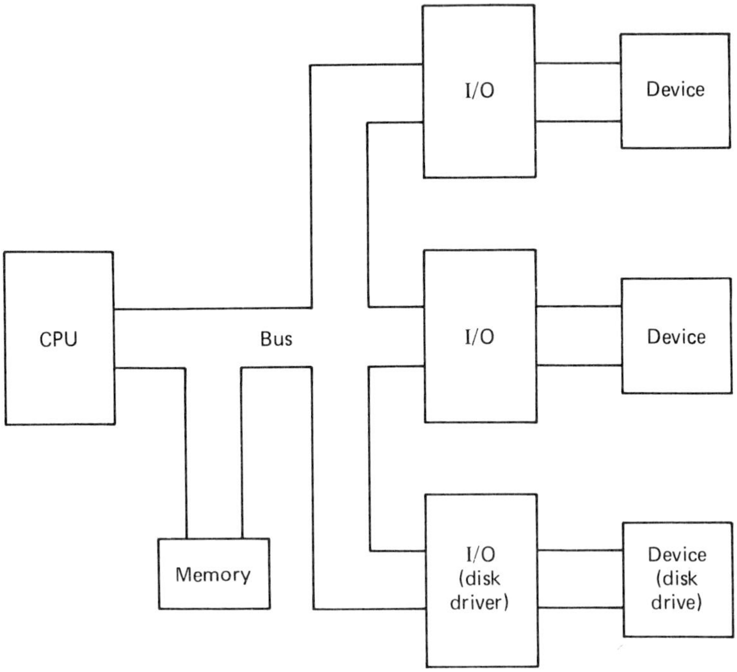

Figure 15-13 Microprocessor system.

into them. Through these internal decoders, the microprocessor controls the activities of the chips. It sends code instructions through *data lines* in the bus.

Most important, the microprocessor is able to access coded instructions stored in the memory section or in the disk drive, as shown in Fig. 15-13. *Memory* in ordinary digital circuits is in the wiring. Every time you turn the circuit on, the wiring will "remember" to do the same things. With these circuits, if you want to sequence and distribute the operations differently, you will have to wire up the circuits differently. These circuits are said to be *hard wired.* The VCR digital control circuit we studied was hard wired.

With microprocessors, if you want a system to work differently, you simply change the sequence of code in memory or the disk. You can have a sequence that starts the motor, stops it, and then reverses it. Later, you might want to change this sequence; you can do so without rewiring the circuits. This is called *programming.*

To the designer, the microprocessor allows a minimum of parts changes when devising new systems. To the troubleshooter, it means that virtually the same circuits will be found in different systems.

Figure 15-14 shows the basic microprocessor lines in the bus paths in Fig. 15-13. The *address lines* carry the address codes used primarily for accessing the instructions that are stored in addressed locations in the memory section. The *data lines* primarily carry instruction codes from memory to control all the actions within and

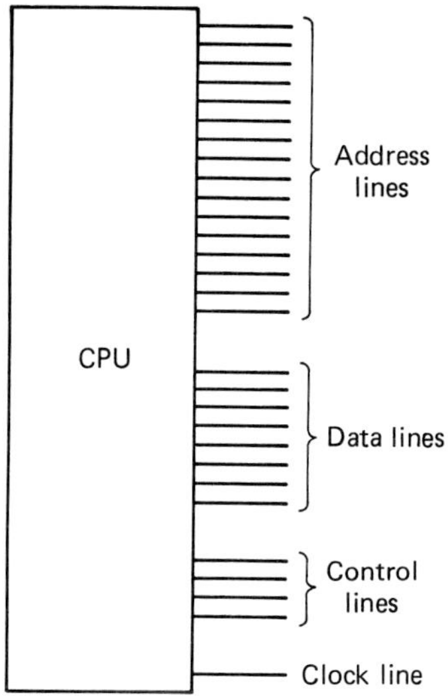

Figure 15-14 Microprocessor and its four types of lines: address lines, data lines, control lines, and clock line.

without the system. If you can recall, the decoder in Fig. 15-3 had two data input lines capable of carrying four different instructions. This microprocessor has eight data lines capable of carrying 255 instructions.

In addition to the lines above, we have the *control lines* that primarily initiate and terminate chip and circuit actions throughout the system. In many cases they activate the enable pins of tri-state buffers (Fig. 14-6) and the clock pins of flip-flops (Fig. 15-7) that are built into many of the chips. The *clock line* may also be connected to these pins. They primarily control the timing of the various actions throughout the system.

Memory chips are an important part of a microprocessor system. They hold the instructions and information necessary for the tasks that the system needs to perform. Figure 15-15 shows two basic types of memory chips: the read-only-memory (ROM) chip and the random-access-memory (RAM) chip. The microprocessor cannot "write" data into the ROM chip. It contains data put into it by the manufacturer. The RAM chip can be both "read" and "written" by the microprocessor. The RAM chips are necessary for storing information coming in from the outside. This enables the microprocessor to respond to the system it is working on. Both RAM and ROM chips have *data pins* (D_0–D_7) for passing instructions and information (data). They also have address pins (A_0–A_7) that input coded data addresses. The chip enable ($\overline{CE}$) or chip select ($\overline{CS}$) pin allows the micoprocessor control line to activate the chip with a lo voltage. In general, the bar over a pin designation indicates that a lo voltage will activate the chip. The output enable ($\overline{OE}$) pin lets the microprocessor read data from the chip with a lo control-line voltage. The RAM chip in Fig. 15-15(b) has a write-enable ($\overline{WE}$) pin that can also be controlled by the microprocessor.

Other chips that the microprocessor works with have similar pins, so it would be helpful if you can remember the pin functions above.

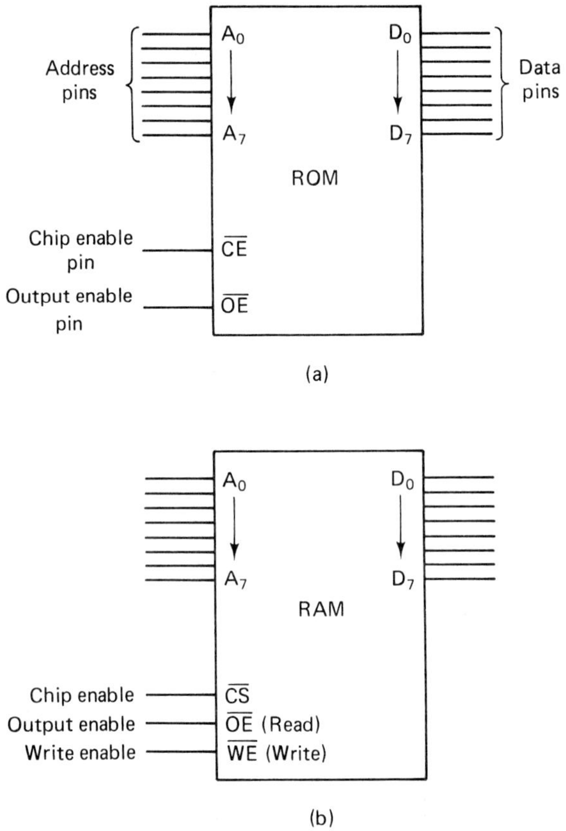

Figure 15-15 The two types of memory chips: (a) read-only memory (ROM) and (b) random-access memory (RAM) or read-and-write memory (RWM).

15-5-2 The Differences

Microprocessor circuits, especially in computer circuits, are difficult to troubleshoot for a number of reasons. First of all, there are only two basic voltage levels, so you will find very few series circuits with more than two parts that will help you isolate shorts and opens.

Another difference lies in the way the various operations are separated from each other and the way they are combined. It is easier to see how analog signals and circuits are separated and combined. You can see separate lines for them. You can see isolation resistors and bypass capacitors. The signals may be combined by simple connections, or by inputting them to the base and emitter terminals of a transistor, or by feeding them to an op-amp mixer circuit as shown in Fig. 14-5(b). You can see all this on a schematic.

In microprocessor circuits, however, the signals are separated by *timing*. You cannot immediately see how this timing works. Normally, no two signals are placed on the same lines at the same instant of time. They alternately share their time on the line. The tri-state buffer circuits in the chips force their output signals to wait until they are enabled. They also control the inputs to various chips, forcing them to wait for an enable pulse before they can input a signal. Flip-flop clock pins work in a similar way.

Another difficulty: Very often, all the lines in microprocessor circuits may be pulsing a million times a second, so that it appears as if they are all working simultaneously. You cannot separate them out by methods that are as simple as the ones we have learned up to now. This makes troubleshooting difficult.

Another big difference is that you cannot always tell what kind of signal is on a line just by looking at it with a scope. A knowledge of programming will enable you to rout signals through various chips and control the operations of certain chips. This would be very helpful in tracking down signals with a scope.

Still another difficulty: The program that runs the microprocessor may be defective. It may be hard to tell whether the defect is in the program or the hardware. Sometimes the defect or "bug" may not show up until years later when a program is run in a way that was not accounted for by the programmer.

A final major difference: The chips and peripheral systems are often connected by many parallel lines that all look alike. The various ways in which the peripheral systems and the microprocessor system control each other are not simple. Simply sending out synchronization signals is often not good enough. You will have to learn the standard ways in which the microprocessor system and the peripheral system work together over these lines.

Needless to say, you will have to gain a lot of experience before you can gain any expertise in troubleshooting digital circuits, especially computer circuits. However, there are some similarities that will help you get started.

15-5-3 The Similarities

Troubleshooting microprocessor and computer boards will be similar to troubleshooting analog circuits in some ways. You will be working your way in through the output and input devices. These devices are often controlled by specifically designated boards or cards. The majority of digital computer systems are repaired by changing cards. These cards bear the names of the devices they control, or "drive." Even the *motherboard,* the one containing the main microprocessor circuit, is exchanged. Also, a large percentage of troubles are corrected by working on the peripheral lines and connections.

The input and output voltages still hold the keys that get you in. You will be tracing the lines from the input–output chip through various chips to the microprocessor. These chips may have names that relate to the output devices they control. Bad chips may not always show you defective voltages; but if they do, it is worth looking into. Of course it's not easy, but let's see what you might be able to do.

Figure 15-16 shows the pins and lines that are commonly found in an input–output chip. Other large chips in the microprocessor system have similar pins. Some may have more pins, and others fewer.

1. The chip select ($\overline{CS}$) pin: The chip goes into action when this pin is activated.
2. Data bus pins (DB_0–DB_7) input or output data, directly or indirectly, to and from the microprocessor.

3. Port data pins (PA_0–PA_7, PB_0–PB_7, PC_0–PC_7, PD_0–PD_7) input or output data to or from the peripheral devices and systems.
5. The address pins (A_0–A_1) determine which port will be used. In this case there are only two; others may have more.
6. The read pin ($\overline{RD}$) enables the microprocessor system to read the peripheral data.
7. The write pin ($\overline{WR}$) enables the microprocessor system to "write" data to the peripherals.

There are other pins you might see. We cannot go into all of them. Just to mention a couple, you might see an *interrupt* pin that allows a peripheral device to stop the microprocessor momentarily from accepting data from the outside; or you might see a pin that inputs and outputs serial code from and to a peripheral.

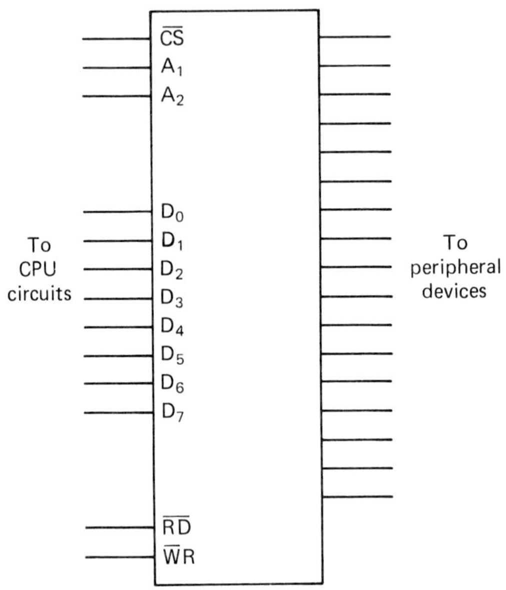

Figure 15-16 The lines going in and out of a typical input–output chip.

15-5-4 Troubleshooting the Signal Paths

After checking the symptoms and the schematic, with some experience, you will have an idea of where the trouble could be. You must now check the voltages in the lines between the input–output chip and the suspected circuit. You should also check the lines leading from the CPU through intermediate chips to the suspected circuit.

It also has output pins that could indicate the status of the chips and circuits connected to these pins.

TIP 15-26

Troubleshooting a board usually begins with the input–output chip. Not only is it first in line, it is also the most likely chip to go bad. Sometimes the only function of an input–output chip is to protect the main circuits.

TIP 15-27

The microprocessor is the least likely to go bad, since it is the farthest away from the output. However, it is always checked because it ultimately handles all signals on the board.

TIP 15-28

To find the bad chip in a microprocessor system, always check the peripheral-device symptoms. This may lead you to the circuit or chip responsible for that symptom. Peripheral devices are often controlled by chips bearing their names or functions.

TIP 15-29

A complete loss of action or defective output can be the result of line conflicts. Check all the enable pins and all chips that output to the lines involved. Check the timing circuits to all the enable pins involved.

TIP 15-30

Data and address lines that begin and end at the same type of pins should have the same resistance to ground when the power is turned off. An ohmmeter check may reveal a faulty line.

TIP 15-31

If two adjacent lines have the same nonpulsing voltage on them, they could be shorted together. Check for a solder short.

WHAT YOU LEARNED TO DO

You learned some simple tips for checking voltages in digital circuits. You learned that some basic troubleshooting rules and procedures can be used when working on these circuits. You learned how to:

1. Identify digital signals.
2. Trace digital signals through the basic logic gates.
3. Check the enable pins of various chips.
4. Check out a digitally controlled VCR cassette-loading circuit. This circuit can be used for various industrial operations.
5. Identify the names and functions of the various logic gates, line buffers, decoders, and multiplexers that helped the CPU work with the main chips.
6. Check voltages on the input, output, and enable pins and lines to isolate the defective pin.
7. See in a general way how the microprocessor works with memory and the peripherals.
8. Identify such control pin functions as chip select, chip enable, read enable, and write enable.

QUESTIONS

1. Draw some sample analog and digital waveforms.

2. What three tasks must a digital circuit perform, and under what conditions are they to be performed?
3. What chip prevented the VCR mechanical switches from interfering with each other?
4. Construct the truth tables for the basic logic gates.
5. Name the chip that selects one of many lines and activates it.
6. Name the chip that selects a voltage from one of many input lines and transmits it over a common output line.
7. Which chip will not transfer a voltage from input to output until it is enabled by a control pin voltage? (It frees the output line when it is not enabled.)
8. Which chip latches a voltage from input to output? (It always has an output voltage; the output line is never free.)
9. What is the difference between the VCR digital control circuit in this chapter and a microprocessor-controlled circuit?
10. Name four types of lines that are connected to the microprocessor.
11. What would lead you to suspect a chip when tracing signal paths?
12. Is a chip likely to be bad if its output is not pulsing when it should be? What else could possibly be bad?
13. What are the chances of a chip developing a dead short between the input pin and the supply pin?
14. How can you find a short from line to ground and from line to supply with a logic pulser?
15. If the enable pin is pulsing, what should the output show on a logic probe?
16. What would you suspect if an enabled output is floating?
17. If the input and output of a noninverting chip are high and the output is enabled, would you suspect the chip?

PRACTICAL TROUBLESHOOTING EXPERIENCE

Obtain a defective digital system or computer card that requires 5 V and an extender-cable kit that will enable you to check the card outside the system. You will need a separate dc supply source. Preferably, use a schematic for this job. You will also need specification sheets for the various chips on the board. The checks below should give you an idea of what is involved when checking voltages on a card or board.

1. With the card not plugged in, check the continuity of the lines between various pins.
2. With the card not plugged in, apply 5 V from a separate source and check the supply voltages to the various chips.
3. Disconnect the 5 V and check the resistances to ground on the data and address lines. Do you find any resistances to ground that are different from the others?

4. If the supply voltages and resistances check okay, plug the card into the extender slot. (*Remember:* The power must be off when changing cards.) Check the voltages on the data and address lines. How can you tell which are address and data lines? Note whether they are hi, lo, floating, or pulsing.
5. If a line has a voltage that is hi, lo, or pulsing, check its continuity with a logic probe or voltmeter. A floating line can also be checked in the same way.
6. Use a logic pulser to apply a pulse to some nonpulsing inputs and see if you can obtain the expected outputs. Note what the output is supposed to be by looking up the specifications for the chip.

16

Intermittents

EXPERIENCE REQUIRED

To understand this chapter, you must know how to find open and shorted parts, conductors, and connections. Knowing how to solder and inspect the work would be helpful. You should know how symptoms point to defective circuits. It would be helpful to know about transistor and IC troubleshooting.

16-1 A CASE HISTORY

Troubleshooting intermittent problems is the toughest job in electronics. It is easier to troubleshoot a system that is not working than it is to troubleshoot one that is working. Before going into the details of how to attack this problem, let us look at a typical case.

16-1-1 A Typical Case

What follows is typical. A VCR comes in for repair: Complaint; stuck cassette. It will not eject. The user heard something running, and then it quit. The cassette is still in the VCR. Easy job. Just clean it up and change the drive belt. You remove the cabinet and turn the cassette front-load pulley by hand to bring out the cassette and return it to the user. You promise to have the job done in a couple of days and the user happily departs. Later you clean the VCR and change the drive belt, making sure that your fingers are clean so that you do not get any oil or grease on the belt or pulleys. Everything works fine. You check it out, now and then, for two days. Just as you thought—an easy job.

Less than a week later, the same VCR is on your bench again. It is not working, and you are going over the mechanical system thoroughly. You turn the front-load pulley by hand and look for signs of drag. You listen for scraping and squeaking sounds. Nothing wrong. You disassemble the cassette holder to inspect the switches, and clean and bend them a little for better contact.

Nothing really wrong there, so your attention is now on the electronic system. You remount the cassette-carriage assembly, so you can work on the circuits. You do not put everything back on, though—only the cassette carriage, so you can see it going down. You insert a see-through cassette that was made by cutting out the top and bottom of an old cassette. With it, you can see the action below. You insert the modified see-through cassette and watch carefully as it goes in and down. Okay, so far. You push the eject button, and watch very carefully. It comes up quickly, smoothly, and it sounds good, too. You fixed it. You feel the satisfaction of doing a good job. Cleaning and bending the reverse-motor switch did the job, you think.

A month later it is on your bench again, and you are working on the reverse-motor amplifier (Fig. 10-8). The see-through cassette is all the way down and sitting on the reverse-motor switch SW_2. You check for 12 V on the reverse terminal of the loading motor. No voltage there. Good sign; you probably will not have to order a new motor, you think. Next, you check pin 6, the reverse-motor output of the big control chip. You have 0.2 V. Good, that is the voltage you need to activate the reverse-motor driver amplifier. Next, the voltmeter probe is moved to the base of transistor Q_5, the input amplifier. Just as soon as the probe touches the base the motor turns and the cassette holder comes up.

You try again. You want to confirm that it is a poor solder connection at the Q_5 base. This time you use a pointed plastic stick. Past experience tells you that perhaps the voltmeter lead will pick up signals or provide a ground return for an open connection. You poke with the plastic stick exactly as you did with the meter probe. It is one of your favorite tools. Yes, the cassette does come up. Next time, you lightly push the lead the other way. Nothing happens. Then you push it the way you did before, and the cassette comes up (Fig. 16-1). Looks as if you got it this time.

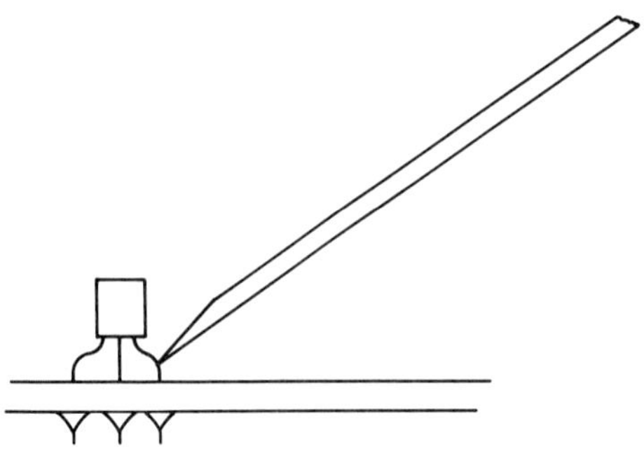

Figure 16-1 Feathering a lead to determine whether it is intermittent.

An inspection of the connection with a magnifying glass reveals nothing. Although the odds are good that it is the connection that is bad, you decide not to take a chance, and put in a new transistor. Past experience has shown that the heat and movement from resoldering the connection can effect a temporary "cure." Also, it is very difficult to distinguish whether the fault is in the external connection or the internal connection of the transistor.

The VCR is returned to the user. It has been working for six years. All it needs now is periodic maintenance.

16-1-2 What Makes Intermittents Such Tough Dogs

Intermittents are known as "tough dogs." They work fine most of the time. There are no bad symptoms and no bad voltages—while they are on the bench. Without symptoms and without bad voltages, there is not a thing you can do—it seems. Worse still, you may find no visible faults, such as burned and cracked parts.

Furthermore, while you are working on an intermittent it will start operating at the worst times. When this happens, you may have to clear the bench for another job. This may not be the best thing to do. You lose concentration and lose your line of thinking. If you do leave it on the bench, you will be losing valuable shop space for an unpredictable length of time. On top of this, troubleshooting methods for intermittents take up more than the usual amount of time. Let's see what we can do to overcome these difficulties.

16-2 WHAT CAUSES INTERMITTENTS?

Intermittent faults come about in several ways. They can be caused by:

1. Poor connections and breaks in the conductor and corroded (oxidized) contacts.
2. Temperature changes.
3. Line voltage changes.
4. Interference from external sources.

It would be very helpful if we examine these causes. We may come up with some valuable troubleshooting information.

16-2-1 Oxidation and Intermittent Opens

Poor solder connections are the most common type of trouble in new equipment. They work okay until the poor connections oxidize (corrode). These bad connections are "fixed" at the factory by the usual jostling on the assembly line. A poor solder connection is jostled all the way to the user facility. It is jostled at the loading docks. It is jostled on the freight train. It is jostled when the dealer takes it out of the carton. It is jostled when it is installed. All this jostling and pounding cleans out the nection and establishes electrical contact. After a period of quiet in the user environment, the connection oxidizes and electrical contact is lost.

It is again cleaned as it is jostled to the shop. It plays perfectly after some more jostling to get it on the bench. Intermittent opens are extremely sensitive. The slightest motion can break or make a contact.

Always look for poor solder connections in new equipment. Look for beady solder connections that do not show a smooth flow onto the foil. Look for insufficient solder. Insufficient solder connections are prone to develop hairline cracks. Inspect these connections with a magnifying glass. Sometimes the soldering is omitted altogether, and all you see is a lead sticking out of a pin hole.

The parts themselves may open up internally. Their leads, which are made of copper, do not make good contact with many unsolderable or unweldable substances within these parts. Capacitors are made of sheets of aluminum foil, which do not take solder, nor are they easily welded. Electrolytics are made of aluminum. Resistors are made of carbon or special wire that is difficult to weld. Semiconductors materials, too, are difficult to connect. Coils are made of enameled wire, which has to be cleaned before it can be soldered to a terminal. Some wires are very thin and difficult to inspect.

The factories turn out millions of capacitors, resistors, semiconductors, and coils. It is no wonder that a few intermittent opens will sneak by. They are jostled when inspected; they are jostled when mounted on the board, and they are jostled all the way to the user installation, where they finally have a period of quiet.

Another form of bad contact can cause detection of radio signals. Sometimes audio amplifiers can be heard with these signals in the background. This can be caused by poor connections that cause an amplifier circuit to rectify and detect a radio signal. Sometimes this is caused by a loose connection that acts like a diode rectifier. The action is similar to that of the old crystal detectors that were operated by tickling a semiconductor silicon or galena crystal with a thin wire (cat whisker). Soldiers on the Anzio Beachhead during World War II used to listen to Axis Sally using razor blades as crystals. All they needed besides this was a pair of field phones and a length of line wire for the antenna.

Printed-circuit boards sometimes have hairline cracks that are difficult to observe. These cause intermittent operation that is difficult to detect. Sometimes the conductors are etched very thinly, almost to the point where the conductive material disappears. Once in a while a tiny hairline piece of foreign matter gets on the board during the manufacturing process and causes thin breaks in the copper. This break is even harder to spot after the board has been coated with solder resist. This is true particularly when there is no crack on the board itself.

Printed-circuit-board ground lines often terminate at the four corner holes of the board. They are grounded to the chassis by screws. Sometimes they are not screwed in tight enough, or the chassis hole is stripped. Oxidation will complete the job of opening up the ground contact.

All in all, when you look at all the parts and lines on a board, you can see why an intermittent is bound to occur once in a while. We can say for sure that intermittents are totally predictable, as predictable as the San Andreas Fault, which is certain to generate the big earthquake in California. ''But when?''

16-2-2 Temperature Changes

Changing temperature conditions will also affect bad connections. Expansion will cause movement, which in turn may reestablish contact. Contraction may cause loss of contact. Or occasionally, just the opposite might happen; expansion might open up a connection and contraction might close it. Expansion and contraction can clean out poor soldering connections and keep the equipment operating forever.

Very often a piece of equipment is kept in storage for a long period of time. During the storage period, the equipment is not turned on and off periodically. There is no warm-up and cool-off, so the bad connection is allowed to develop a good oxide insulation. As you know, oxide is used as insulation in electrolytic capacitors. A very thin coating does the job. When the equipment is brought out of storage and turned on, it does not work. On top of that, all the mechanical contacts have oxidized, or their lubrication has dried out to form a nice film of insulation. This time, the equipment is sent to permanent storage in the junk pile. It could have been repaired.

Temperature can cause slow changes in defective-part resistances. Many faults show up only after the equipment has warmed up for a while. In some cases the parts are so sensitive to temperature changes that their action is unpredictable. All parts, even good ones, are sensitive to temperature changes. They are all designed to operate within certain temperature limits. Resistors normally tend to increase resistance with temperature. Most capacitors normally tend to increase capacitance with temperature. Polystyrene types decrease capacitance with temperature increase. Transistor output paths tend to lose resistance with temperature; that is, they pass more current with temperature increases.

Defective resistors tend to greatly increase resistance with temperature. Among defective resistors, the old $\frac{1}{2}$-W types with high value are the worst. Leaky capacitors also get worse with temperature increase. Leaky transistors are worst of all in this respect. Dirty boards with high resistance leakage also respond to temperature changes.

Temperature changes cause faults in two ways. They can slowly bring about changes, or they can cause sudden changes. Defective connections change suddenly. Parts can have internal opens that occur suddenly; otherwise, they usually change gradually. However, in some cases, they may not produce any changes until a breakover point is reached. Feedback-controlled circuits, for instance, may hold things in line until a breakover point is reached. Transistor operations often are not affected by defective parts until the limits of their operating points are reached. Most circuits are designed to operate under worst-case conditions. They are okay until the worst-case conditions are reached. Digital circuits can work with extremely distorted, low-level pulses. Digital 5-V systems will work with as low as 2.4 V for their logic hi's. They will continue operating with bad parts until finally the voltage goes under 2.4 V.

TIP 16-1

Sudden intermittents are associated with poor connections and breaks in the conductors and parts. They respond primarily to mechanical changes. They can respond secondarily to random temperature changes and nonrandomly after a given warm-up period.

Also look for borderline voltages that can go over the brink when the circuit is disturbed. Note the position of the controls, and note what faults are being compensated for.

TIP 16-2

Slowly developing faults are associated with the material in the defective parts. They respond to temperature changes. Monitor suspected points for creeping voltage changes.

Note the position of the controls, and note what faults are being compensated for.

16-2-3 Voltage Changes

We saw that temperature changes were associated with changes in operating voltages that resulted in intermittents. Let's look at other types of voltage-related intermittents. Open high-voltage connections, 100 V or more, may heal themselves intermittently by arcing and self-welding. This type of trouble reveals itself in stored equipment. Not recommended is the practice of pounding them and letting it go at that. It will come back to you in more ways than one.

In the old vacuum-tube days, technicians used to jump electrolytics with good ones, accompanied by loud snapping sparks. They found that they could cure the bad electrolytics that way. Foreign matter, or solder splash shorts, are often cleared by high-voltage surges. You cannot snap and spark around transistors or ICs this way. They cannot withstand high-voltage surges.

It is a widely held belief, which seems to be borne out by experience, that transistors can be healed by slight voltage spikes called *transients.* It is more likely that the transistor contacts are cleaned by slight contact rather than the transistor itself being healed. True or not, technicians handle intermittents very gingerly. They do not want to cause a voltage that will heal a transistor or IC pin unless they know which one it is. When checking voltages, they feel that even their test prod might pick up or cause slight voltages that bring the circuit back to life.

Voltage spikes on the lines are more likely to cause damage than healing. These spikes are of short duration, and in some places they have been measured as high as 1000 V. This is okay for most equipment because of the short duration, but it is not good for transistors and IC chips.

Some localities have problems with excessive ac power-line variations. These variations may cause intermittent operation. This is especially true of equipment with marginally defective parts. Line-voltage changes usually occur at certain hours of the day, usually during times of peak or low usage in the locality.

> TIP 16-3
>
> Defective operation at certain periods of the day can be caused by ac power-line variations. Faults associated with these changes include borderline parts defects. Check for borderline voltages and resistances.

16-2-4 Interference

Problems caused by interference from the outside are sometimes mistaken for internal intermittents. This can result in vague information from the user as to the exact nature of the trouble. Interference is picked up by the ac power line, receiver antenna, or the equipment circuits themselves. Static interference from nearby motors or defective automotive ignition systems might be the source of this interference. This is especially true in areas where the signal strength is poor and the receiver AGC is operating at maximum gain. At maximum gain the receiver picks up all kinds of interference.

New building construction in the area can cause poor reception in equipment that had previously worked fine in the area. Moving constructon cranes can cause all sorts of strange reception problems. Sometimes such problems occur during certain times of the day when the crane is stopped in a certain position and the operator steps out for lunch or for the day.

> TIP 16-4
>
> Interference symptoms may occur at certain times of the day. Check for interference sources that go into operation at these hours.

Generally speaking, electronic systems are designed to reject most types of interference and line-voltage variations. Intermittents related to these sources are usually due to a fault in the circuits themselves. Look for borderline faults. Of course, these are the hardest to detect.

16-2-5 Sloppy Work

Another source of intermittents should be mentioned. Very often it is the sloppy work of a troubleshooter that is the cause of an intermittent. If the complaint is intermittent operation, inspect the circuits for bad soldering. Look for opens and solder splashes. Make sure that your own work is clean.

Besides sloppy soldering, indiscriminate banging and wiggling of parts can also cause intermittents. Tugging and bending the leads can cause the internal connections of parts to open. The PC-board connections can also open. Look for insufficient solder on a PC connection.

Look out for heavy parts that are mounted on vertical PC boards. Transporting and banging the chassis around can loosen up or break the lead connections of the big electrolytics and coils.

Very often you will have to move aside a lot of wires to get at the parts. Dress these leads back into their original position after you are through with the job. The circuits may work okay with the leads moved around, but they may not work okay under different conditions at another time and place.

16-3 TROUBLESHOOTING PROCEDURE FOR INTERMITTENTS

Troubleshooting intermittents is just that—shooting trouble. You have to approach it systematically.

> TIP 16-5
>
> When troubleshooting intermittents, do not start out by banging the chassis. It may start operating and work well for several days before you have a chance to work on it again. Banging is done only when you have no other way of establishing that the intermittent is caused by physical motion.

16-3-1 Questioning the User

You have to know the symptoms. When the equipment is working well, you will never find out what's wrong unless you ask the user. To question effectively, you have to understand people. They do not like trouble. They tend either to underplay it or to exaggerate it. You may have to ask specific questions:

1. Does the trouble appear when the set is first turned on or later? If later, how much later? In 15 minutes? An hour? See Tip 16-1, Tip 16-2, and the solutions in this section.
2. Do the symptoms gradually or suddenly change? How long does the trouble last when it does occur? See Tip 16-1, Tip 16-2, and the procedures in this section.
3. Has the user made any adjustments? What kind of adjustments were made? Did the user hit the equipment or jiggle the controls to make it work? See Tip 16-1, Tip 16-2, Tip 16-3, Tip 16-4, and what follows.
4. Does the trouble occur during certain hours of the day? See Tip 16-3, Tip 16-4, and the following procedure.
5. Make sure that you are fixing the right fault. The real complaint could be an intermittent. Many users will live with faulty symptoms until a new one shows up. Then they'll bring it in.

16-3-2 Reproducing Symptoms with Heat

The fundamental strategy for troubleshooting intermittents is to reproduce the symptom. The information you have may lead you to suspect a specific area where the trouble might be. If the trouble occurs only after warm-up, you can hasten the process by applying heat with a lamp or hair dryer over the suspected area. If you have no specific area under suspicion, you can cover the equipment up with a blanket, canvas, or a cardboard box. Do not leave the covered equipment unattended.

If you have the space and the time, it may be best to let the equipment warm up on its own, while you go on to some other job. The idea behind this is to stay as close as you can to the way the user does things. If you do things differently, the trouble might not show up.

If you suspect a specific part, apply heat to it carefully with a soldering iron without touching it. Heat all the leads going into the suspected part, but not excessively. An intermittent will respond to heat very quickly. Even good parts will respond to heat if you apply enough of it.

TIP 16-6

When troubleshooting intermittents, you might reproduce the symptoms by applying heat. You can heat the general area by using a heat lamp or a hair-drying blower. You can heat specific parts with a soldering iron without touching them. Do not apply excessive heat.

16-3-3 Reproducing Symptoms with Spray Coolant

After warm-up you can apply spray coolant to the suspected area. If the symptom shows up, the intermittent is in that area. Let things warm up again until operation is restored. You can apply heat to hasten restoration. Then carefully apply coolant to the individual parts and connections to regain the symptoms. This procedure requires considerable skill and experience.

If the symptoms reappear when you spray a part, that part—if you sprayed it correctly—is defective. Correct spraying means first of all that you do not get the spray on any other part. Second, it means that you should not spray too much. All it takes is a touch of spray. If you spray too much, even a good part will react to it. You might also get spray on the other parts. Third, it means that you must give it time to take effect. If you spray very little, it takes time for the coldness to penetrate through the outer material. At the same time, the outer materials of the surrounding parts are even less penetrated. So less spraying isolates the suspected part even more. Spraying with coolant is an art.

If the trouble comes or goes suddenly, the trouble can be caused by an open that is inside a part. Spray the leads to isolate the fault.

Do not forget that the cooling procedure applies to intermittent connections as well as to parts. Sometimes you can get better results by spraying the solder side of the board.

Old and dirty boards also respond to spray cooling. Do not be fooled into thinking you found a bad part when it was actually the board. Scraping and cleaning the board with special cleaning fluids will usually do the job.

TIP 16-7

Intermittent symptoms may be reproduced by spraying the general area with coolant. Next, the specific parts, connections, or conductors can be isolated by pinpointing them with a touch of coolant. Avoid spraying the other parts.

TIP 16-8

Sudden changes caused by parts having internal opens may be isolated by spraying the leads with coolant.

16-3-4 Troubleshooting Voltage Problems

If the information you received suggests that line voltage variations may be the problem, "cook" the equipment for a while at high ac input voltage (130 V ac) with a variable transformer. If the trouble shows up, check the transistor voltages to find the intermittent part. Similarly, operate the equipment at low voltage (105 V ac) if necessary. Monitor key voltages if the trouble does not show up for a long time (see Tips 16-1 and 16-2).

The information you have with regard to the symptoms can be helpful. Frequent circuit-breaker cutouts or fuse blowouts suggest that high line voltage is the problem. Narrow raster, lack of brightness, and lack of sharpness would suggest low line voltage. You may know of certain localities in your area that have trouble with low or high line voltages. Take this into account.

> **TIP 16-9**
>
> When troubleshooting intermittents, you might be able to reproduce the symptoms by operating the equipment at high or low ac line voltages.

You can also introduce substitute voltages from an external dc power source or through a jumper resistor from the circuit supply line. This voltage is usually applied to the base.

> **TIP 16-10**
>
> You can reproduce an intermittent symptom by introducing substitute voltages into the circuit. Use an external variable power source, or jump a potentiometer from the circuit supply line. The insertion is usually at the base of a transistor.
>
> You can also reproduce the symptoms by shorting a part or opening up a signal or voltage path. You can short out a path to ground. Usually, the *BE* terminals of a transistor can be shorted without damaging anything.

16-3-5 Reproducing Symptoms by Touching and Bending

If the trouble occurs intermittently, but not necessarily after warm-up, you may have an intermittent connection. Gently touch the parts in the suspected circuit. Use a plastic stick. If nothing happens, tap the parts and the board around the suspected circuit. If still nothing happens, bend the board by pressing it slightly with a plastic stick (see Fig. 16-2; the bend in the board is exaggerated for illustrative purposes). Push from both sides in various places. Finally, tap around the entire board. Then

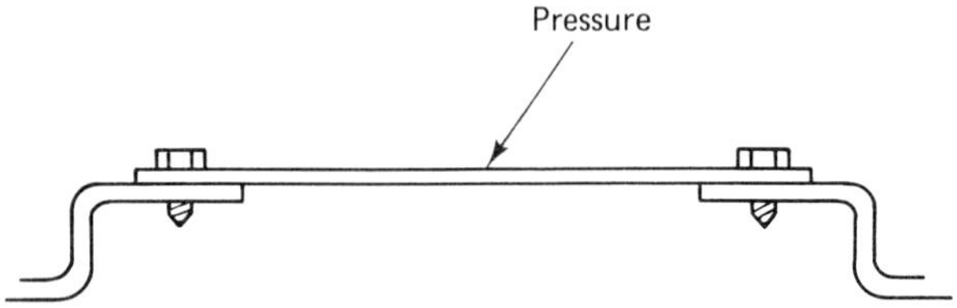

Figure 16-2 Bending a board to reveal a hairline crack or loose connection.

bang the whole chassis—not too hard—you're not angry; you're just trying to establish whether movement is going to do the trick.

Bending boards has its own peculiarities. Sometimes you bend one way and nothing happens. Bend it the other way and the symptom shows up. Sometimes the trouble shows up when you bend one part of the board and not elsewhere. Sometimes the trouble shows up when you push a part and not when you push another part. Sometimes no matter what, how, or where you bend, the trouble shows up. An intermittent can be very touchy. You have to develop a good sense for it. The idea is to work around the area and narrow in on the kill.

TIP 16-11

When looking for intermittent connections or cracks in the board by bending it, the trick is to start out with just slight touches all around the board and parts. Then gradually increase the pressure each time you work around the board.

If at some point in the process the circuit responds to tapping or bending, you must then try to narrow in on the fault. Touching a part, even if lightly, can be very deceptive. If touching a part brings the circuit back to life, touch another nearby part. Chances are high that this part, too, will respond. Touch other parts ever more lightly. The parts that respond to the lightest touch are closest to the bad part or connection. Keep touching the parts ever more lightly until you are sure you have the right part. Then you can look at it. Examine the solder connection. Resolder the suspected connection. Tap the area to see if you get a reaction. Cool off the area. Perhaps the heat from the soldering restored operation. Check the circuit out for a couple of days before putting it back into the cabinet and sending it out. This touching–bending–tapping–banging sequence takes a lot of skill and experience. A few successes will help you gain the ''touch.''

Many circuits have boards that can be plugged into sockets. The contacts will corrode and cause intermittent troubles. When pulling these boards out, pull them straight out or wiggle them in the lengthwise direction so as not to break the socket solder connections. These solder connections cause intermittent troubles that are hard to pinpoint. Resolder them as well as cleaning and tightening the contacts. Another thing: Lugs on the board are prone to poor soldering.

TIP 16-12

When troubleshooting intermittents, the symptoms might be restored by tapping the suspected areas. The suspected parts may be isolated just by touching them lightly.

TIP 16-13

Many intermittents are caused by poor solder connections and corroded board or cable connector contacts. Also look for poor lug connections on the boards.

16-3-6 Restoration with Heat, Cold, and Touch

Sometimes the circuit does not operate normally for a while and then kicks into operation often enough to hamper your voltage checking. In this case you can touch the suspected area or part with a plastic stick to find the defective part, or at least to get back the bad voltage you are looking for. You can also apply heat or cold for the same reasons. Of course, you have to use the same techniques and take the same precautions.

Old and dirty boards also respond to spray cooling. Do not be fooled into thinking you found a bad part when it was actually the board.

TIP 16-14

Old or dirty boards can cause intermittent operation. They respond to heat and cold treatments.

16-3-7 Checking Voltages to Locate Intermittents

Concentrate on measuring and recording as many voltages as you can between the time you turn the equipment on and the time the trouble appears. Many symptoms do not show up until a certain threshold voltage is reached. Look for these voltages.

Check these voltages as you apply heat or cold to the suspected parts. Also record as many voltage readings as you can when the equipment is not working normally and bad symptoms are showing. You can then compare the two sets of voltages, provided that you did not forget to record them.

If you find any borderline voltages, relate them to the symptoms. If a voltage seems high, it could be on the verge of making the circuit go bad. Just a little shove might make the circuit go over the brink. The same thing can be done to a lower-than-normal voltage. Borderline voltages are caused by partial opens and shorts.

In other words, if you see any borderline voltages, change them by jumping the base to emitter, or the base to collector, with a 10-kΩ resistor, or by applying variable voltages at strategic points. For instance, if the symptoms show a frequency problem, see if you can re-create it by changing the base voltage of the oscillator. You should have a variable dc power supply for this purpose. In some cases you can change the borderline voltages by changing the ac line voltages with a variac.

Do not overlook the the position of the controls. If they are at one extreme, rotate them back toward normal and see if you can reproduce the symptoms. Figure out what defective part is being compensated for. Look at the schematic and compare the voltages.

TIP 16-15

When troubleshooting intermittents, look for controls that are adjusted to the extreme positions. The symptoms might be reproduced by extending their effects on the circuit voltages. Look for the parts that could cause them to be adjusted to the extreme.

Certain makes are known to have bad runs on certain parts. If you have one on your bench, replace the troublesome parts without question. On vertically mounted boards, look for broken solder around the leads of the larger parts on the board. They often break loose on account of their weight.

TIP 16-16

When troubleshooting intermittents, change those parts in certain makes that are known to have had bad runs.

Also, if you cannot find anything wrong and if it is not going to take too long, change all the small electrolytics, all high-value resistors that have high voltage on them, and all low-voltage capacitors in the suspected circuit. Change all transistors relating to the symptom. If the parts are identified on the conductor side of the board, just resolder the leads in the suspected circuit. Otherwise, resolder the more suspicious-looking connections all over the board. Clean and bend or straighten all contacts.

16-3-8 "Cooking" the Equipment at High Voltage

After working on an intermittent, you may not be sure that you changed the right part or soldered the right connection. For this reason you have to ''cook'' or operate the equipment for a day or so. If the problem is related to line voltage, cook it for a while at 130 or 105 V ac, whichever applies. Do not disturb the equipment while you let it cool off overnight. In the mornings, when you first turn it on, note the operation very carefully. Too often, technicians have failed to do this. Many symptoms occur when the instrument is cold. Very often, technicians change parts after they have worked on the chassis for a while. When they see that the equipment is working after the replacement, they think they have effected a repair. The truth is: The warming up and the jostling did the job. Always check the job the next morning when things are cooler. It takes a long time for a circuit to cool off.

It takes a while for oxidation to set in, too. Preferably, the equipment should be allowed to rest quietly for a week or two, but this is not usually feasible.

If feasible, duplicate the user environment with the shop air conditioner or by covering the equipment partially or wholly with a sheet of cloth or plastic.

TIP 16-17

After troubleshooting an intermittent, cook the job for a few days to make sure the trouble does not show up again. If possible, duplicate the user environment by increasing or decreasing the line voltage with a variac. Check the operation first thing in the mornings when the equipment has had a chance to cool off and settle down.

16-4 AN ALTERNATIVE

This method works when no expensive parts are involved. Make all the preliminary touch and wiggle, heat, and cool tests. Check all the voltages. Then resolder all connections and replace all parts that have anything to do with the trouble. The advantages of this method are:

1. Less time is wasted in thinking, replacing, and testing with each part you think is bad. It is faster.
2. The systematic approach makes for a faster and neater job. It is psychologically less stressful, and this will improve your work in general.
3. A large number of alternatives are removed from consideration at one crack.
4. The overall statistical probability may favor this approach. This may be especially true if you have found no bad voltages. (Do not forget that board exchange and replacement may be wiser.)

WHAT YOU LEARNED TO DO

In this chapter you learned to troubleshoot intermittents:

1. Caused by poor connections, contacts, and oxidation.
2. Caused by partial shorts and opens that produced borderline voltages.
3. That respond to temperature, voltage, and mechanical changes.
4. By reproducing the symptoms.
5. By heating, cooling, and tapping the suspected area to reproduce the symptoms. You could also reproduce the symptoms by changing the ac line voltage.
6. By skillfully isolating defective parts and connections by heating or cooling, or by touching the suspected part, board slot contact, or solder connection. You learned how to bend boards skillfully to find a hairline crack or intermittent connection.
7. By checking for borderline voltages and pushing them over the brink. You learned that you could do this by jumping the base connections to change the voltage.
8. By cooking the equipment after troubleshooting it.

QUESTIONS

1. What is the most basic thing you must do to locate an intermittent fault?
2. Cite the things you should not do when looking for an intermittent.
3. What questions should you ask the user?
4. What should you look for primarily if the intermittent occurs suddenly and randomly?
5. What should you look for primarily if the intermittent occurs suddenly after a period of warm-up?
6. What should you look for primarily if the trouble comes on gradually?
7. What can cause trouble during certain periods of time?

8. How can you reproduce a trouble after warm-up?
9. How can you reproduce a trouble that goes away after warm-up?
10. How can you reproduce troubles that are related to faulty voltages?

PRACTICAL TROUBLESHOOTING EXPERIENCE

Get hold of a piece of discarded electronic equipment. It is likely to be an intermittent "dog" that some pro technician wisely turned down. You may have the time to work on it. Look for signs of previous work, such as resoldered connections and disconnected leads. If you see any signs like these, you are in for real trouble. However, it is a challenge and a real learning experience, even if you fail. Eyeball the boards for burned or broken parts. If you find an expensive part that is burned out or otherwise damaged, forget it unless you have an old replacement part available. Anyway, we assume that you have a chassis to work on.

Replace all visibly burned resistors and reconnect all loose leads. Then apply reduced voltage, through a series test lamp or a variac. Check the power-supply and power-output sections. Jump the regulator if necessary. If nothing is shorted, run the chassis at full power. If the set works, leave it on for two hours. If you're lucky, it should go bad after a while. If it does not, tap the chassis lightly, then hit it hard. Do not hit CRTs. If this does not get a response, cover it with a box or cloth. Do not leave it unattended. Operate it at high and low voltages. Cool it off by spraying it over wide areas. Go through several cycles of heating and cooling. If you cannot get it to go bad, let it cook all day. Try again the next morning.

If no trouble shows up, just let it set in a safe place and turn it on for a few hours a day for several days or weeks. Do not leave the area while it is on. Stick around and work on something else.

If you do see the intermittent trouble, note the symptoms. Try to reproduce the symptoms by pinpoint heating, cooling, wiggling, and by varying the ac input voltage. Look for borderline voltages. Monitor the key voltages, usually at the collectors, at all times.

Index